T0351078

SPACE TELESCOPE SCIENCE INSTITUTE

SYMPOSIUM SERIES: 20

Series Editor S. Michael Fall, Space Telescope Science Institute

MASSIVE STARS: FROM POP III AND GRBs TO THE MILKY WAY

This volume is based on a meeting held at the Space Telescope Science Institute on May 8–11, 2006.

This collection of review papers, written by world experts in the many aspects of massive stars, provides an invaluable resource, both to professional astronomers and astrophysicists, and for students. The topics covered range from the formation of massive stars, to their role in the early universe, and from stellar winds to pair-production supernovae.

Other titles in the Space Telescope Science Institute Series.

1 Stellar Populations
Edited by C. A. Norman, A. Renzini and M. Tosi 1987 978-0521-33380-1
2 Quasar Absorption Lines
Edited by C. Blades, C. A. Norman and D. Turnshek 1988 978-0521-34561-3
3 The Formation and Evolution of Planetary Systems
Edited by H. A. Weaver and L. Danly 1989 978-0521-36633-5
4 Clusters of Galaxies
Edited by W. R. Oegerle, M. J. Fitchet and L. Danly 1990 978-0521-38462-9
5 Massive Stars in Starbursts
Edited by C. Leitherer, N. R. Walborn, T. M. Heckman and C. A. Norman 1991 978-0521-40465-5
6 Astrophysical Jets
Edited by D. Burgarella, M. Livio and C. P. O'Dea 1993 978-0521-44221-3
7 Extragalactic Background Radiation
Edited by D. Calzetti, M. Livio and P. Madau 1995 978-0521-49558-5
8 The Analysis of Emission Lines
Edited by R. E. Williams and M. Livio 1995 978-0521-48081-9
9 The Collision of Comet Shoemaker-Levy 9 and Jupiter
Edited by K. S. Noll, H. A. Weaver and P. D. Feldman 1996 978-0521-56192-1
10 The Extragalactic Distance Scale
Edited by M. Livio, M. Donahue and N. Panagia 1997 978-0521-59164-5
11 The Hubble Deep Field
Edited by M. Livio, S. M. Fall and P. Madau 1998 978-0521-63097-9
12 Unsolved Problems in Stellar Evolution
Edited by M. Livio 2000 978-0521-78091-9
13 Supernovae and Gamma-Ray Bursts
Edited by M. Livio, N. Panagia and K. Sahu 2001 978-0521-79141-0
14 A Decade of *Hubble Space Telescope* Science
Edited by M. Livio, K. Noll and M. Stiavelli 2002 978-0521-82459-0
15 The Dark Universe: Matter, Energy, and Gravity
Edited by M. Livio 2003 978-0521-82227-5
16 Astrophysics of Life
Edited by M. Livio, I. N. Reid and W. B. Sparks 2005 978-0521-82490-3
17 The Local Group As an Astrophysical Laboratory
Edited by M. Livio and T. M. Brown 2006 978-0521-84759-9
18 Planets to Cosmology: Essential Science in *Hubble*'s Final Years
Edited by M. Livio and S. Casertano 2006 978-0521-84758-2
19 A Decade of Extrasolar Planets around Normal Stars
Edited by M. Livio, K. Sahu and J. Valenti 2008 978-0-521-89784-6

Massive stars: From Pop III and GRBs to the Milky Way

Proceedings of the
Space Telescope Science Institute Symposium,
held in Baltimore, Maryland
May 8–11, 2006

Edited by
MARIO LIVIO
Space Telescope Science Institute, Baltimore, MD 21218, USA
EVA VILLAVER
Space Telescope Science Institute, Baltimore, MD 21218, USA

Published for the Space Telescope Science Institute

Shaftesbury Road, Cambridge CB2 8EA, United Kingdom

One Liberty Plaza, 20th Floor, New York, NY 10006, USA

477 Williamstown Road, Port Melbourne, VIC 3207, Australia

314–321, 3rd Floor, Plot 3, Splendor Forum, Jasola District Centre, New Delhi – 110025, India

103 Penang Road, #05–06/07, Visioncrest Commercial, Singapore 238467

Cambridge University Press is part of Cambridge University Press & Assessment,
a department of the University of Cambridge.

We share the University's mission to contribute to society through the pursuit of
education, learning and research at the highest international levels of excellence.

www.cambridge.org
Information on this title: www.cambridge.org/9780521762632

First published 2009

A catalogue record for this publication is available from the British Library

ISBN 978-0-521-76263-2 Hardback

Contents

Participants

Aloisi, Alessandra	Space Telescope Science Institute/ESA
Attard, Michael	University of Western Ontario
Balsara, Dinshaw	University of Notre Dame
Beckwith, Steven	Space Telescope Science Institute
Belczynski, Chris	New Mexico State University
Blair, William	The Johns Hopkins University
Bonanos, Alceste	Carnegie Institution of Washington
Bonnell, Ian	University of St. Andrews
Bromm, Volker	University of Texas
Chen, Hsiao-Wen	University of Chicago
Chevalier, Roger	University of Virginia
Crowther, Paul	University of Sheffield
Dessart, Luc	Steward Observatory
Disney, Mike	Cardiff University
Duncan, Douglas	University of Colorado
Edwards, Michelle	University of Florida
Eldridge, John	Queen's University Belfast
Elmegreen, Bruce	IBM Watson Research Center
Fabbiano, Giuseppina	Harvard-Smithsonian Center for Astrophysics
Figer, Donald	Rochester Institute of Technology
Fox, Derek	The Pennsylvania State University
Fruchter, Andrew	Space Telescope Science Institute
Galliano, Frédéric	NRC/NPP, NASA Goddard Space Flight Center
Garcia, Miriam	Instituto de Astrofisica de Canarias
Gardner, Jonathan P.	NASA Goddard Space Flight Center
Gehrels, Neil	NASA Goddard Space Flight Center
Godon, Patrick	Villanova University
Gouliermis, Dimitrios	Max-Planck Institute for Astronomy
Gull, Theodore	NASA Goddard Space Flight Center/EUD/EP&SA
Hasan, Hashima	NASA Headquarters
Hauser, Michael	Space Telescope Science Institute
Hillier, D. John	University of Pittsburgh
Hunter, Ian	Queen's University Belfast
Johnson, Kelsey	University of Virginia
Kashlinsky, Alexander Sasha	NASA Goddard Space Flight Center
Kaspi, Victoria	McGill University
Koekemoer, Anton	Space Telescope Science Institute
Kratter, Kaitlin	University of Toronto
Krumholz, Mark	Princeton University
Kudritzki, Rolf	Institute for Astronomy, University of Hawaii
Langer, Norbert	Utrecht University
Lawlor, Timothy	The Pennsylvania State University, Wilkes-Barre
Leistra, Andrea	University of Arizona
Leitherer, Claus	Space Telescope Science Institute
Lennon, Daniel	The Isaac Newton Group of Telescopes
Livio, Mario	Space Telescope Science Institute
Loeb, Abraham	Harvard University

Lopez Merino, Antonio	Niels Bohr Institute, University of Copenhagen
Margon, Bruce	Space Telescope Science Institute
Meixner, Margaret	Space Telescope Science Institute
Mikles, Valerie	University of Florida
Mirabel, Felix	European Southern Observatory, Chile
Mitalas, Romas	University of Western Ontario
Mountain, Matt	Space Telescope Science Institute
Muñoz-Marin, Victor Manuel	Instituto de Astrofísica de Andalucía (IAA-CSIC)
Nazé, Yaël	Institut d'Astrophysique et de Geophysique (ULg)
Nordlund, Aake	JILA, University of Colorado
Norman, Michael	University of California, San Diego
O'Shea, Brian	Los Alamos National Lab
Oey, Sally	University of Michigan
Patel, Nimesh	Harvard-Smithsonian Center for Astrophysics
Pellerin, Anne	Space Telescope Science Institute
Puzia, Thomas H.	Space Telescope Science Institute
Robberto, Massimo	Space Telescope Science Institute
Rothberg, Barry	Space Telescope Science Institute
Sahu, Kailash	Space Telescope Science Institute
Scannapieco, Evan	Kavli Institute for Theoretical Physics
Schaerer, Daniel	Geneva Observatory
Scowen, Paul	Arizona State University
Smartt, Stephen	Queen's University Belfast
Smith, Nathan	University of Colorado
Smith, Britton	The Pennsylvania State University
Sonneborn, George	NASA Goddard Space Flight Center
Sota, Alfredo	Space Telescope Science Institute
Tej, Anandmayee	Tata Institute of Fundamental Research
Tilley, David	University of Notre Dame
Tominaga, Nozomu	University of Tokyo
Townsley, Leisa	The Pennsylvania State University
Tumlinson, Jason	Yale University
Villaver, Eva	Space Telescope Science Institute
Walborn, Nolan	Space Telescope Science Institute
Whalen, Daniel	Los Alamos National Lab
Whitmore, Brad	Space Telescope Science Institute
Whitney, Barbara	Space Science Institute, Boulder
Wiseman, Jennifer	NASA Headquarters
Yao, Lihong	University of Toronto
Young, David	Queen's University Belfast
Young, Teresa	Amateur Astronomer
Young, Timothy	University of North Dakota
Zinnecker, Hans	Astrophysikalisches Institut Potsdam
van der Kruit, Colly	Kapteyn Astronomical Institute
van der Kruit, Pieter	Kapteyn Astronomical Institute

Preface

The Space Telescope Science Institute Symposium on *Massive Stars: From Pop III and GRBs to the Milky Way* took place during May 8–11, 2006.

These proceedings represent only a part of the invited talks that were presented at the symposium. We thank the contributing authors for preparing their manuscripts.

Traditionally, massive stars played the important roles of being responsible for supernova explosions, of being the progenitors of stellar-mass black holes, and of producing heavy elements. In recent years, massive stars have gained additional importance in our understanding of cosmic history. Very massive stars (Population III) are now recognized as constituting the first population of stars in the universe, and massive stars have been identified as being the progenitors of the long-duration Gamma-ray Bursters. In addition, very massive stars may produce supernova explosions by a new mechanism—pair instability—that has been anticipated theoretically, but has never been unambiguously detected (the recent SN 2006gy may have been such an event).

The ST ScI symposium on Massive Stars attempted to capture all the aspects involved in the astrophysics of massive stars.

We thank Sharon Toolan of ST ScI for her help in preparing this volume for publication.

Mario Livio
Eva Villaver
Space Telescope Science Institute
Baltimore, Maryland

High-mass star formation by gravitational collapse of massive cores

By MARK R. KRUMHOLZ†

Department of Astrophysical Sciences, Princeton University, Princeton, NJ 08544-1001, USA

The current generation of millimeter interferometers have revealed a population of compact ($r \lesssim 0.1$ pc), massive ($M \sim 100\,M_\odot$) gas cores that are the likely progenitors of massive stars. I review models for the evolution of these objects from the observed massive-core phase through collapse and into massive-star formation, with particular attention to the least well-understood aspects of the problem: fragmentation during collapse, interactions of newborn stars with the gas outside their parent core, and the effects of radiation-pressure feedback. Through a combination of observation, analytic argument, and numerical simulation, I develop a model for massive-star formation by gravitational collapse in which massive cores collapse to produce single stars or (more commonly) small-multiple systems, and these stars do not gain significant mass from outside their parent core by accretion of either gas or other stars. Collapse is only very slightly inhibited by feedback from the massive star, thanks to beaming of the radiation by a combination of protostellar outflows and radiation-hydrodynamic instabilities. Based on these findings, I argue that many of the observed properties of young star clusters can be understood as direct translations of the properties of their gas-phase progenitors. Finally, I discuss unsolved problems in the theory of massive-star formation, and directions for future work on them.

1. Introduction

Massive-star formation occurs in the densest, darkest parts of molecular clouds. These clumps of gas have masses of thousands of $M_\odot$, radii $\lesssim$1 pc, volume densities of $\sim$$10^5$ H atoms cm^{-3}, column densities of $\sim$1 g cm^{-2}, visual extinctions of hundreds, and velocity dispersions of several km s^{-1}. Observations often reveal indicators of massive-star formation such as maser emission and infrared point sources within them, but the majority of their mass appears to be dark and cold. Due to their high extinctions and low temperatures, these regions are only accessible to observation through millimeter emission, either in molecular lines (e.g., Plume et al. 1997; Shirley et al. 2003; Yonekura et al. 2005; Pillai et al. 2006b) or dust continuum (e.g., Carey et al. 2000; Mueller et al. 2002), or through infrared absorption (e.g., Egan et al. 1998; Menten et al. 2005; Rathborne et al. 2005; Simon et al. 2006; Rathborne et al. 2006). They are likely the progenitors of the rich clusters in which massive stars form.

In the last few years, observations using millimeter interferometers to obtain still higher resolution have identified "cores" within these dense clumps, objects small enough that they approach the stellar-mass scale. Cores are distinguished by even higher volume densities than the massive clumps around them, 10^6 H cm^{-3} or more, and smaller radii, $r \lesssim 0.1$ pc. Some show mid-IR (MIR) point sources in their centers (e.g., Pillai et al. 2006a), while others show no MIR emission, or even MIR absorption, indicating that they are starless or contain only very low-mass stars (e.g., Sridharan et al. 2005). In some cases they show signs of molecular outflows, but not MIR emission, indicating that the extremely massive core contains a very low-mass protostar, and thus is near the onset of star formation (Beuther et al. 2005).

The characteristic mass, size, and density of massive cores make them appealing candidates to be the progenitors of massive stars (e.g., Garay 2005). Moreover, as I discuss

† Hubble Fellow

in more detail in Section 2, the observed mass and spatial distribution of protostellar cores is quite similar to that of stars in young clusters. If massive cores are the direct precursors of massive stars, then one can explain many of the properties of newborn clusters directly from the observed properties of their gas-phase progenitors. The goal of this review is to construct a rough scenario based on this idea by following observed cores through collapse, fragmentation, accretion, and feedback, to the final formation of massive stars. In Section 2, I briefly review observations of the properties of massive cores to provide initial conditions for this scenario. In Sections 3–5, I discuss three major questions about how these cores turn into stars: Do they fragment into many stars, or only a handful? Do the stars they form accrete significant mass from outside the parent core? Does feedback significantly inhibit accretion? Finally, in Section 6, I discuss some outstanding problems in the modeling of massive-core evolution, and suggest directions for future work.

2. Massive cores: Initial conditions for massive-star formation

We know disappointingly little about massive cores, despite great observational effort. Due to their small sizes and large distances, massive cores are only marginally resolved, even in observations with the highest resolution telescopes available. Nonetheless, observations do allow us to determine some gross properties of individual massive cores, and of the massive-core population as a whole. Observations show that cores are centrally concentrated, although the exact density profile is difficult to determine with interferometer measurements, and fairly round, with aspect ratios of roughly 2:1 or less. They are cold, $T \approx 10$–40 K, except near stellar sources, so their observed velocity dispersions of ~ 1 km s^{-1} imply the presence of highly supersonic motions (e.g., Reid & Wilson 2005; Beuther et al. 2005, 2006). At the characteristic density of $\sim 10^6$ H cm^{-3} found in these cores, the free-fall time is only $\sim 10^5$ yr, so the implied accretion rate when a massive core collapses is 10^{-4}–$10^{-3}\,M_\odot$ yr^{-1}.

McKee & Tan (2003) propose a simple self-similar model of massive cores in which the core density and velocity dispersion are power-law functions of radius, such that at every radius, turbulent motions provide enough ram pressure to marginally support the core against collapse. The central idea is that, at the high pressures found in massive-star-forming regions, massive cores must be supported by internal turbulent motions. While a self-similar spherical model is obviously a significant simplification of a turbulently supported gas cloud, it provides a reasonably good fit to the available observations, and makes it possible to calculate quantities such as the timescale for star formation and the relationship between core mass, column density, pressure, and velocity dispersion. It also provides a good starting point for simulations and more detailed analytic work.

For the massive-core population as a whole, we know somewhat more, and the observations bolster the idea that massive cores might really be the progenitors of massive stars. Several authors, using different techniques and observing different regions, find that the mass distribution of massive cores matches the stellar initial mass function, shifted to higher mass by a factor of a few, with a Salpeter slope of $\Gamma \approx -2.3$ at high masses, and a flattening at lower masses (Beuther & Schilke 2004; Reid & Wilson 2005, 2006a,b). This extends earlier observational work indicating that in low- and intermediate-mass star-forming regions the core-mass function resembles the stellar initial mass function (IMF) (e.g., Motte et al. 1998; Testi & Sargent 1998; Johnstone et al. 2001; Onishi et al. 2002), and suggests that the stellar IMF may simply be set by the mass distribution of prestellar cores, reduced by a factor of a few due to mass ejection by protostellar outflows (Matzner & McKee 2000). Simulations and analytic arguments, can, in turn, explain the core-mass distribution as arising naturally from the supersonic turbulence present in star-forming clumps (Padoan & Nordlund 2002; Tilley & Pudritz 2004; Li et al. 2004).

Clark & Bonnell (2006) argue that the mass distribution of *bound* cores in simulations does not have a Salpeter slope and thus is unlikely to be the origin of the stellar IMF. However, this misses a critical point: the Salpeter slope is only observed for stars significantly above the peak of the IMF. The full IMF is closer to a broken power law (Kroupa 2002) or a lognormal (Chabrier 2003), with the break or peak at $\sim 0.5\, M_\odot$. This is roughly the Jeans mass in star-forming clumps, and indeed the simulations of Clark & Bonnell (2006) do show something like a lognormal distribution, with a peak at roughly the Jeans mass of their simulations. (The simulations are scale-free, since they include only hydrodynamics and gravity and use an isothermal equation of state.)

Furthermore, recent observations focusing on the spatial distribution of cores have shown that cores are mass segregated (Stanke et al. 2006) in much the same manner as stars in very young clusters: the core-mass function has the same lognormal or broken power-law form everywhere in clumps, with the exception that the most massive cores—those larger than a few $M_\odot$ in size—are found only in the very center. The stellar population of the Orion Nebula Cluster (ONC) follows the same pattern (Hillenbrand & Hartmann 1998; Huff & Stahler 2006), indicating that the observed mass segregation in stars may simply be an imprinting of the prestellar-core spatial distribution. At least some of the mass segregation must be primordial rather than a result of dynamical evolution (Bonnell & Davies 1998), although recent evidence that cluster formation takes several crossing times (Tan et al. 2006) suggests that evolution may be important too. Nonetheless, it is quite suggestive that both the IMF and the spatial distribution of stars in a cluster seem to be explicable solely from the observed distribution of gas from which star clusters form. However, the origin of the mass segregation of cores is at present unknown.

3. Fragmentation of massive cores

It is only possible to explain the properties of stars in terms of the properties of cores if there is a more or less direct mapping from core mass to star mass. Such a mapping exists only if cores do not fragment too strongly, i.e., if massive cores typically produce one or a few massive stars, rather than many low-mass stars. Fragmentation to a few objects does not present a problem, since observationally constructed mass functions are generally uncorrected for multiple systems, but fragmentation to many objects does.

One might expect massive cores to fragment because they contain many thermal Jeans masses of gas. At the densities of $\sim 10^6$ H cm^{-3} and temperatures of ~ 10 K typical of massive cores, the Jeans mass is only $\sim M_\odot$, so one might expect a $50\, M_\odot$ core to form tens of stars. Dobbs et al. (2005) simulate the collapse and fragmentation of massive cores with initial conditions based on the McKee & Tan (2003) model, using a code that includes hydrodynamics and gravity. They try both isothermal and barotropic equations of state. (Barotropic here means that the gas is assumed to be isothermal at densities below some critical density, chosen to be 10^{-14} g cm^{-3} in the Dobbs et al. simulations, and adiabatic at higher densities). Dobbs et al. find that the cores fragment, forming anywhere from a few to several tens of objects, depending on the assumed initial conditions and equation of state. In no case do their simulations form a massive star.

However, the Dobbs et al. (2005) calculation omits the critical effect of radiation feedback from the forming star. The high densities in massive cores produce high accretion rates, so that the first protostar to condense within a core will immediately produce a large accretion luminosity as the gas that falls onto it radiates away its potential energy.

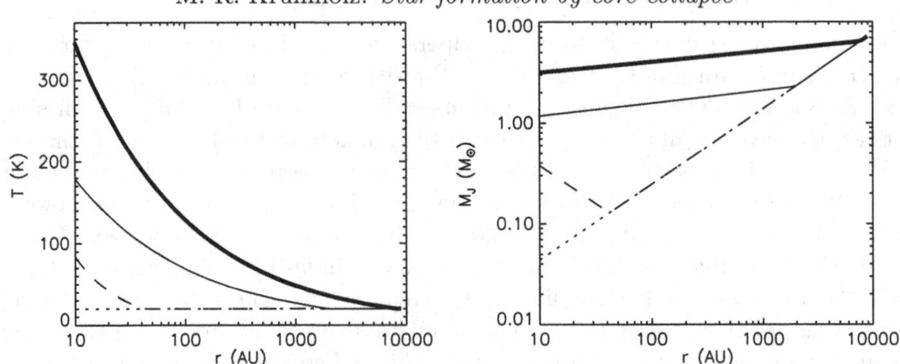

FIGURE 1. Temperature (left panel) and Jeans mass (right panel) versus radius in a core with a mass of $50\,M_\odot$, a column density of 1 g cm^{-2}, and a density profile $\rho \propto r^{-1.5}$, taken from the models of Krumholz et al. (2006a). The lines show the result from a radiative transfer calculation when the central protostar is $0.05\,M_\odot$ (thin solid line) and when it is $0.8\,M_\odot$ (thick solid line), and from using the barotropic approximation (dashed line) or an isothermal equation of state (dotted line). The Jeans mass is computed using the density and temperature at each radius, and is defined as $M_J = \frac{4}{3}\pi^{5/2}[k_B T/(G\mu)]^{3/2}\rho^{-1/2}$, where ρ is the density, T is the temperature, and $\mu = 2.33 m_H$ is the mass per particle for a gas of molecular hydrogen and helium in the standard interstellar abundance.

For a typical accretion rate of $10^{-4}\,M_\odot$ yr^{-1} at massive core densities, a $0.1\,M_\odot$, $1\,R_\odot$ star releases approximately $300\,L_\odot$ of accretion power. Because the core is very optically thick, the radiation is trapped within it and heats the gas as it diffuses out. As a result, the densest, inner parts of the core, where fragmentation is most likely to take place, are subject to rapid heating, which suppresses fragmentation. Isothermal or barotropic approximations completely ignore this effect.

Krumholz et al. (2006a) examines how feedback heating affects fragmentation in the context of a simple model of core accretion using a high-accuracy analytic radiative-transfer approximation (Chakrabarti & McKee 2005). Figure 1 shows a sample result, the radial temperature profile and Jeans mass versus radius for a McKee & Tan (2003) core with a mass of $50\,M_\odot$ and a column density of 1 g cm^{-2} accreting onto a protostar in its center. The figure compares the results using a radiative transfer approach to what one would find by neglecting radiative transfer and simply using a barotropic or isothermal equation of state. As the plot shows, both an isothermal equation of state and the barotropic approximation make order-of-magnitude errors in the temperature and Jeans mass.

One might worry whether this analytic treatment done in spherical symmetry applies to more realistic massive cores with complex density structures (e.g., Bonnell et al. 2007). The natural way to address this question is with radiation-hydrodynamic simulations of massive core evolution. Krumholz et al. (2007) simulate the collapse and fragmentation of massive cores to examine this effect. The simulations use an adaptive mesh-refinement radiation code to solve the Euler equations of gas dynamics coupled to gray radiation transport and radiation pressure force in the flux-limited diffusion approximation (Truelove et al. 1998; Klein 1999; Howell & Greenough 2003). They use the adaptive mesh capability to guarantee that the local Jeans length is always resolved by at least eight cells (Truelove et al. 1997), and that the radiation energy density changes by no more than 25% per cell, so radiation gradients are well resolved. The code uses Eulerian sink particles to represent stars (Krumholz et al. 2004), and the sink particles are, in turn, coupled to a simple protostellar evolution model (McKee & Tan 2003), which computes

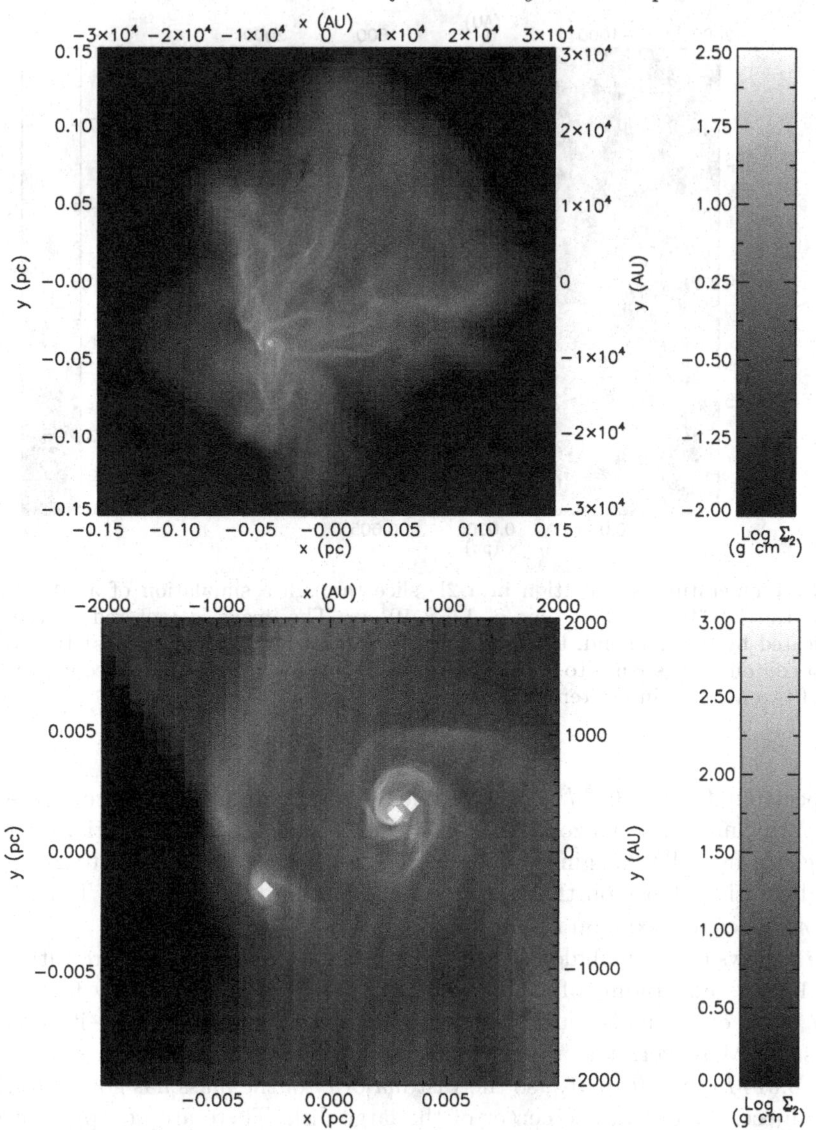

FIGURE 2. Column density of the entire core (upper panel) and zoomed in on the forming stars (lower panel) for a simulation of a 100 $M_\odot$, 0.1 pc McKee & Tan (2003) core at a time of 2.0×10^4 yr. The positions of the stars are indicated by the diamonds. Their masses are, from left to right, 0.31 $M_\odot$, 5.33 $M_\odot$, and 0.16 $M_\odot$.

the instantaneous stellar luminosity, including the effects of accretion, Kelvin-Helmholtz contraction, deuterium burning, and hydrogen burning. This luminosity becomes a source term in the radiation equation. Further details on the code are given in Krumholz et al. (2005a).

The simulations begin with cores following the model of McKee & Tan (2003). The initial density profile is chosen with $\rho \propto r^{-1.5}$, to a maximum density of $\rho = 10^{-14}$ g cm^{-3}, corresponding roughly to the density of the inner, thermally supported zone of McKee & Tan cores. The temperature is 20 K throughout the core. There are initial turbulent velocities chosen from a Gaussian random distribution (Dubinski et al. 1995) with a

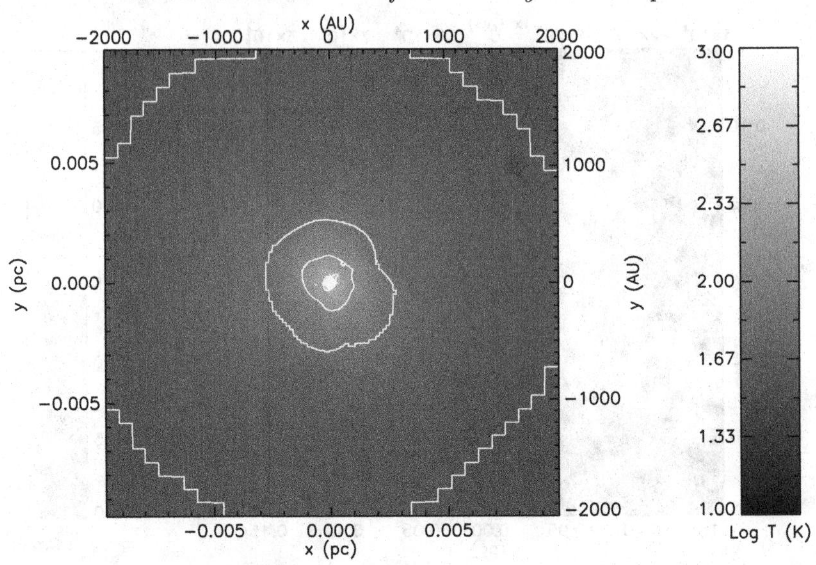

FIGURE 3. Temperature distribution in a 2D slice through a simulation of a $100\,M_\odot$, 0.1 pc McKee & Tan (2003) core at a time of 1.6×10^4 yr. The image is centered on the $3.83\,M_\odot$ star, indicated by the diamond, the most massive present in the simulation at that time. The outermost contour corresponds to a temperature of 40 K, and each subsequent contour represents a factor of two increase in the temperature.

power spectrum $P(k) \propto k^{-4}d^3k$ over wavelengths ranging from the size of the core to the size of the inner thermal zone, subject to the constraint that the initial velocity field be divergence free. The magnitude of the velocity field is normalized to give approximate hydrostatic balance on the largest scale (equation 6 of McKee & Tan 2003). The simulations reach a maximum resolution of 10 AU.

Figure 2 shows the column density distribution in a simulation of a core with an initial mass of $100\,M_\odot$ and radius of 0.1 pc, 2.0×10^4 yr (0.37 mean-density free-fall times) after the start of a simulation. The core is not forming many stars—it is forming a triple system. Moreover, it is a highly unequal triple: the masses of the three stars are $5.33\,M_\odot$, $0.31\,M_\odot$, and $0.16\,M_\odot$, so the vast majority of the mass has gone into the most massive object, the one at the center of the large disk. There are no apparent signs of further fragmentation, so unless feedback disrupts this system, it seems destined to form a massive star incorporating a significant fraction of the initial core mass, rather than dozens of small stars. The weak fragmentation we find from simulations provides strong support for the idea that the core-mass function directly sets the stellar-mass function.

To understand the origin of the weak fragmentation, it is helpful to examine the temperature distribution in the core. Figure 3 shows the temperature distribution in the simulation at $t = 1.6 \times 10^4$ yr, when the central star is only $3.83\,M_\odot$. At this point, the star has not yet begun hydrogen burning, and the luminosity of a few thousand $L_\odot$ is entirely due to accretion. This accretion power has doubled the initial temperature of the gas out to more than 2000 AU from the central star, and increased the temperature to more than 100 K over a radius of many hundreds of AU. This heating strongly suppresses fragmentation in the densest gas, where it is most likely to occur. Of the two stars that do form in addition to the most massive, one does so at an initial separation of several thousand AU, far enough that it can condense, and the other does so inside the protostellar disk, where the high column density provides shielding against the stellar

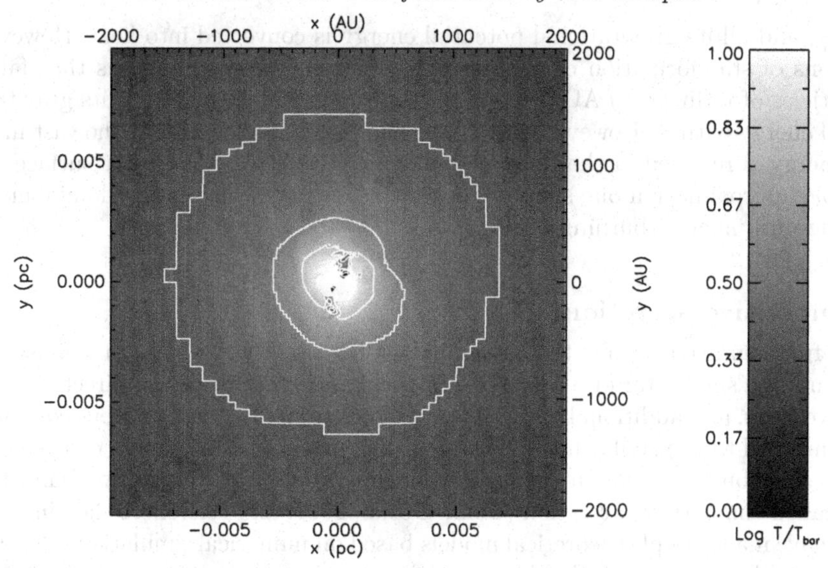

FIGURE 4. Ratio of the temperature distribution shown in Figure 3 to the temperature one would infer using the barotropic equation of state of Dobbs et al. (2005). Note that the color scale starts at a ratio of 1.0, so any color other than black indicates that the true temperature is higher than the barotropic temperature. The outermost contour corresponds to a ratio of 0.4 dex, increasing by 0.2 dex with each subsequent contour.

radiation and produces a lower temperature. In examining a movie of the simulation, one clearly sees many overdense clumps that look like they might collapse, but do not do so because they are bathed in the radiation field of the central star. Rather than forming stars of their own, they fall onto the central star and accrete.

Part of the reason suppression of fragmentation is effective is because, despite the complex density structure shown in Figure 2, the temperature distribution is relatively round and smooth. The only significant deviation from sphericity is in the protostellar disk. Thus, the gas around the protostar is heated quite uniformly, and outside the disk, where shear suppresses most fragmentation, there are no cold spots favorable to fragmentation. This is to be expected: the entire core is very optically thick, so radiation diffuses outward rather than free streaming. As a result, there is very little shadowing, and clumps that are only starting to collapse are not sufficiently overdense to exclude the radiation field and remain cooler than their surroundings. As I discuss in Section 5, only when gas reaches the densities typical of accretion disks, or when optically thick structures begin to form, can there can be significant temperature anisotropies due to collimation.

The weak fragmentation shown in simulations with radiation is strikingly different from what one obtained without it, where the number of fragments generally approaches the number of thermal Jeans masses in the initial cloud. Figure 4 shows why: the barotropic equation of state severely underestimates the temperature, making fragmentation far easier than it should be. The magnitude of the underestimate ranges from factors of a few at distances of thousands of AU to orders of magnitude in the central hundreds of AU. Since the Jeans mass depends on temperature to the 1.5 power, the error in the critical mass for fragment growth is larger still. It is easy to intuitively understand why the barotropic approximation fails so badly: the physical assumption underlying the barotropic approximation is that above some critical density the gas cannot radiate efficiently, and all its gravitational potential energy is converted into heat. However, 3D simulations of star formation cannot resolve stellar surfaces, so any gas that falls into sink particles of radius ~10 AU disappears from the simulation, taking its gravitational potential energy with it. However, since potential energy varies as r^{-1}, the vast majority of the energy is released in the final plunge from ~10 AU to the stellar surface. In the barotropic approximation one simply ignores this energy, which is the dominant source of heating until nuclear burning begins.

4. Competitive accretion

Weak fragmentation means that a significant fraction of the mass in a massive core will end up in a single star or a few stars. However, for cores to be the direct progenitors of massive stars, any additional mass a star accretes from outside its parent core must be small compared to the stellar mass. The idea that most of a star's mass comes not from a parent core, but from gas in the cluster-forming clump not originally bound to that star, is called competitive accretion (Bonnell et al. 2007, and references therein). Several authors have made simple theoretical models based on numerical simulations that appear to show exactly this process. In these models, all stars are born from cores at roughly equal masses, with the initial mass ranging from brown dwarf masses (Bate & Bonnell 2005) to as much as $\sim 0.5\,M_\odot$—the peak of the IMF—in the most recent models (Bonnell & Bate 2006). In these models, most of the seeds do not accrete much mass in addition to that in their parent core, but the special location of a few stars allows them to undergo rapid accretion, reaching high masses. This process determines the IMF above the peak.

4.1. *Under what conditions does competitive accretion occur?*

Competitive accretion definitely occurs in some simulations, and there is no reason to doubt those simulations produce the correct result for the physics they include. However, in order to determine whether the properties of real star-forming clumps are accurately reflected, it is necessary to investigate the physics behind the competitive accretion process. Krumholz et al. (2005d) defines the fractional mass change $f_M \equiv \dot{M}_* t_{\rm dyn}/M_*$ as the fractional mass change that a star of mass M_* undergoes per dynamical (crossing) time $t_{\rm dyn}$ of its parent clump, where $\dot{M}_*$ is understood to refer to the accretion rate after the star has consumed its initial bound core. Competitive accretion models require $f_M \gg 1$. Accretion of gas that is not initially bound to a star can occur in one of two forms: either the star may capture other gravitationally bound cores and then accrete them, as proposed for example by Stahler et al. (2000), or it may accrete gas that is not organized into bound structures.

The former process is reasonably easy to understand, since it is simply an extension of standard calculations of collision rates in stellar dynamics. The only significant complication is that collisions between stars and cores that occur at too large a relative velocity do not result in capture, since the star will simply plough through the core without dissipating enough energy for the two to become bound. Even with this complication, the calculation is relatively straightforward, and Krumholz et al. (2005d) show that the fractional mass change due to captures of cores with mass comparable to the stellar mass M_* in a star-forming clump of mass M is

$$f_{M\text{-cap}} \approx 0.4\phi_{\rm co}\left[4 + 2u^2 - (4 + 7.32u^2)\exp(-1.33u^2)\right] \quad , \tag{4.1}$$

where $\phi_{\rm co}$ is the fraction of the clump mass that is in bound cores, $u \approx 10\alpha_{\rm vir}^{-1}(M_*/M)^{1/2}$ is the ratio of the escape velocity from the surface of a core to the velocity dispersion in the clump, and $\alpha_{\rm vir}$ is the virial parameter for the core, roughly its ratio of turbulent

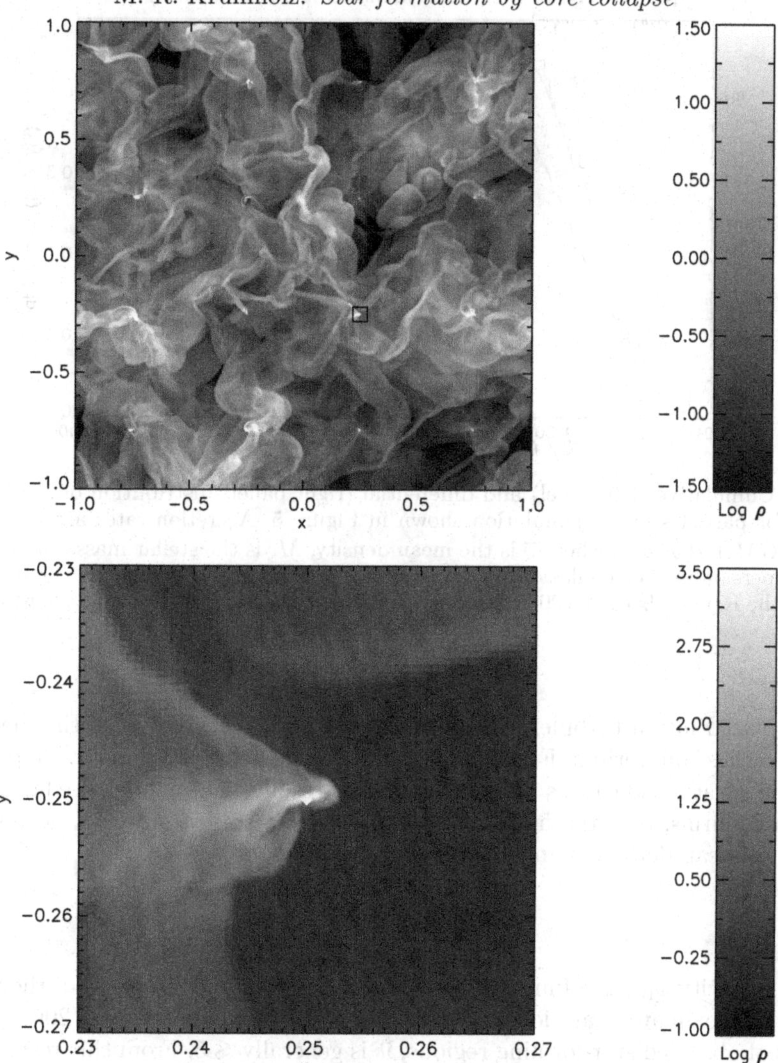

FIGURE 5. Slices through a simulation of Bondi-Hoyle accretion in a turbulent medium by Krumholz et al. (2006b), showing density in the entire simulation box (upper panel) and in a small region (indicated by the black box) around one of the accreting particles (lower panel). The position of the particle is indicated by the small white diamond. The density and length are in dimensionless units where the mean density in the box is unity, and the box extends from -1 to 1. The maximum resolution of the simulation is 8192^3.

kinetic energy to its gravitational potential energy. The significant thing to notice about this expression is that it does not approach unity unless u is quite large, which in turn only happens for virial parameters $\alpha_{\text{vir}} \ll 1$, i.e., for clumps where the turbulent velocity dispersion is small compared to that needed to prevent collapse.

Accretion of unbound gas is somewhat more complex, since to determine the accretion rate one requires a theoretical model for Bondi-Hoyle accretion in a turbulent medium. Krumholz et al. (2005b, 2006b) have developed such a theory and shown that it reproduces the results of simulations quite well. Figure 5 shows a sample of an adaptive mesh-refinement simulation in which a grid of 64 Eulerian sink particles (Krumholz et al.

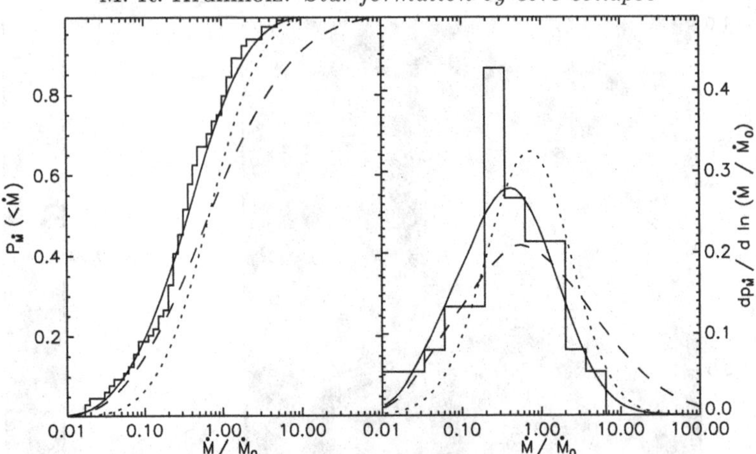

FIGURE 6. Cumulative (left panel) and differential (right panel) distribution of accretion rates measured for particles in the simulation shown in Figure 5. Accretion rates are normalized to $\dot{M}_0 \equiv 4\pi\overline{\rho}(GM_*)^2/(\sqrt{3}\sigma)^3$, where $\overline{\rho}$ is the mean density, M_* is the stellar mass, and σ is the 1D velocity dispersion in the simulation box. The histogram shows the simulation results, the solid line shows the Krumholz et al. (2006b) model, and the dashed and dotted lines show alternative models.

2004) are placed into a turbulent medium and allowed to accrete until the mean accretion rate reaches equilibrium. Figure 6 compares the model prediction for the probability distribution of accretion rates to the simulation results. The model predicts, and the simulation confirms, that the mean accretion rate for a star of mass M_* accreting from a medium of mean density $\overline{\rho}$ and 1D velocity dispersion σ is

$$\dot{M}_* 4\pi\phi_{\rm BH} \approx \overline{\rho}\frac{(GM_*)^2}{(\sqrt{3}\sigma)^3} \quad , \tag{4.2}$$

where the quantity $\phi_{\rm BH}$ is a function of the Mach number and size scale of the turbulent region, an approximate analytic form which is given in Krumholz et al. (2006b). For the properties of observed star-forming regions, it is generally $\lesssim 5$. From this result, one can compute f_M due to accretion of unbound gas in a star-forming clump of mass M:

$$f_{M-\rm BH} \approx 10\phi_{\rm BH}\alpha_{\rm vir}^{-2}(M_*/M) \quad . \tag{4.3}$$

Again, for cluster-clumps hundreds to thousands of $M_\odot$ in mass, $f_{M-\rm BH}$ can be of order unity only for $\alpha_{\rm vir} \ll 1$.

Combining the two potential sources of mass, one can derive an approximate criterion that a star-forming gas clump must satisfy in order for competitive accretion to occur within it. For seed stars of mass $0.5\,M_\odot$, this condition is

$$\alpha_{\rm vir}^2 M \lesssim 50\,M_\odot \quad . \tag{4.4}$$

Straightforward application to observed star-forming clumps shows that they are nowhere near meeting this condition, since their typical masses are many hundreds to thousands of $M_\odot$, and their observed virial parameters generally are near unity. From this, Krumholz et al. (2005d) conclude that competitive accretion does not occur in real clumps. It occurs in simulations only because there either was very little turbulence present in the initial conditions (e.g., Klessen & Burkert 2000, 2001), or the initial turbulence has decayed away (e.g., Bonnell et al. 2003), leaving the clumps sub-virial.

In light of this claim, Bonnell & Bate (2006) have re-analyzed the simulations of Bonnell et al. (2003). They argue that competitive accretion occurs for a few stars that sit at the center of collapsing regions. These regions are stagnation points of the larger turbulent flow, where the velocity dispersion is much smaller than the mean and the density much larger. Thus, although the clump as a whole is turbulent, the regions where massive stars form are effectively decoupled from the large-scale flow. Since these decoupled regions have masses much smaller than that of the entire clump, and their virial parameters are smaller than unity due to their state of global collapse, they satisfy the competitive accretion condition (4.4), and stars within them gain mass via competitive accretion.

I discuss whether such decoupled regions exist in real clumps in Section 4.2. However, one potential difficulty with the idea that competitive accretion occurs in locally detached parts of the flow is the assumption that such detached regions would fragment and still produce many stars to compete, as it does in non-radiative simulations. However, these collapsing regions bear a striking resemblance to the objects we observe as massive cores: they are bound, high-density regions with coherent velocity structures. As shown in Section 3, such objects do not fragment strongly when one includes radiative transfer. Indeed, Bate & Bonnell (2005) estimate the typical size scale of these collapsing regions as $\sim$1000 AU, well inside the effective heating radius of a single rapidly accreting star. Instead of competitive accretion, the true behavior might simply be monolithic collapse of a massive core to a single system. In this case, competitive accretion would be nothing more than core accretion simulated without a sufficiently accurate treatment of radiative transfer. Investigating that possibility will have to wait for future simulations.

4.2. *Competitive accretion and global collapse*

The Bonnell & Bate (2006) analysis of competitive accretion concurs with Krumholz et al. (2005d) that competitive accretion cannot occur under typical conditions in a star-forming clump. Instead, the possibility of competitive accretion depends on the existence of long-lasting stagnation points within which the gas and stars move together with gravity as the only significant force. The existence or non-existence of such stagnation points is something that is at least indirectly subject to observational determination. As Bonnell & Bate (2006) point out, if any source adds a significant amount of turbulent kinetic energy to the star-forming clump, such that the rate of energy injection is comparable to the rate at which energy is lost through radiative shocks, then ram pressure from flowing gas will push on gas, but not on stars. No long-lived stagnation points will exist, and gas and star velocities will become decoupled. In this case, stars will randomly sample the density and velocity field of the clump, consistent with the treatment of Krumholz et al. (2005d) and Krumholz et al. (2006b), and competitive accretion will not occur.

The most likely candidate for significant energy injection on the size scales of cluster-forming molecular clumps is feedback from protostellar outflows (Norman & Silk 1980). Observational efforts to estimate whether the kinetic energy added by outflows is significant tentatively conclude that it is (Williams et al. 2003; Quillen et al. 2005). While these results are preliminary, if they are confirmed, then competitive accretion cannot occur. One can also look for signs of large-scale collapse onto stagnation points. While there are a few examples of apparent infall signatures (Motte et al. 2005; Peretto et al. 2006) on the scale of thousand $M_{\odot}$ star-forming clumps (as opposed to onto individual protostars, which is expected whether competitive accretion occurs or not), the majority of searches for infall signatures have turned up negative (Garay 2005). Competitive accretion predicts that infall should be ubiquitous. It is, however, possible that these non-detections are due to observational confusion or lack of resolution.

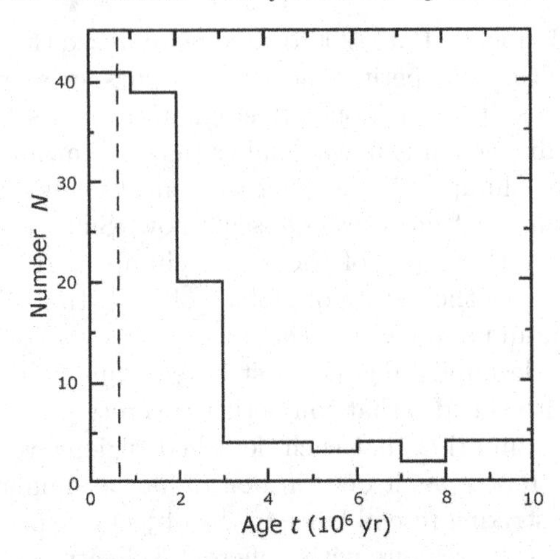

FIGURE 7. Number of stars out of a sample of 162 versus age in IC 348 (histogram), and dynamical time in the cluster (dashed vertical line). The histogram is reprinted from Palla & Stahler (2000) with permission from the authors.

A potentially more promising approach is to look for indirect signs of significant energy injection from feedback. If not, then the turbulence in protocluster gas clumps should decay rapidly (e.g., Stone et al. 1998), leading to global collapse in roughly a crossing time. Over this time, a substantial fraction of the gas should be converted into stars. We can therefore test whether competitive accretion can occur by observationally estimating the time scale of star-cluster formation and the star-formation rate in pre-cluster gas.

On the question of time scale, Elmegreen (2000) has argued that star formation proceeds in a crossing time, but a more careful examination of much of the same data by Tan et al. (2006) shows that the typical time scale is closer to 3–5 crossing times (6–10 free-fall times). There are several lines of evidence for this, including the fairly round morphologies of most protocluster clumps, the lack of strong sub-clustering apparent in young clusters such as the Orion Nebula Cluster, the estimated momentum carried by the combined protostellar outflows of forming clusters, the age spreads of stars determined by fitting to pre-main sequence evolutionary tracks, and the ages of dynamical ejection events. In contrast, in the simulation of Bonnell et al. (2003) where feedback is neglected, star formation has completely ended, and more than 50% of the gas has been accreted in one crossing time, $\lesssim 5 \times 10^5$ yr. Figure 7 shows an example of how this prediction for the time scale of star-cluster formation compares to observations: the solid line shows the observed ages of stars in IC 348, a cluster whose most massive member is a mid-B star, computed by Palla & Stahler (2000). The dashed vertical line shows the crossing time of ≈ 0.6 Myr, computed by Tan et al. (2006). If the cluster formed in global collapse, all the stars should be to the left of the line. In this case and all the others analyzed by Tan et al. (2006), the observed age spreads and formation times of clusters are strongly incompatible with the idea that star-forming clumps are in a state of global collapse, and hence are strongly incompatible with competitive accretion.

One can also examine the star-formation rate and compare to populations of dense gas clumps, as in the recent analysis by Krumholz & Tan (2006). As an example, Gao & Solomon (2004b) and Wu et al. (2005) have shown that there is a strong correlation between luminosity in the HCN(1-0) line and infrared luminosity that extends from individual cluster-forming gas clumps in the Milky Way up to entire ultraluminous infrared galaxies. Since the infrared emission traces the star-formation rate (e.g., Kennicutt

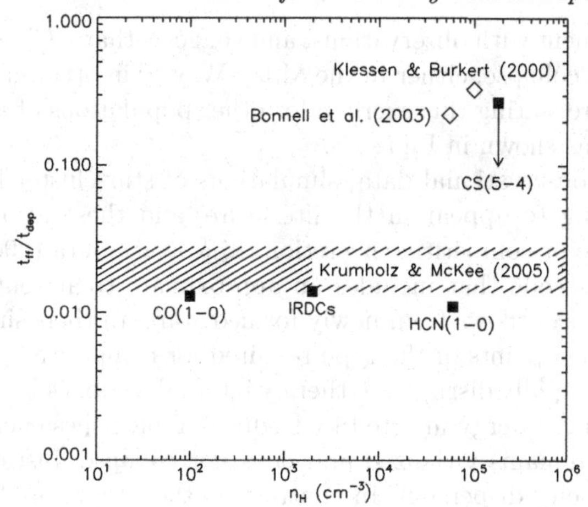

FIGURE 8. Ratio of the free-fall time to the depletion time (defined as the time required to convert all gas into stars at the current star-formation rate), estimated for gas at a given characteristic density. The plot shows observationally determined points (solid squares) for gas traced by CO(1-0) emission (giant molecular clouds), infrared dark clouds taken from the Simon et al. (2006) catalog, gas traced by HCN(1-0) emission (Gao & Solomon 2004b; Wu et al. 2005), and gas traced by CS(5-4) emission (Shirley et al. 2003). The CS(5-4) point is an upper limit. The plot also shows points from simulations (open diamonds) by Klessen & Burkert (2000) and Bonnell et al. (2003), and the Krumholz & McKee (2005) model (hatched region) applied under the assumption that star formation occurs in clouds with virial parameters $\alpha_{\rm vir} = 1$–2 and 1D Mach numbers $\mathcal{M} = 20$–40. The plot is adapted from Krumholz & Tan (2006).

1998), and HCN(1-0) emission traces the gas mass at densities $n \approx 6 \times 10^4$ H cm^{-3} (Gao & Solomon 2004a), a typical density for protocluster gas, this correlation is a direct measure of the star-formation rate in protocluster gas. Wu et al. (2005), based on observationally calibrated conversions from IR light to star-formation rate and HCN(1-0) emission to gas mass find that their correlation corresponds to a star-formation law $\dot{M}_* \approx M_{\rm HCN}/(80\ \rm Myr)$, where $M_{\rm HCN}$ is the mass of gas emitting HCN(1-0). Including some corrections to those calibrations, Tan et al. (2006) estimate a star-formation law

$$\dot{M}_* \approx \frac{M_{\rm HCN}}{17\ \rm Myr} \quad . \tag{4.5}$$

We can compare this observational law to the predictions of theoretical models. Bonnell et al. (2003) simulate a gas clump with a mean density of $n = 5.5 \times 10^4$ H cm^{-3}, almost exactly the same as the density of observed HCN clumps. They find that clusters form in a free-fall collapse in which 58% of the gas is incorporated into stars after 2.6 free-fall times. At the HCN density of 6×10^4 H cm^{-3}, the free-fall time is 0.18 Myr, so the star-formation law one would infer from the simulations is

$$\dot{M}_* \approx \frac{M_{\rm HCN}}{0.8\ \rm Myr} \quad , \tag{4.6}$$

inconsistent with the observed relation. In contrast, if one assumes that HCN-emitting gas clumps are virialized, $\alpha_{\rm vir} \sim 1.5$, and turbulent, Mach number $\mathcal{M} \sim 25$, then the Krumholz & McKee (2005) estimate for the star-formation rate predicts

$$\dot{M}_* \approx \frac{M_{\rm HCN}}{10\ \rm Myr} \quad . \tag{4.7}$$

This is in good agreement with observations, and suggests that HCN-emitting gas cannot be in a state of global collapse, either in the Milky Way or in other galaxies. Krumholz & Tan (2006) find that repeating this exercise for other populations of star-forming objects gives similar results, as shown in Figure 8.

In addition to the observational data, simulations of star-cluster formation including feedback are beginning to appear in the literature, and these also cast doubt on the idea of global collapse or competitive accretion. Li & Nakamura (2006) simulate a star-forming clump with a code that includes an approximate treatment of the energy and momentum injected by outflows from newly formed stars. In their simulations, there are no long-lived stagnation points of the type required for competitive accretion. Any such stagnation points are rapidly disrupted either by internal feedback from stars within them or by external shocking. Energy injected by feedback replenishes energy lost in radiative shocks, so the cloud remains virialized and does not collapse. Instead, it approaches a roughly constant velocity dispersion. In contrast to the results of Bonnell et al. (2003) without feedback, Li & Nakamura (2006) find that only ~6% of the mass collapses into stars per free-fall time, in reasonable agreement with both observational estimates of the cluster-formation time scale and the star-formation rate in dense gas.

In summary, the question of whether competitive accretion occurs reduces to the question of whether cluster-forming gas clumps are in a state of global collapse, and both observational estimates and simulations including feedback seem to rule out this possibility. The mass in a protostellar core is all the mass that a star will ever have.

5. Feedback and accretion

As shown in the previous sections, massive cores do not fragment strongly, nor do the stars that form within them gain significant mass from outside their parent core. There remains, however, the problem of getting mass from the core onto a star. This is potentially difficult because massive protostars have short Kelvin-Helmholtz times that enable them to reach the main sequence while they are still forming from their parent clouds (Shu et al. 1987). Once nuclear burning begins, the star will have the immense luminosity of a main sequence star of the same mass. Radiation might inhibit accretion in two different ways: first, it exerts a direct radiation pressure force on dust grains suspended in the incoming gas, and this force could be stronger than gravity. Second, ionizing photons can create an H II region around the star, within which the temperature is 10^4 K and the sound speed is 10 km s^{-1}. The H II region is at very high pressure, and may therefore be able to expand and unbind the accreting gas. If either of these feedback processes is capable of halting accretion, then the mapping from core mass to star mass will not be simple at the high mass end, and it will not be possible to explain the properties of stars in terms of the properties of cores.

5.1. *Radiation pressure feedback*

Early spherically symmetric calculations of the effects of radiation pressure on accretion flows onto massive stars found that radiation becomes stronger than gravity, and halts accretion at masses of 20–40 $M_\odot$ (Kahn 1974; Wolfire & Cassinelli 1987) for typical Galactic metallicities. More recent work has relaxed this limit by considering the effects of accretion disks, which allow most of the mass to accrete from within the optically thick disk, where it is shielded from radiation pressure (Nakano 1989; Nakano et al. 1995; Jijina & Adams 1996). Furthermore, rotational flattening can cause the radiation field itself to become asymmetric, as more radiation escapes in the lower-density polar direction than through the higher-density equatorial plane. Cylindrically symmetric

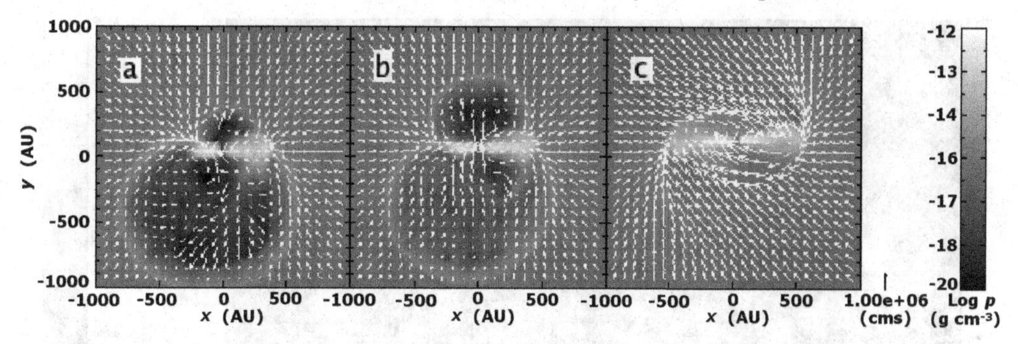

FIGURE 9. Slices in the XZ plane showing the density (grayscale) and velocity (arrows) in a simulation of the collapse of a 100 $M_{\odot}$ core by Krumholz et al. (2005a). The panels show times of (a) 1.5×10^4 yr, (b) 1.65×10^4 yr, and (c) 2.0×10^4 yr after the start of the simulation. The corresponding stellar masses at those times are 21.3 $M_{\odot}$, 22.4 $M_{\odot}$, and 25.7 $M_{\odot}$. Note that these masses are considerably larger than those of the turbulent runs discussed in Section 3 at the same times, primarily due to the steeper initial density profile assumed in these simulations.

radiation-hydrodynamic simulations show that this flashlight effect allows accretion to slightly more than 40 $M_{\odot}$ (Yorke & Sonnhalter 2002). These simulations also reveal that forming a massive star is easier in simulations that include a multifrequency treatment of the radiation field than in those that use a gray approximation, probably because the collimation of the radiation field is reduced by the gray approximation.

More recent 3D simulations show that the flashlight effect persists into 3D, and is actually enhanced by a qualitatively new effect. Figure 9 shows time slices from a radiation-hydrodynamic simulation of the collapse of a 100 $M_{\odot}$, 0.1 pc-radius core with an initial density profile $\rho \propto r^{-2}$, performed by Krumholz et al. (2005a, 2007). The initial core has a temperature of 40 K and no turbulence, just overall solid-body rotation at a rate such that the rotational kinetic energy is 2% of the gravitational potential energy, a typical rotation rate observed for low-mass cores (Goodman et al. 1993). The simulation shows that, at stellar masses $\lesssim 17\,M_{\odot}$, stellar radiation is too weak to reverse inflow, and the gas falls onto an accretion disk and accretes in a steady, cylindrically symmetric flow. At about 17 $M_{\odot}$, stellar radiation begins to reverse inflow and drive radiation bubbles above and below the accretion disk. The interiors of these bubbles are optically thin and very low density, while the walls reach densities $\gtrsim 10^{10}$ H cm^{-3}. Gas reaching the bubbles ceases to move radially inward. Instead, it travels along the bubble wall until it falls onto the accretion disk. At that point, it is shielded from stellar radiation and is able to accrete onto the star. As shown in Figure 10, the bubbles also collimate radiation, increasing the flux in the polar direction and decreasing it in the equatorial plane. At the time shown in Figure 10a, the radiation flux just outside the radiation bubble in the polar direction exceeds that in the equatorial plane at the same radius by more than an order of magnitude.

The bubbles expand asymmetrically with respect to both the polar axis and the equatorial plane due to an instability. The direction in which a bubble has expanded farthest and cleared out a low optical depth region represents the path of least resistance for radiation leaving the star to escape to infinity. As a result, radiation is collimated in whichever direction the bubble has expanded farthest. In turn, this causes the radiation pressure force to be largest in that direction, which amplifies the rate of bubble expansion. Figure 9a shows this asymmetric expansion phase. As the simulation progresses, the instability becomes more violent and the bubbles start to collapse, as shown in Figure 9b. The collapse is due to radiation Rayleigh-Taylor instability: the gas is a heavy

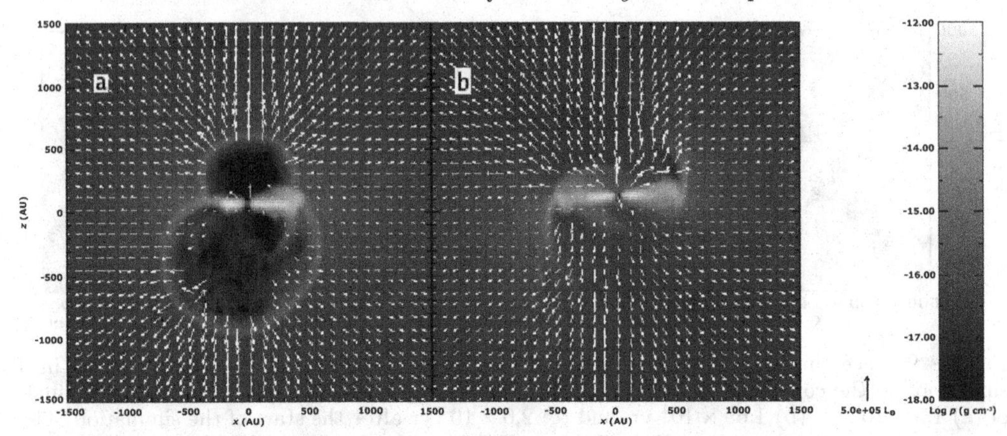

FIGURE 10. Slices in the XZ plane showing the density (grayscale) and radiation flux multiplied by $4\pi r^2$ (arrows) in the simulation shown in Figure 9. For clarity, the flux vectors in the optically thin bubble interior in panel (a) have been omitted. The times shown in panels (a) and (b) are the same as those shown in panels (b) and (c) of Figure 9.

fluid, and the radiation is a light fluid trying to hold it up, an unstable configuration. Once the bubbles collapse, as shown in Figure 9c, infalling gas is deflected by radiation onto the dense remnants of the bubble walls, which are optically very thick, and therefore self-shielding against radiation. The gas travels along the walls onto the disk and then accretes. In this configuration the remnant walls collimate the accretion, as shown in Figure 10b. The radiation flux behind the walls is very small, so gas accreting from that direction feels very little radiation pressure force.

Simulations to date have reached $\approx 34\,M_\odot$ in this manner, with 5–10 $M_\odot$ more in the accretion disk and no sign of a reversal of infall. Note that these calculations use gray radiative transfer, for which case Yorke & Sonnhalter (2002) found an upper limit of about $20\,M_\odot$ for the stellar mass. This suggests that multifrequency 3D calculations would show even easier accretion.

Krumholz et al. (2005c) point out a second mechanism for overcoming radiation-pressure feedback. Massive protostars have outflows that appear to be simply scaled-up versions of the outflows from lower-mass stars (see Beuther & Shepherd 2005 for a recent review), or that are somewhat wider in angle for the most massive, O-type stars. These outflows can have a very strong effect on the radiation field. Outflows are launched from the inner accretion disk around the star, where the gas is so hot that any dust grains present within it sublime. As a result, outflows contain almost no dust when they are launched, and simple calculations show that due to the velocities of $\gtrsim$500 km s^{-1} with which they flow, there is insufficient time for large dust grains to reform before the outflowing gas reaches distances $\gtrsim$0.1 pc from the source star. This makes the outflow cavities optically thin, so they can collimate radiation and very effectively carry it out of the optically thick accreting envelope. Krumholz et al. (2005c) use Monte Carlo radiative-transfer calculations to explore this effect, and show that it can reduce the radiation pressure force on infalling gas by an order of magnitude. This can in turn shift the radiation pressure force from a regime where it is stronger than the gravitational force to one where it is weaker than gravity, allowing accretion to continue where it might have been halted without the protostellar outflow.

It is unclear how this effect will interact with radiation pressure bubbles, since the simulations in which the bubbles appear do not yet include a model for protostellar outflows. Nonetheless, the overall conclusion one may draw from the two effects is that radiation pressure cannot easily halt accretion. There are at least two mechanisms which are capable of rendering the radiation field aspherical, allowing accretion over the range of solid angle where the radiation pressure force is reduced. Gravity is very effective at channeling gas into those solid angles, and then onto the star.

5.2. *Ionization*

The second form of radiative feedback that accretion onto a massive star must overcome is ionization, first recognized as a problem by Larson & Starrfield (1971) and Yorke & Kruegel (1977). If the ionizing photons from a massive protostar are able to reach gas at a distance from the star where the escape speed is $\lesssim$10 km s^{-1}, then the ionized gas will be unbound and may escape to infinity. However, Walmsley (1995) shows that sufficiently rapid accretion can avoid this problem by keeping the ionized region quenched against the stellar surface. For spherically symmetric infall, if the accretion rate is larger than

$$\dot{M}_{\rm crit} = \sqrt{\frac{4\pi G M_* S m_H^2}{\alpha^{(2)}}} \approx 2\times 10^{-5}\left(\frac{M_*}{10\,M_\odot}\right)^{1/2}\left(\frac{S}{10^{49}\ \mathrm{s}^{-1}}\right)^{1/2} M_\odot\ \mathrm{yr}^{-1}\ , \quad (5.1)$$

where S is the ionizing luminosity of the star in photons s^{-1} and $\alpha^{(2)}$ is the recombination coefficient to excited levels of hydrogen, then all ionizing photons will be absorbed near the stellar surface. The escape velocity there is much larger than 10 km s^{-1}, so this ionized gas will be confined by stellar gravity and accretion may continue unimpeded.

Observational estimates of H II region lifetimes indeed appear to require confinement of this sort. Ultracompact H II regions, roughly those $\lesssim$0.1 pc in radius, have a dynamical expansion time $\lesssim$0.1 pc/10 km s$^{-1} \approx 10^4$ yr, but the number of observed ultracompact H II regions implies that their lifetimes must be closer to 10^5 yr (Wood & Churchwell 1989; Kurtz et al. 1994). Moreover, some observed systems show signs of both accretion and the presence of an ultracompact H II region (e.g., Hofner et al. 1999; Sollins et al. 2005), providing direct evidence that accretion can continue past H II region formation.

Hoare et al. (2007) gives a recent review of ultracompact H II regions, and Keto (2007) attempts to construct a unified theoretical framework within which one can understand the relationship between H II regions and accretion (though see Beuther et al. 2007). In this model, the behavior of an H II region is controlled by the relative sizes of the ionization radius within which the ionization and recombination rates balance and all ionizing photons are absorbed, the gravitational radius where a spherical accretion flow accelerates from subsonic to supersonic relative to the sound speed in ionized gas, and the disk radius, at which angular momentum in accreting gas causes infall to become aspherical and a disk to form. This model can roughly explain observed H II region morphologies. One of its consequences is that if the ionization radius is smaller than the gravitational radius, the H II region will remain confined and accretion will continue unimpeded. For the accretion rates of $10^{-4} - 10^{-3}\,M_\odot$ yr^{-1}, one expects that given the conditions in massive cores, massive protostars will remain in this regime until they have exhausted essentially all of their available gas. Thus, ionization feedback does not present a significant barrier to accretion.

6. Conclusions and future directions

6.1. *Massive-star formation from massive cores*

The results summarized in the previous sections combine to give a unified scenario of massive-star formation: massive stars form by the gravitational collapse of massive cores.

We observe these cores, from a few tenths to a few hundred $M_{\odot}$ in mass, in the dense molecular clumps where star clusters form. Their mass distribution matches the stellar IMF, and their spatial distribution is roughly consistent with the spatial distribution of massive stars in clusters. When massive cores collapse, they do not fragment strongly, because radiation feedback from the high accretion rates they produce warms their inner regions, raising the Jeans mass and suppressing fragmentation. As a result, all of their mass falls onto one or a few stars. This occurs even though the cores are turbulent and contain complex density and velocity structures.

The stars that form in massive cores do not gain significant mass from outside their parent cores, because gas that is not gravitationally bound to them at birth is too turbulent for significant accretion to occur. The mostly likely agent for keeping the gas turbulent is feedback from protostellar outflows, and both observational estimates and simulations suggest that this mechanism works. Regardless of the agent, though, observations of the age spread of young clusters and the star-formation rate in pre-cluster gas unambiguously require that star clusters form from gas that is not in free-fall collapse.

The gas falling onto a massive protostar at the center of a massive core must, in the face of significant radiation, accrete feedback in the form of radiation pressure on dust grains suspended in the inflowing gas, and ionizing radiation that generates high-pressure ionized gas. However, neither mechanism substantially impedes accretion. Radiation escapes through low optical-depth channels, which either form spontaneously due to radiation-hydrodynamic instabilities, or are created by protostellar outflows. Gas accretes through high optical-depth channels onto an accretion disk, where it is shielded from protostellar radiation. At the same time, the accretion flow is capable of absorbing all the ionizing stellar photons near the star, where the escape speed is larger than the ionized gas sound speed. As a result, formation of an H II region is either suppressed entirely, or the ionized region is kept close to the star where gravity confines it and prevents it from expanding. Accretion continues through the ionized region.

Because gas from a massive core accretes onto one or two stars with no significant impedance from feedback, and those stars do not gain a significant amount of mass from outside their parent cores, there is a direct mapping from the properties of cores to the properties of stars. Consequently the mass and spatial distributions of young clusters are direct imprints of the mass and spatial distributions of cores in their gas-phase progenitors, possibly with some additional redistribution of mass due to dynamical mass segregation over the several crossing times that it takes for a cluster to assemble. Observables in the gas phase connect directly to observables in the stellar phase.

6.2. *Problems for the future*

While this scenario explains a great deal of what we observe about massive stars, it also omits a number of physical effects that a more complete picture should include, and it needs more rigorous testing against observation. In this section, I discuss the prospects for improvement in these areas in the next few years.

6.2.1. *Magnetic fields*

None of the models of massive-star formation proposed to date have included the effects of magnetic fields on the dynamics of the collapsing gas. This is partly due to a lack of observational information about magnetic fields in massive-star-forming regions. Crutcher (2005) summarizes the state of observations, and concludes that magnetic fields are marginally dynamically significant in massive-star-forming cores. However, this determination is extremely uncertain due to the difficulty of interpreting observational indicators of magnetic-field strength. Part of this uncertainty comes from geometry. The critical magnetic flux required to hold up a core against gravity depends on the core shape and mass distribution, and differing assumptions about the shape can produce qualitatively different conclusions about how the magnetic-field strength compares to that required for it to be dynamically significant (Bourke et al. 2001).

Beyond the geometric effect, measurements of field strengths themselves are quite uncertain. The two most commonly used techniques for determining magnetic-field strengths in molecular gas are Zeeman splitting of OH or CN lines and the Chandrasekhar-Fermi method, in which one uses the dispersion of polarization vectors seen in dust emission to estimate how the ratio of magnetic to kinetic energy density. Both of these techniques suffer from major systematic errors. Zeeman splitting averages the magnetic field along the line of sight, potentially washing out the signal, and is also only sensitive to regions where the observed species are present. Since many species freeze out onto dust grains in the densest parts of cores (Tafalla et al. 2002), Zeeman measurements may reveal more about magnetic field strengths in the diffuse gas around cores than in cores themselves. For Chandrasekhar-Fermi measurements, one also faces an uncertainty about where along the line of sight the observed polarized light is being emitted, and whether one is really measuring field in the core or in its outskirts. This uncertainty is made even worse by the fact that the magnetic field one deduces for a given dispersion of polarization vectors depends on the gas density, so uncertainty in the gas properties in the emitting region translates directly into uncertainty about the magnetic field strength.

The level of uncertainty is driven home by the result that, in at least some regions where both have been used, the inferred magnetic field strengths differ by a factor of $\sim$100 (e.g., Crutcher 1999; Lai et al. 2001, 2002). While in principle this is possible because Zeeman splitting measures the field along the line of sight and Chandrasekhar-Fermi measures the field in the plane of the sky, it seems implausible that the magnetic field in multiple sources could be so perfectly aligned perpendicular to the line of sight. A more likely explanation is that there are large systematic errors in one or both techniques, or that the two techniques are simply measuring the field in different regions.

In the absence of more conclusive observational data, it is hard to know whether it is important to include magnetic fields in massive-star-formation models or not. In principle, one could make models both with and without dynamically significant magnetic fields to explore their effects. However, at present magnetohydrodynamics codes using either smoothed particle hydrodynamics or adaptive mesh refinement—one of which is required to achieve the requisite dynamic range to simulate star formation—are still in their infancy (e.g., Price & Monaghan 2004; Crockett et al. 2005). There are no codes past the experimental stage that include both MHD and radiative transfer, which the discussion above demonstrates is absolutely crucial. The question of the influence of magnetic fields will therefore have to wait for progress in both code development and observations.

6.2.2. *Improved simulation physics*

In addition to magnetic fields, there are numerous other physical effects that the current generation of simulations ignore. It is chastening to realize that for almost every new process in physics that has been added to simulations of massive-star formation, a qualitatively new and unexpected phenomenon has been revealed:

- adding radiative transfer to simulations of turbulent cores shows that fragmentation is much weaker than one might suppose from purely hydrodynamic calculations (Section 3);

• moving from gray to multifrequency radiative transfer in two-dimensional simulations doubles the maximum stellar mass formed due to enhancement of radiation beaming (Yorke & Sonnhalter 2002);

• moving from two to three dimensions with gray radiative transfer reveals the formation of unstable, asymmetric radiation bubbles (Krumholz et al. 2005a, and Section 5);

• including protostellar outflows in models greatly reduces the efficacy of radiation feedback on the scale of individual cores (Krumholz et al. 2005c); and

• changes the dynamics of star-cluster formation from rapid collapse and competitive accretion to quasi-equilibrium behavior where turbulent decay is balanced by injection of kinetic energy by outflows (Li & Nakamura 2006).

This pattern suggests that there may be new effects yet to be discovered as simulation physics improves. An important corollary here is that we have very likely reached the limit of what pure gravity plus hydrodynamics simulations with no feedback can teach us. Major work in the future will not come from simply running more and bigger simulations of the same type, but from adding new physics to the problem.

The next logical steps in modeling massive-star formation include extending multifrequency radiative transfer to three dimensions, using radiation-hydrodynamic codes to simulate the formation of star clusters instead of just individual stars, and including protostellar outflows in radiation-hydrodynamic simulations of massive cores. These are all relatively straightforward extensions of existing techniques. Their primary limit is computer power, since gray radiation-hydrodynamic simulations in three dimensions take months on available supercomputers. However, Moore's Law will help in this regard, and these simulations should become feasible in the next few years. More difficult are the problems for which not only computer power, but also our current simulation methods are inadequate, and for which we will have to develop new techniques. One problem in this category is putting more realistic dust physics into three-dimensional models, including coagulation, shattering, and differential motion of grains and gas, which may substantially change opacities (Preibisch et al. 1995; Sonnhalter et al. 1995; Suttner et al. 1999). Another is radiative transfer that goes beyond flux-limited diffusion (e.g., Hayes & Norman 2003). Improvements in these areas are likely to take somewhat longer, perhaps becoming reasonable on the time scale of five or more years.

6.2.3. *Connecting models to observations*

A third area in which substantial progress must be made is connecting simulation and theoretical models to observations. There has been a fair amount of work on relating simulation results to statistical diagnostics of young clusters such as the IMF, and binary and brown-dwarf properties. However, these indicators are quite difficult to use as a means of distinguishing theories because they focus on the outcome only after the star cluster is fully formed, and there appear to be multiple mechanisms capable of producing the same outcome (e.g., Padoan & Nordlund 2002 and Bonnell et al. 2003 on the origin of the IMF, or Bonnell & Bate 2005 and Krumholz et al. 2006a on the binarity properties of massive stars). In addition, observations of the results of star formation are difficult to connect to models because we do not know the properties of the progenitor gas system for a given stellar system. It is often possible to reproduce a given stellar system simply by tuning the properties of the assumed progenitor.

Better diagnostics are likely to come from comparing the results of simulations and theoretical models to observations of systems that are still actively forming stars. Masers provide one potential connection, since they make it possible to trace gas properties and kinematics on very small scales (e.g., Torrelles et al. 2001, who find expanding bubbles traced by masers, and Greenhill et al. 2004 who detect a disk and an outflow cavity traced by masers). These may be particularly useful for tracing radiation-driven bubbles, since the density of 10^9–10^{10} cm^{-3} and temperature of a few hundred K of bubble walls is favorable for maser emission. As discussed elsewhere in this volume, observational searches for accretion disks around massive stars, which are expected for massive-star formation by core collapse, but probably not for collisional formation, are another example. Morphologies of warm gas, as seen in high-resolution observations of "hot cores" (e.g., Garay 2005), provide another potential point of comparison to simulations. One can carry such comparisons even further by post-processing the simulation results with radiative transfer codes to produce simulated line profiles and intensity maps. One can also look for telltale signs of star formation by collisions, such as infrared flares (Bally & Zinnecker 2005) or embedded clusters in which the stellar density reaches the $\sim 10^7$ pc^{-3} required for efficient collisions—such high densities should produce distinctive spectral energy distributions that peak at wavelengths approaching 100 μm (Chakrabarti & McKee 2005, 2008). Yet another technique is to use extragalactic observations, in which one can observe the entire galactic population of a given sort of object (e.g., HCN-emitting clouds Gao & Solomon 2004b), and then make arguments based on population statistics (e.g., Krumholz & Tan 2006). Any or all off these are likely to be far more definitive than comparisons to stellar systems.

Prospects for improvement in this area of massive-star formation theory are quite good. Current and planned millimeter instruments such as PdBI, the SMA, CARMA, and ALMA will allow higher resolution examination of more distant, embedded star-forming systems than has heretofore been possible, and surveys or both Galactic and extra-Galactic star-forming regions with *Spitzer* are providing large databases that we can use to test models on the level of populations. These sources of data have already begun to let us distinguish between models of massive-star formation, and will provide even more powerful diagnostics in the next few years.

I thank H. Beuther, R. I. Klein, Z.-Y. Li, C. F. McKee, F. Nakamura, J. M. Stone, and J. C. Tan for helpful discussions. This work was supported: NASA through Hubble Fellowship grant HSF-HF-01186 awarded by the Space Telescope Science Institute, which is operated by the Association of Universities for Research in Astronomy, Inc., for NASA under contract NAS 5-26555; the Arctic Region Supercomputing Center; the National Energy Research Scientific Computing Center, which is supported by the Office of Science of the US Department of Energy under contract DE-AC03-76SF00098, through ERCAP grant 80325; the NSF San Diego Supercomputer Center through NPACI program grant UCB267; and the US Department of Energy at the Lawrence Livermore National Laboratory under contract W-7405-Eng-48.

REFERENCES

Bally, J. & Zinnecker, H. 2005 *AJ* **129**, 2281.

Bate, M. R. & Bonnell, I. A. 2005 *MNRAS* **356**, 1201.

Beuther, H., Churchwell, E. B., McKee, C. F., & Tan, J. C. 2007. In *Protostars and Planets V* (eds. B. Reipurth, D. Jewitt, & K. Keil). p. 165. University of Arizona Press.

Beuther, H. & Schilke, P. 2004 *Science* **303**, 1167.

Beuther, H. & Shepherd, D. 2005. In *Cores to Clusters: Star Formation with Next Generation Telescopes* (eds. M. S. Nanda Kumar, M. Tafalla, & P. Caselli), reprinted from *Astrophysics and Space Science Library*, Vol. 324, p. 105. Springer.

Beuther, H., Sridharan, T. K., & Saito, M. 2005 *ApJ* **634**, L185.

Beuther, H., Zhang, Q., Sridharan, T. K., Lee, C.-F., & Zapata, L. A. 2006 *A&A* **454**, 221.

Bonnell, I. A. & Bate, M. R. 2005 *MNRAS* **362**, 915.
Bonnell, I. A. & Bate, M. R. 2006 *MNRAS* **370**, 488.
Bonnell, I. A., Bate, M. R., & Vine, S. G. 2003 *MNRAS* **343**, 413.
Bonnell, I. A. & Davies, M. B. 1998 *MNRAS* **295**, 691.
Bonnell, I. A., Larson, R. B., & Zinnecker, H. 2007. In *Protostars and Planets V* (eds. B. Reipurth, D. Jewitt, & K. Keil). p. 149. University of Arizona Press.
Bourke, T. L., Myers, P. C., Robinson, G., & Hyland, A. R. 2001 *ApJ* **554**, 916.
Carey, S. J., Feldman, P. A., Redman, R. O., Egan, M. P., MacLeod, J. M., & Price, S. D. 2000 *ApJ* **543**, L157.
Chabrier, G. 2003 *PASP* **115**, 763.
Chakrabarti, S. & McKee, C. F. 2005 *ApJ* **631**, 792.
Chakrabarti, S. & McKee, C. F. 2008 *ApJ* **683**, 693.
Clark, P. C. & Bonnell, I. A. 2006 *MNRAS* **368**, 1787.
Crockett, R. K., Colella, P., Fisher, R. T., Klein, R. I., & McKee, C. F. 2005 *JCP* **203**, 422.
Crutcher, R. M. 1999 *ApJ* **520**, 706.
Crutcher, R. M. 2005. In *Massive Star Birth: A Crossroads of Astrophysics* (eds. R. Cesaroni, M. Felli, E. Churchwell, & M. Walmsley). IAU Symposium 227, p. 98. Cambridge University Press.
Dobbs, C. L., Bonnell, I. A., & Clark, P. C. 2005 *MNRAS* **360**, 2.
Dubinski, J., Narayan, R., & Phillips, T. G. 1995 *ApJ* **448**, 226.
Egan, M. P., Shipman, R. F., Price, S. D., Carey, S. J., Clark, F. O., & Cohen, M. 1998 *ApJ* **494**, L199.
Elmegreen, B. G. 2000 *ApJ* **530**, 277.
Gao, Y. & Solomon, P. M. 2004a *ApJS* **152**, 63.
Gao, Y. & Solomon, P. M. 2004b *ApJ* **606**, 271.
Garay, G. 2005. In *Massive Star Birth: A Crossroads of Astrophysics* (eds. R. Cesaroni, M. Felli, E. Churchwell, & M. Walmsley). IAU Symposium 227, p. 86. Cambridge University Press.
Goodman, A. A., Benson, P. J., Fuller, G. A., & Myers, P. C. 1993 *ApJ* **406**, 528.
Greenhill, L. J., Reid, M. J., Chandler, C. J., Diamond, P. J., & Elitzur, M. 2004. In *Star Formation at High Angular Resolution* (eds. M. Burton, R. Jayawardhana, & T. Bourke). IAU Symposium 221, p. 155. Cambridge University Press.
Hayes, J. C. & Norman, M. L. 2003 *ApJS* **147**, 197.
Hillenbrand, L. A. & Hartmann, L. W. 1998 *ApJ* **492**, 540.
Hoare, M. G., Kurtz, S. E., Lizano, S., Keto, E., & Hofner, P. 2007. In *Protostars and Planets V* (eds. B. Reipurth, D. Jewitt, & K. Keil). p. 181. University of Arizona Press.
Hofner, P., Peterson, S., & Cesaroni, R. 1999 *ApJ* **514**, 899.
Howell, L. H. & Greenough, J. A. 2003 *JCP* **184**, 53.
Huff, E. M. & Stahler, S. 2006 *ApJ* **644**, 355.
Jijina, J. & Adams, F. C. 1996 *ApJ* **462**, 874.
Johnstone, D., Fich, M., Mitchell, G. F., & Moriarty-Schieven, G. 2001 *ApJ* **559**, 307.
Kahn, F. D. 1974 *A&A* **37**, 149.
Kennicutt, R. C. 1998 *ARA&A* **36**, 189.
Keto, E. 2007 *ApJ* **666**, 976.
Klein, R. I. 1999 *J. Comp. App. Math.* **109**, 123.
Klessen, R. S. & Burkert, A. 2000 *ApJS* **128**, 287.
Klessen, R. S. & Burkert, A. 2001 *ApJ* **549**, 386.
Kroupa, P. 2002. In *Modes of Star Formation and the Origin of Field Populations* (eds. E. K. Grebel & W. Brandner), ASP Conf. Ser. 285, p. 86. Astronomical Society of the Pacific.
Krumholz, M. R., Klein, R. I., & McKee, C. F. 2005a. In *Massive Star Birth: A Crossroads of Astrophysics* (eds. R. Cesaroni, M. Felli, E. Churchwell, & M. Walmsley). IAU Symposium 227, p. 231. Cambridge University Press.
Krumholz, M. R., Klein, R. I., & McKee, C. F. 2007 *ApJ* **656**, 959.
Krumholz, M. R., Matzner, C. D., & McKee, C. F. 2006a *ApJ* **653**, 361.
Krumholz, M. R. & McKee, C. F. 2005 *ApJ* **630**, 250.

Krumholz, M. R., McKee, C. F., & Klein, R. I. 2004 *ApJ* **611**, 399.
Krumholz, M. R., McKee, C. F., & Klein, R. I. 2005b *ApJ* **618**, 757.
Krumholz, M. R., McKee, C. F., & Klein, R. I. 2005c *ApJ* **618**, L33.
Krumholz, M. R., McKee, C. F., & Klein, R. I. 2005d *Nature* **438**, 332.
Krumholz, M. R., McKee, C. F., & Klein, R. I. 2006b *ApJ* **638**, 369.
Krumholz, M. R. & Tan, J. C. 2006 *ApJ* **654**, 304.
Kurtz, S., Churchwell, E., & Wood, D. O. S. 1994 *ApJS* **91**, 659.
Lai, S., Crutcher, R. M., Girart, J. M., & Rao, R. 2001 *ApJ* **561**, 864.
Lai, S., Crutcher, R. M., Girart, J. M., & Rao, R. 2002 *ApJ* **566**, 925.
Larson, R. B. & Starrfield, S. 1971 *A&A* **13**, 190.
Li, P. S., Norman, M. L., Mac Low, M., & Heitsch, F. 2004 *ApJ* **605**, 800.
Li, Z.-Y. & Nakamura, F. 2006 *ApJ* **640**, L187.
Matzner, C. D. & McKee, C. F. 2000 *ApJ* **545**, 364.
McKee, C. F. & Tan, J. C. 2003 *ApJ* **585**, 850.
Menten, K. M., Pillai, T., & Wyrowski, F. 2005. In *Massive Star Birth: A Crossroads of Astrophysics* (eds. R. Cesaroni, M. Felli, E. Churchwell, & M. Walmsley). IAU Symposium 227, p. 23. Cambridge University Press.
Motte, F., Andre, P., & Neri, R. 1998 *A&A* **336**, 150.
Motte, F., Bontemps, S., Schilke, P., Lis, D. C., Schneider, N., & Menten, K. M. 2005. In *Massive Star Birth: A Crossroads of Astrophysics* (eds. R. Cesaroni, M. Felli, E. Churchwell, & M. Walmsley). IAU Symposium 227, p. 151. Cambridge University Press.
Mueller, K. E., Shirley, Y. L., Evans, N. J., & Jacobson, H. R. 2002 *ApJS* **143**, 469.
Nakano, T. 1989 *ApJ* **345**, 464.
Nakano, T., Hasegawa, T., & Norman, C. 1995 *ApJ* **450**, 183.
Norman, C. & Silk, J. 1980 *ApJ* **238**, 158.
Onishi, T., Mizuno, A., Kawamura, A., Tachihara, K., & Fukui, Y. 2002 *ApJ* **575**, 950.
Padoan, P. & Nordlund, Å. 2002 *ApJ* **576**, 870.
Palla, F. & Stahler, S. W. 2000 *ApJ* **540**, 255.
Peretto, N., André, P., & Belloche, A. 2006 *A&A* **445**, 979.
Pillai, T., Wyrowski, F., Carey, S. J., & Menten, K. M. 2006a *A&A* **450**, 569.
Pillai, T., Wyrowski, F., Menten, K. M., & Krügel, E. 2006b *A&A* **447**, 929.
Plume, R., Jaffe, D. T., Evans, N. J., Martin-Pintado, J., & Gomez-Gonzalez, J. 1997 *ApJ* **476**, 730.
Preibisch, T., Sonnhalter, C., & Yorke, H. W. 1995 *A&A* **299**, 144.
Price, D. J. & Monaghan, J. J. 2004 *MNRAS* **348**, 123.
Quillen, A. C., Thorndike, S. L., Cunningham, A., Frank, A., Gutermuth, R. A., Blackman, E. G., Pipher, J. L., & Ridge, N. 2005 *ApJ* **632**, 941.
Rathborne, J. M., Jackson, J. M., Chambers, E. T., Simon, R., Shipman, R., & Frieswijk, W. 2005 *ApJ* **630**, L181.
Rathborne, J. M., Jackson, J. M., & Simon, R. 2006 *ApJ* **641**, 389.
Reid, M. A. & Wilson, C. D. 2005 *ApJ* **625**, 891.
Reid, M. A. & Wilson, C. D. 2006a *ApJ* **644**, 990.
Reid, M. A. & Wilson, C. D. 2006b *ApJ* **650**, 970.
Shirley, Y. L., Evans, N. J., Young, K. E., Knez, C., & Jaffe, D. T. 2003 *ApJS* **149**, 375.
Shu, F. H., Adams, F. C., & Lizano, S. 1987 *ARA&A* **25**, 23.
Simon, R., Jackson, J. M., Rathborne, J. M., & Chambers, E. T. 2006 *ApJ* **639**, 227.
Sollins, P. K., Zhang, Q., Keto, E., & Ho, P. T. P. 2005 *ApJ* **624**, L49.
Sonnhalter, C., Preibisch, T., & Yorke, H. W. 1995 *A&A* **299**, 545.
Sridharan, T. K., Beuther, H., Saito, M., Wyrowski, F., & Schilke, P. 2005 *ApJ* **634**, L57.
Stahler, S. W., Palla, F., & Ho, P. T. P. 2000. In *Protostars and Planets IV* (eds. V. Mannings, A. P. Boss, & S. S. Russell). p. 327. University of Arizona Press.
Stanke, T., Smith, M. D., Gredel, R., & Khanzadyan, T. 2006 *A&A* **447**, 609.
Stone, J. M., Ostriker, E. C., & Gammie, C. F. 1998 *ApJ* **508**, L99.

Suttner, G., Yorke, H. W., & Lin, D. N. C. 1999 *ApJ* **524**, 857.

Tafalla, M., Myers, P. C., Caselli, P., Walmsley, C. M., & Comito, C. 2002 *ApJ* **569**, 815.

Tan, J. C., Krumholz, M. R., & McKee, C. F. 2006 *ApJ* **641**, L121.

Testi, L. & Sargent, A. I. 1998 *ApJ* **508**, L91.

Tilley, D. A. & Pudritz, R. E. 2004 *MNRAS* **353**, 769.

Torrelles, J. M., Patel, N. A., Gómez, J. F., Ho, P. T. P., Rodríguez, L. F., Anglada, G., Garay, G., Greenhill, L., Curiel, S., & Cantó, J. 2001 *Nature* **411**, 277.

Truelove, J. K., Klein, R. I., McKee, C. F., Holliman, J. H., Howell, L. H., & Greenough, J. A. 1997 *ApJ* **489**, L179.

Truelove, J. K., Klein, R. I., McKee, C. F., Holliman, J. H., Howell, L. H., Greenough, J. A., & Woods, D. T. 1998 ApJ **495**, 821.

Walmsley, M. 1995. In *Circumstellar Disks, Outflows and Star Formation.* Revista Mexicana de Astronomia y Astrofisica Serie de Conferencias, Vol. 1, p. 137.

Williams, J. P., Plambeck, R. L., & Heyer, M. H. 2003 *ApJ* **591**, 1025.

Wolfire, M. G. & Cassinelli, J. P. 1987 *ApJ* **319**, 850.

Wood, D. O. S. & Churchwell, E. 1989 *ApJS* **69**, 831.

Wu, J., Evans, N. J., Gao, Y., Solomon, P. M., Shirley, Y. L., & Vanden Bout, P. A. 2005 *ApJ* **635**, L173.

Yonekura, Y., Asayama, S., Kimura, K., Ogawa, H., Kanai, Y., Yamaguchi, N., Barnes, P. J., & Fukui, Y. 2005 *ApJ* **634**, 476.

Yorke, H. W. & Kruegel, E. 1977 *A&A* **54**, 183.

Yorke, H. W. & Sonnhalter, C. 2002 *ApJ* **569**, 846.

Observations of massive-star formation

By NIMESH A. PATEL

Harvard-Smithsonian Center for Astrophysics, 60 Garden Street, Cambridge, MA 02138, USA

High-mass stars form in deeply embedded cores with very high visual extinction. Such star-forming regions are typically located at distances >1 kpc from the Sun. Radio interferometric observations are hence vital for studying such regions at spatial resolutions of <1000 AU. I will review radio observations of high-mass young stellar objects in our Galaxy, with emphasis on recent results from the Submillimeter Array. There now exists a large sample of sources which represent the earliest stages of high-mass star formation. Radio observations of these sources in dust continuum and molecular line emission have shown that they share many characteristics with low-mass star formation. Stars with masses up to $\sim 20\,M_{\odot}$ may form via the disk-accretion mechanism instead of merging of lower-mass stars. Several questions regarding masses and stability of such disks still remain outstanding, such as driving mechanisms of the outflows, and multiplicity of sources. Detailed observations of higher-mass stars, which are at >2 kpc, will be possible with the next generation of radio interferometers, such as the Atacama Large Millimeter Array, which will help address these questions.

---

1. Introduction

Massive stars are important due to their role in the creation of heavy nuclei and the chemical enrichment of our Galaxy, and their dynamical effects on the surrounding interstellar medium via stellar winds, photoionization and supernovae shocks, which could even trigger further star formation (Patel et al. 1998). Yet, our understanding of the formation of massive stars is still rather insecure, compared to low-mass star formation. Theoretical models that have to rely upon observational facts and observations of massive-star formation are challenging because of several well-recognized factors: Inherently short timescales of evolution $\sim 10^4$–10^5 years (Stahler, Palla, & Ho 2000); rarity compared to low-mass stars (e.g., for every $20\,M_{\odot}$ star, there are about a hundred thousand low-mass stars in the Galaxy; Massey 2003); multiplicity of sources, and typically greater distances and high visual extinction.

We are now in a period of remarkable developments in the instrumentation required for radio observations of massive-star formation, particularly with the commissioning of the Submillimeter Array (SMA; Ho et al. 2004), the Expanded Very Large Array, and the new arrays on the horizon: the California Radio Millimeter Array and the Atacama Large Millimeter Array (ALMA). Here, I will briefly review radio observations of massive-star formation, with emphasis on recent results from the SMA.

2. Identification of earliest stages

Searches for sites of massive-star formation began with the use of the IRAS colors (Wood & Churchwell 1989), resulting in a catalog of Ultra-compact H II (UCHII) regions. While UCHIIs are among the most luminous objects at far-IR wavelengths and help locate the regions of massive-star formation, they do not represent the earliest stages, since at this stage, the protostar has already started ionizing the surrounding gas (see review by Churchwell 2000a). Precursors to UCHIIs are revealed in later surveys based mostly on the Midcourse Space Experiment (MSX) catalog and follow-up studies of mm-wave continuum emission. The *Spitzer*'s GLIMPSE survey is likely to revolutionize such a catalog of early stages in massive-star formation (Benjamin et al. 2003). Besides

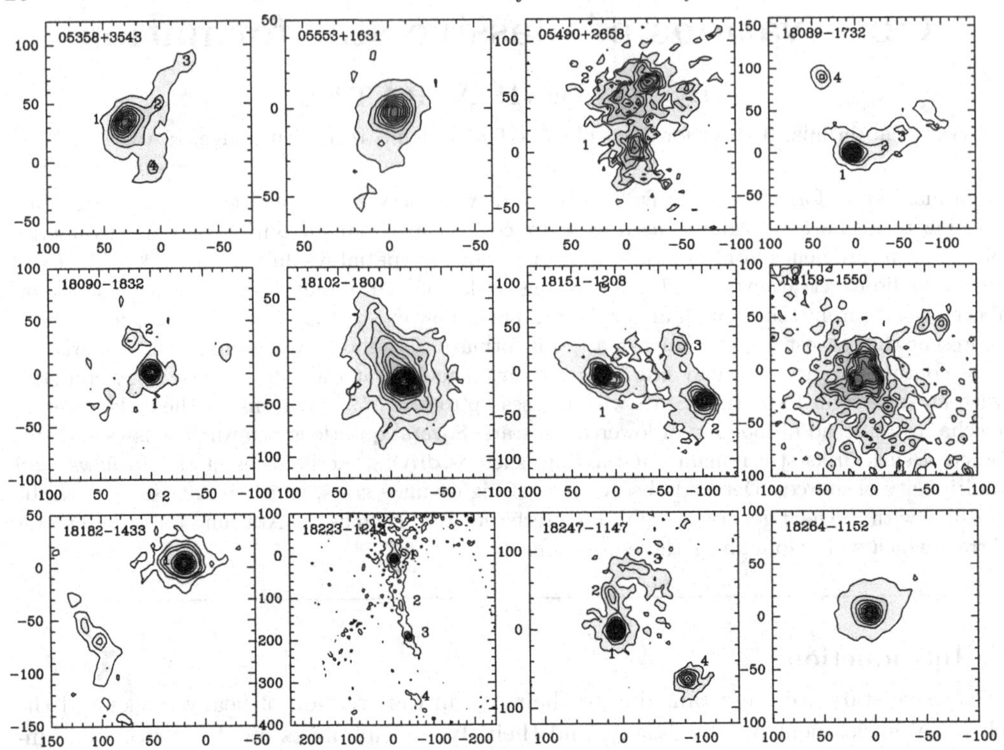

FIGURE 1. This figure is an excerpt from Figure 1 of Beuther et al. (2002a) showing 1.2 mm continuum emission mapped using the IRAM 30-m telescope. The x and y axes are position offsets in arcseconds along east-west and north-south, respectively, from the position of the IRAS sources (Sridharan et al. 2002). A wide range of morphologies are seen in the HMPOs, with several sources showing multiplicity.

providing a list of targets for interferometric studies, such combination of near-IR, mid-IR and mm-wave continuum observations are important in providing initial conditions for theoretical models of massive-star formation (see paper by Krumholz, this volume).

2.1. *High-mass protostellar objects and starless cores*

Sridharan et al. (2002) identified 69 IRAS sources following a modified set of criteria as previously suggested by Wood & Churchwell (1989; see also Molinari et al. 1996 and Ramesh & Sridharan 1997). These sources have a higher mass-to-luminosity ratio compared to the UCHIIs, and a comparison of the 1.2 mm continuum emission from these sources (Beuther et al. 2002a), and the MSX images, which show either no detectable emission associated with these sources, or show absorption features, suggest that these sources are younger in the evolution towards UCHIIs (see Figure 1). Sridharan et al. (2005a) further studied this sample of sources and found that many fields had 1.2 mm continuum with either absorption in mid-IR images or lack of mid-IR emission. Using this technique, they identified 56 candidates for High-Mass Starless Cores (HMSCs). From the 1.2 mm fluxes, they estimate the masses of these cores in the range of 10^2–10^3 $M_\odot$. When compared with similar observations of HMPOs and UCHIIs, NH_3 observations of these cores show smaller internal motions, implying that these cores are indeed likely to be prestellar and preceding in evolution compared to the HMPOs.

One of the HMPOs, IRAS 18089−1732 has been followed up with observations with the SMA by Beuther et al. (2004, 2005). Continuum emission at 1.3 mm and 850 μm shows a single-peaked elongated source suggesting a compact core with halo emission. The central core mass is estimated to be $40\,M_\odot$. Tentative detection of an outflow in SiO $J = 5$–4 emission and disk emission in $HCOOCH_3$ are also reported. This source is at a distance of 3.6 kpc and has a luminosity of $10^{4.5}\,L_\odot$. It has a massive core of $\sim 2000\,M_\odot$ and is an example of an early stage of a hot core. Beuther et al. (2005) detected more than 50 molecular lines from 18 molecules/isotopomers in the 850 μm band. A slightly elongated structure with a velocity gradient in $HCOOCH_3$ emission is seen as a tentative indication of the presence of a disk, but not in CH_3CN emission, perhaps due to optical depths effects. The dynamical mass of the enclosed gas within the outer radius of the disk is estimated to be $\sim 20/\sin^2(i)\,M_\odot$. The kinematics of $HCOOCH_3$ emission does not show any signature of Keplerian rotation. In other sources, CH_3CN has been found to be a good tracer of disk emission in other high-mass young stars: IRAS 20126+4104 (Cesaroni et al. 2005) and Cepheus–A HW2 (Patel et al. 2005, see Section 5.2). In the case of IRAS 18182–1433, another HMPO source from the sample of 69 sources (Sridharan et al. 2002), both CH_3CN and $HCOOCH_3$ show lack of coherent velocity structure. The question of which molecular line tracer is most suitable for searches for disks associated with massive protostars remains elusive.

2.2. *Infrared dark clouds*

Infrared dark clouds (IRDCs) were first identified from the mid-infrared images from the Midcourse Space Experiment (MSX) and *Infrared Space Observatory* (*ISO*) images, as silhouettes against the background mid-IR emission in the Galactic plane. The kinematic distances of these clouds are estimated to be between 2.2 and 9 kpc (Carey et al. 1998). Simon et al. (2006) have recently identified 10,931 IRDCs using a contrast enhancing algorithm on the MSX (8 μm) images. These clouds typically show a filamentary structure with sizes ranging from 0.4 to 15 pc. From observations of H_2CO emission, Carey et al. (1998) find these cloud cores to be dense and cold ($n > 10^5$ cm^{-3}, $T < 20$ K). The filaments are most likely representing cylindrical rather than sheet-like structures (Menten et al. 2005), based on available measurements of densities and H_2 column densities. For a recent review, see Menten et al. (2005).

Rathborne et al. (2006) studied a sample of 38 IRDCs identified from the larger survey by Simon et al. (2006) based on extinction features seen on MSX images. Continuum emission at 1.2 mm was detected in all of these sources with the IRAM 30-m telescope. Twenty-two of these sources appear to be compact, with the most massive source of $210\,M_\odot$; 16 of the sources appear filamentary, appearing to have higher mass cores of up to $660\,M_\odot$. The IRDCs have masses in the range of 120–16,000 $M_\odot$, with a median value of $940\,M_\odot$ and sizes of a few pc (see Figure 2).

Further evidence for evolutionary progression in the cores associated with IRDCs come from NH_3 observations carried out by Pillai et al. (2006). These authors observed 10 IRDCs in NH_3 (1,1) and (2,2) lines with the 100-m Effelsberg telescope. The sources were selected from a sample by Carey et al. (1998). The IRDCs have a size scale of 1–10 pc and densities $>10^6$ cm^{-3}. The mean linewidths and kinetic temperatures appear to have a progression in various types of cores as summarized below (from Pillai et al. 2006):

	IRDC	→	HMPO	→	UCHII
Δv (km s^{-1})	1.51		2.05		2.52
T (K)	13.9		18.3		22.6

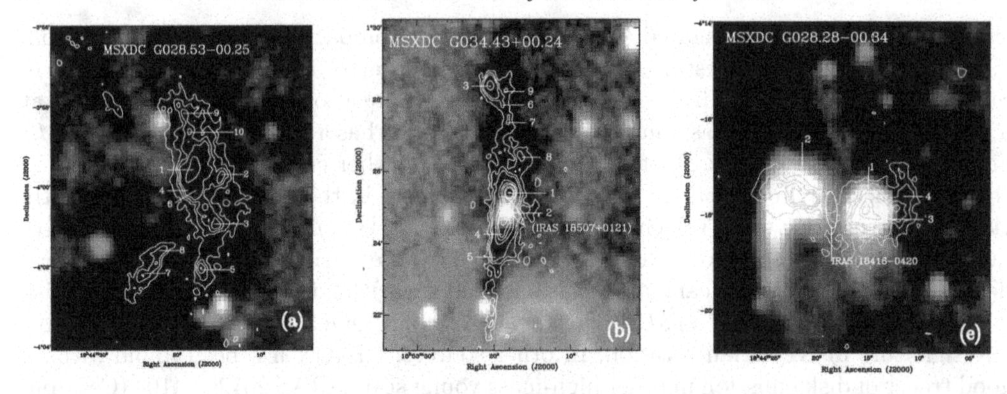

FIGURE 2. This figure is adopted from Figures 4a, 5b and 6e of Rathborne et al. (2006), showing 1.2 mm continuum emission mapped using the IRAM 30-m telescope, overlaid on MSX 8.3 μm emission. The three panels from left to right represent the evolutionary progression from the earliest, intermediate, and later stages (respectively) of the massive cores in the IRDCs.

3. Outflows

Outflows have been used as signposts of star formation, but for massive stars outflows have a special significance. According to theoretical models, outflows can assist the formation of stars more massive than $30\,M_\odot$ (Krumholtz et al. 2005; Yorke 2004; Yorke & Sonnhalter 2002). Krumholtz et al. (2005) have shown that an outflow can provide an optically thin channel through which radiation can escape, allowing protostellar accretion to proceed along the equatorial plane, hence increasing the final mass of the star.

Outflow searches have been carried out in high-mass star-forming regions by Shepherd & Churchwell (1996), Zhang et al. (2001) and Beuther et al. (2002a). Outflows appear to be as common for high-mass young stars as for low-mass YSOs. Outflow energetics are correlated with mass over six orders of magnitude (Churchwell 2000b). Interferometeric observations following up on these single-dish surveys have revealed some sources which were previously thought to be single outflows to be multiple outflows (Beuther et al. 2003), and the orientation to change from small to large scales (Shepherd et al. 2000). Typically, however, the driving source is not clearly identified, making it difficult to link the outflow directly with an accretion mechanism.

A large fraction of surveyed HMPOs in CO 2−1 emission have shown molecular outflows with varying morphologies (Beuther et al. 2002b; Wu et al. 2004; Zhang et al. 2005, see Figure 3). In all cases, the physical characteristics of the outflows—such as mechanical force and energy—appear to be about one to a few orders of magnitude greater than those associated with low-mass YSOs. Massive outflows may also be a significant contributor to the turbulent energy in the surrounding clouds (Zhang et al. 2005). The large detection rate of outflows in the sample of HMPOs implies that accretion may play a critical role in the formation of massive stars. It is not clear, however, if these massive outflows are generated in the same way as the low-mass outflows (Keto et al. 2006; Shang et al. 2004).

Outflows in massive YSOs are generally found to be poorly collimated compared to the low-mass YSOs. Shepherd (2005) has reviewed observational properties of outflows associated with young OB stars and compared them with outflows in low-mass star formation. Some of the most highly collimated outflows associated with low-mass YSOs were recently mapped with the SMA. The HH−211 outflow was mapped with the SMA in the SiO $J = 5$–4 emission (Hirano et al. 2006) and in SiO $J = 8$–7 emission (Palau et al. 2006). These observations show a highly collimated jet of dense ($>10^6$ cm^{-3}) and hot

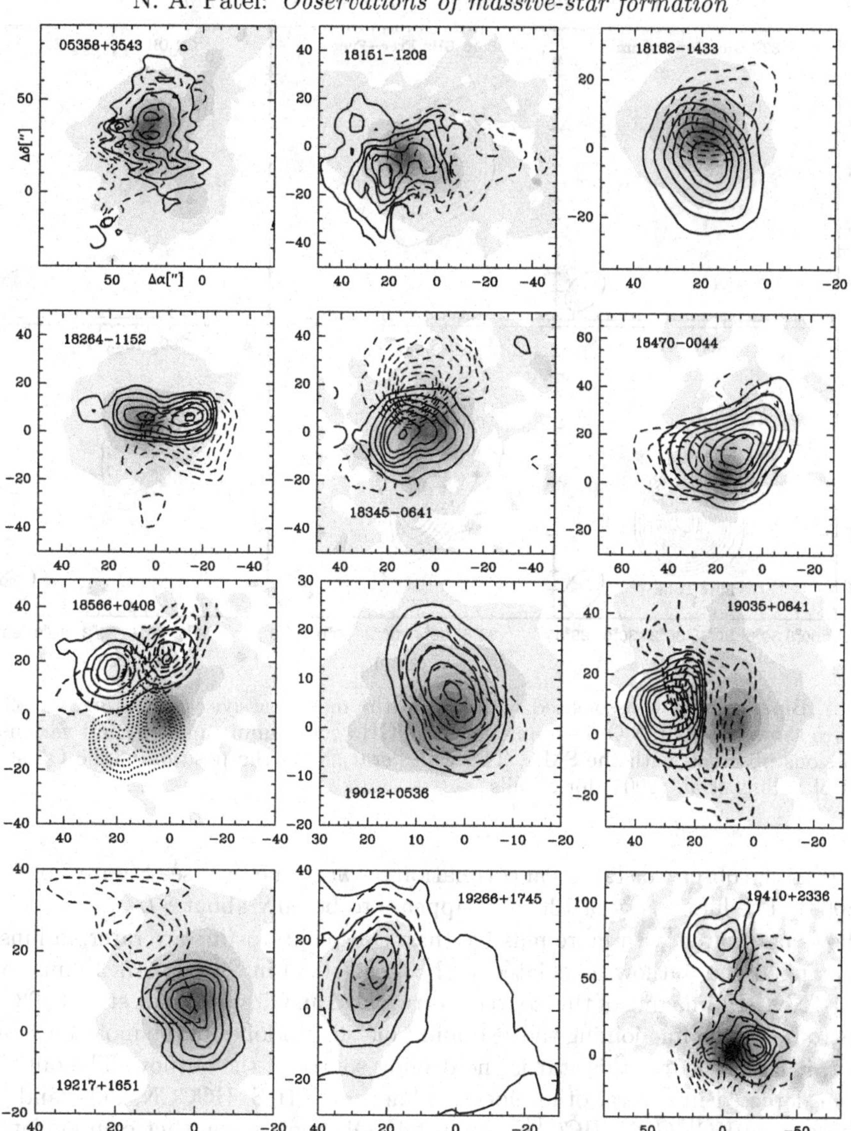

FIGURE 3. Molecular outflows associated with HMPOs, mapped in CO 2−1 emission by Beuther et al. (2002b). The halftone images show the 1.2-mm continuum emission (from Beuther et al. 2002a). Position offsets along east-west and north-south are with respect to the IRAS positions given in Sridharan et al. (2002).

(300–500 K) gas. The higher excitation SiO line emission appears closer to the protostar. The ratio of (8–7)/(5–4) emission reveals density and temperature gradients along the jet, with the innermost knots there about an order of magnitude denser compared to the outer knots (Palau et al. 2006). Similar results were found in another low-mass protostellar outflow studied by Zapata et al. (2005) with the SMA, in CO 2–1 emission, in the OMC–1 South region. This highly collimated outflow appears to be driven by the source 136−359 detected in 1.3-mm continuum emission, based on positional coincidence with respect to the outflow lobes. However, the mechanical luminosity of the outflow of $0.3\,M_\odot$ km s^{-1} yr^{-1} suggests the driving source to have a high luminosity of $\sim 10^4\,L_\odot$,

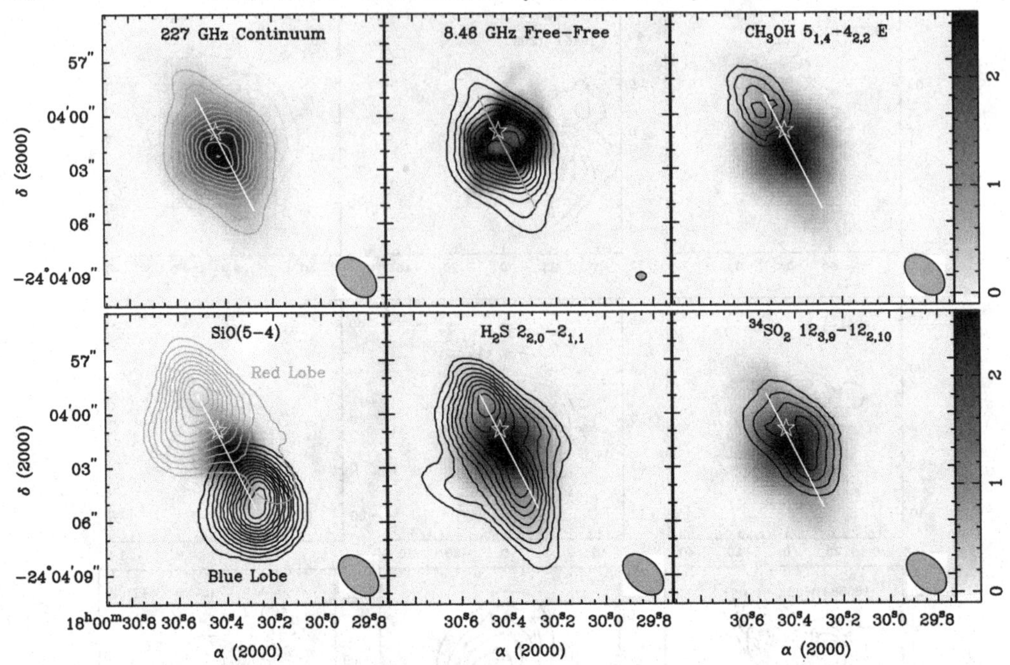

FIGURE 4. Bipolar outflow associated with one of the most massive young stars, G5.89−0.39. This figure shows maps in SiO 5−4 emission, 227 GHz continuum emission, and various other line emissions observed with the SMA. The white star shows the position of the O5 star. See Figure 1 of Sollins et al. (2004) for details.

according to the observed statistical correlation between the two (see Figure 3 of Beuther et al. 2003). The luminosity of 136−359 appears to be only about $8\,L_{\odot}$.

Outflows associated with more massive than B-type protostars are rarer. Sollins et al. (2004) mapped the outflow associated with G5.89−0.39 in SiO J = 5–4 emission (see Figure 4). Near-IR imaging of this source has revealed an O5 star (Feldt et al. 2003) which appears to be on the line joining the red- and blue-shifted lobes of the molecular outflow. However, it is not clear if this star is the driving source of the outflow. The outflow has also been detected in several other spectral lines, e.g., H_2S, HCCCN, SO_2, and $^{34}SO_2$, while lines from $HC^{13}CCN$, $HCC^{13}CN$ and CH_3OH show a compact component at the ambient velocity. The total mass associated with the dust emission from the 1.3-mm continuum measurements with the SMA is $83\,M_{\odot}$. The mass of the gas associated with the outflow is estimated to be $18\,M_{\odot}$. Since the O5 star is not symmetrically centered between the two lobes of the outflow (though it is in the same line), it is not clear if the O5 star is driving the outflow. The inclination angle of the outflow is largely unknown, which would help clarify the geometry and the association of this outflow with the UCHII region or the O5 star.

The most well-studied outflow associated with massive-star formation and the nearest source is the Orion KL region (Greenhill et al. 2005). The spatio-kinematic structure of SiO and H_2O masers obtained from multi-epoch proper motion measurements suggest the driving source of this outflow is believed to be *source I*. The nature of source I was not well known, and there was a hypothesis that it was a late-type star. From the SED characterization, the submillimeter data are useful in proving that source I is a protostellar object with an SED fitted by proton-electron free-free emission below 100 GHz and a dust component contributing in the submillimeter band (Beuther et al.

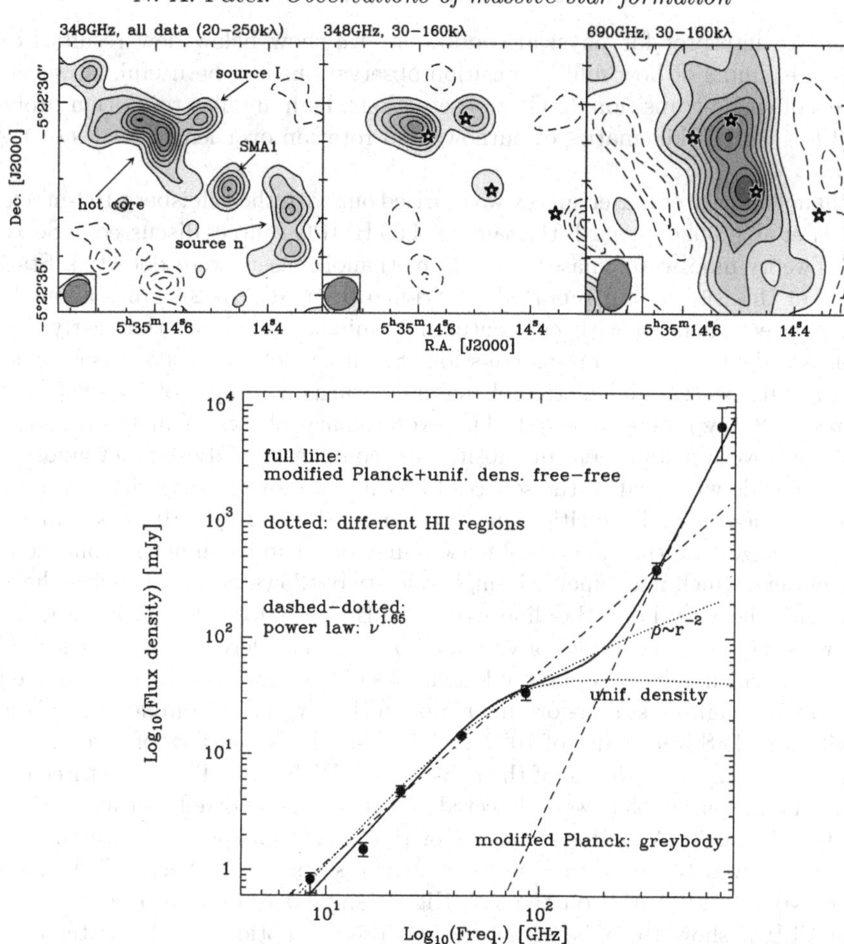

FIGURE 5. Top: Submillimeter continuum images of Orion KL. Top left and middle panels: 348 GHz continuum data (Beuther et al. 2004). Top right panel: New 690 GHz continuum data covering. Bottom: SED of source I. The measurement at 690 GHz helps constrain the model and shows that source I is likely to be a young protostar with continuum emission modeled as free-free and greybody dust emission. This figure is adapted from Figures 3 and 5 of Beuther et al. (2006).

2006, see Figure 5). A new submillimeter continuum source, SMA−1, was also detected, and this may be the youngest source in the region since there is no evidence of any corresponding ionizing radiation at cm wavelengths. This source is also not detected at IR wavelengths due to deep extinction.

4. Masers

While radio observations of thermal line emission from outflows have helped identify regions of massive-star formation and allowed statistical studies, the angular resolution is generally inadequate to probe the regions where the outflows are launched. Maser line emission in several molecules has tremendous potential in probing the physical conditions at very high spatial resolutions. Due to their intrinsically bright emission, they are easier to observe to carry out surveys with single-dish observations, achieving high-angular resolution in selected sources with interferometers. Observations of just a single transition is not easy to interpret for physical conditions—we show below new results from the SMA as an example of how multi-transition observations can be useful. However, even in the case of single-transition 22 GHz observations, high-angular resolution probes can be useful to map the kinematics of outflows and rotation over length scales of 1 AU to 1000 AU.

A methanol and water maser survey was carried out with the Effelsberg 100-m telescope by Beuther et al. (2002c) towards the sample of 69 HMPO sources discussed in Section 2.1 (above). Twenty-nine water masers and 26 methanol masers were detected. Statistical properties of this survey are reported by Sridharan et al. (2002). Only about 20% of the sources are associated with cm continuum emission, implying that early stages in evolution are delineated by maser emission. Szymczak et al. (2005) also carried out a survey in OH, water, and methanol masers towards a sample of 79 sources. Forty-one sources (28 new) were detected. The evolutionary phases of methanol and water masers largely overlap, and their luminosities are comparable. OH and methanol centroid velocities coincide with that of the sources to within 5% for the majority of the sources. Water-maser line-centroid velocities, on the other hand, differ by >10 km s^{-1} in $\sim$40% of the sources, suggesting that methanol masers may occur in the inner regions, compared to water masers, which may emerge in high-velocity outflows. Selected sources have been observed with the Very Long Baseline Array (VLBA) by Moscadelli et al. (2005).

We now discuss some examples of very long baseline interferometric studies of 22 GHz water masers, which help us probe the kinematics of gas very close to the massive young stars. W75N is a high-mass star-forming region in the Cygnus-X complex at a distance of 2 kpc, with an IRAS luminosity of $10^5\ L_\odot$. A large-scale CO outflow of several parsecs is known to be associated with one of the sub-regions: W75N(B). There are three compact radio-continuum sources that were detected with the VLA (Torrelles et al. 1997); two of these, VLA$-$1 and VLA$-$2, have clusters of H_2O masers associated with them. These sources are believed to be excited by early B-type stars. VLA$-$1 and VLA$-$2 are separated by $\sim$0.″7 ($\sim$1400 AU) on the sky. Higher-angular-resolution observations, made with the VLBA, show that the structure and proper motions of the water masers in VLA$-$1 and VLA$-$2 show strikingly different geometries. In VLA$-$1, the masers appear to be tracing a highly collimated outflow with an opening angle of $\sim$10° and almost in the plane of the sky. The linear structure traced by these masers has a spatial extent of 200×2000 AU and it shows proper motion of $\sim$2 mas/yr (19 km s^{-1}) parallel to the thermal radio jet. The masers associated with VLA$-$2, on the other hand, are distributed on a shell of radius of $\sim$160 AU with proper motion of $\sim$3 mas/yr ($\sim$28 km s^{-1}). Torrelles et al. (2003) suggest the different morphologies of the outflows represent different evolutionary stages, with the spherical geometry being the younger of the sources. However, further observations at 0.″1 resolution, in submillimeter wavelengths to characterize the SEDs of these sources, are required to better understand the nature of these sources.

Nearly isotropic outflows traced by water masers over scales of one to hundreds of AU are known in three other sources: IRAS 21391+5802 in the IC1396 region (Patel et al. 2000), a group of masers named R5 in the Cepheus-A HW2 region (Torrelles et al. 2001) and in G24.78+0.08 (Moscadelli 2005). IRAS 21391+5802 is a 400 $L_\odot$, intermediate-mass YSO embedded in a globule associated with the IC1396 H II region complex (Patel et al. 1995). The loop of water masers is only 1 AU in size, with no clear signature of expansion, but a velocity spread of $\sim$2 km s^{-1}. In Cepheus$-$A HW2, the R5 structure is most remarkable, with deviation from a fit to a circle to within 0.1% in all three epochs—each separated by about a month. This structure is of radius 62 AU and the proper motions show nearly perfect radial expansion by 9 km s^{-1}, implying a dynamical time of only 33 years. Presumably, this is a limb-brightened shocked shell of gas representing

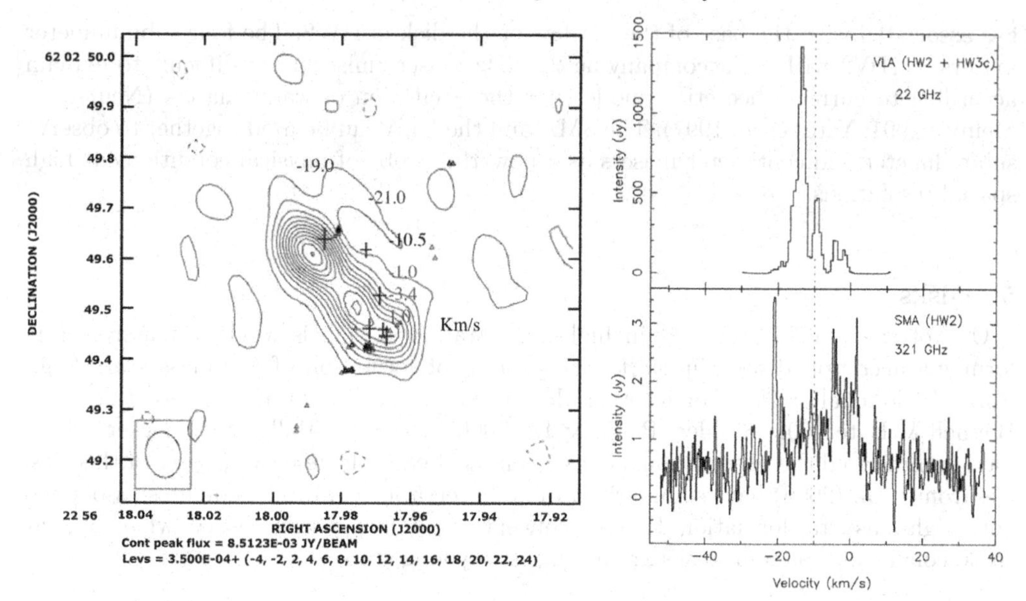

FIGURE 6. 321 GHz water maser emission from Cepheus–A HW2. Left: Contours show 1.3 cm continuum emission and triangles show the 22 GHz water masers mapped with the VLA. The positions of the submillimeter water masers are shown with error bars and labels denoting V_{LSR} velocity (the systemic velocity of HW2 is ~ -10 km s^{-1}). See Figs. 1 & 2 from Patel et al. (2007). Right: Comparison of spectra of 22 GHz (upper) and 321 GHz masers (lower). Note that the strongest 22 GHz emission occurs at relatively smaller velocities, while the strongest 321 GHz emission occurs at relatively higher velocities.

an episodic event of ejection of material from a young star at the center. At the center of the implied circle, a continuum source was later identified by Curiel et al. (2002). This source is likely to be a young B3 star. Interestingly, some of the masers in this R5 loop were observed about eight years later with the VLBA, and their predicted positions are in very good agreement with the newly observed positions according to the previously measured proper motions (Vlemmings et al. 2006). Linear polarization was also detected by these authors, implying a magnetic field strength of about 30–100 mG, with direction along the shell expansion. The nature of these nearly isotropic expanding shells traced by water masers is not yet clearly understood.

While the 22 GHz water masers have been extremely useful in diagnosing the spatial structure and kinematics of the gas to within several AU to 1000 AU scale as discussed above, this single maser transition alone cannot probe the physical condition of the gas—particularly its temperature—except by indicating that the gas has to be dense (10^8–10^9 cm^{-3}) and hotter than 500 K. Nearly simultaneous observations of several masing transitions in water can be helpful. Using the SMA, we have begun such a study with observations of the 10_{29}–9_{36} transition at 321.226 GHz (discovered previously by Menten et al. 1990) along with VLA observations of the 22 GHz masers in the Cepheus–A HW2. Preliminary results from this work (Patel et al. 2007) are shown in Figure 6. The submillimeter masers appear preferentially along the jet, with the blue- and red-shifted masers consistent with the larger scale molecular outflow (Gomez et al. 1999). On the other hand, the cm masers are distributed along the jet, as well as along the disk. Three out of the seven detected submillimeter masers are well correlated with the centimeter masers spatially as well as kinematically, while there are several 22 GHz maser spots without corresponding submillimeter masers, presumably arising in relatively cooler (~500 K) gas associated with the base of the outflow or the disk in HW2. The four submillimeter masers in HW2 without accompanying 22 GHz maser emission are difficult to explain according to current theoretical models for the excitation of water masers (Neufeld & Melnick 1991; Yates et al. 1997). The SMA and the VLA can be used together to observe submillimeter and centimeter masers as a powerful probe of physical conditions at high spatial resolutions.

5. Disks

One of the central questions in high-mass star formation is whether massive stars form via accretion disks, similar to the process of formation of low-mass stars (e.g., Yorke & Sonnhalter 2002) or whether they form via mergers of lower-mass stars (e.g., Bonnell & Bate 2002; Stahler, Palla, & Ho 2000). Recently, Bally & Zinnecker (2005) have summarized the observational consequences of each of these two classes of theories. Cesaroni et al. (2006) have summarized radio observations of disk candidates associated with high-mass star formation. Here we concentrate on two specific sources which appear to be convincing cases for disks around massive young stars.

5.1. *IRAS 20126+4104*

IRAS 20126+4104 (I20126) is a $10^4\,L_\odot$ high-mass young star at a distance of 1.4 kpc, first identified by Wood & Churchwell (1989) in their catalog of UCHII sources. Zhang et al. (1998) discovered a flattened rotating structure in NH_3 emission exhibiting a Keplerian velocity gradient consistent with an enclosed mass of $\sim$20 $M_\odot$ over a radius of $\sim$10,000 AU. Since then, this source may be the most well-studied object in the context of investigation of disk-outflow systems associated with massive-star formation. Cesaroni et al. (2005) have carried out a more detailed interferometric study with the IRAM PdBI in $C^{34}S$ (2–1) and (5–4) and in several lines of CH_3OH and CH_3CN emission. The disk is detected in CH_3CN and $C^{34}S$ emission. CH_3OH appears to be tracing the disk, as well as the outflow. The new $C^{34}S$ observations show kinematics that are consistent with a Keplerian disk of radius $\sim$7,600 AU rotating about a 7 $M_\odot$ star. From available continuum data in the range of 40 to 230 GHz, the mass of the disk is estimated to be 1–4 $M_\odot$ within a radius of $\sim$5,000 AU from the young star. The larger value of mass ($\sim$20 $M_\odot$) obtained by Zhang et al. earlier may be due to a sampling of a larger region, including a substantial portion of the mass of the inner disk. This structure of $\sim$10,000 AU scale is most likely a large-scale flattened envelope rather than a circumstellar disk. The outflow in I20126 is discussed in detail by Cesaroni et al. (2005), who present evidence of precession based on molecular line maps and near-IR H_2 images. Sridharan et al. (2005b) obtained high-angular resolution, near-IR images of I20126 which show a dark lane in the K-band image, separating emission peaks in L', M' and mm-wave continuum bands, lying close to the dark lane. This structure is interpreted to be a nearly edge-on disk of 850 AU thickness for radius $<$1000 AU. A new point source was found at a separation of 1000 AU in line with the disk, interpreted as a possible binary companion.

5.2. *Cepheus–A HW2*

Cepheus–A is a well-known high-mass star-forming region at a distance of 725 pc (nearest, after Orion BNKL), with a bolometric luminosity of $2.5 \times 10^4\,L_\odot$. Several hot cores have been identified in this region from VLA continuum observations (Hughes & Wouterloot 1984), with HW2 being the dominant source of B0.5 spectral type and about 15 $M_\odot$. The second-most powerful molecular outflow in our Galaxy (after Orion) is in HW2

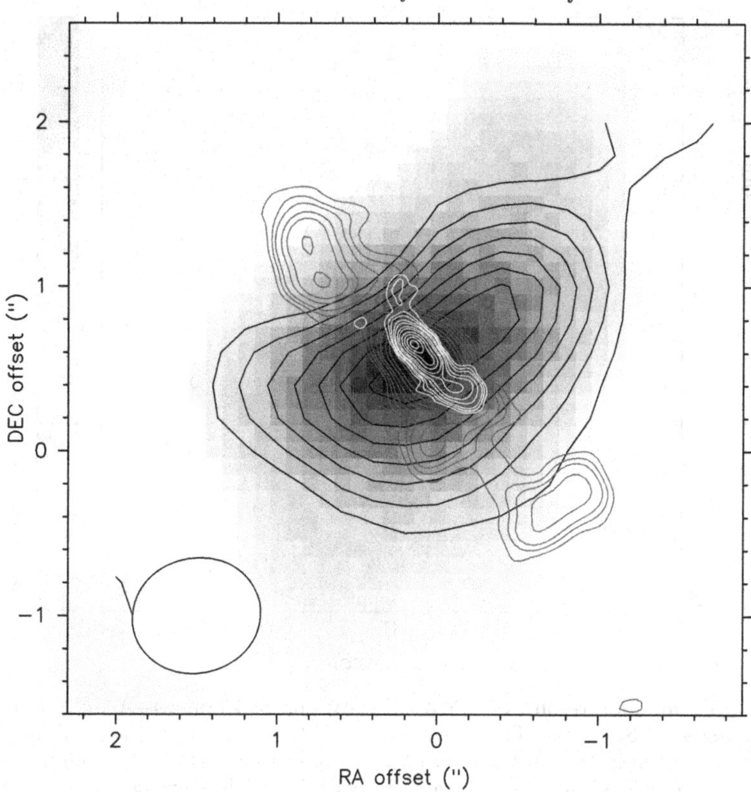

FIGURE 7. Dust continuum emission at 327 GHz from Cepheus−A HW2, shown as halftone ranging linearly from 0 to 1.5 Jy beam^{-1}. Integrated CH_3CN $J = 18$–17, $K = 0, 1, 2, 3$ line emission is overlaid as black contours with levels from 5 to 40 Jy beam^{-1} km s^{-1} in steps of 5 Jy beam^{-1} km s^{-1}. The SMA beam size was $0\farcs8 \times 0\farcs7$ with a position angle of $-78\fdg6$ (shown in lower left corner). Also shown as gray and white contours is the continuum emission from the thermal jet observed with the VLA at 3.6 cm and 1.3 cm wavelength, respectively (Torrelles et al. 1996; Curiel et al. 2006).

(Rodriguez, Ho, & Moran 1980). Gomez et al. (1999) presented mm-wave interferometric maps of this outflow with a tentative indication of an associated protostellar disk.

Perhaps the strongest evidence of a circumstellar disk associated with the highest-mass young star of 15–20 $M_\odot$ is provided by recent SMA observations of HW2. Patel et al. (2005) recently reported direct submillimeter detection and imaging of the dust and gas emission from Cepheus−A HW2. The continuum emission at 900 μm and CH_3CN $J = 18$–17 ($K = 0, 1, 2, 3$) line emission both show an elongated structure that is nearly perpendicular to the previously observed thermal jet (Torrelles et al. 1996; Curiel et al. 2006), strongly supporting the disk hypothesis (see Figure 7). The disk has a radius of 330 AU and the mass is estimated to be 1–8 $M_\odot$ from submillimeter continuum emission. CH_3CN emission shows a velocity gradient of 6 km s^{-1} over $0\farcs5$, implying an enclosed binding mass of $19 \pm 5\,M_\odot$. Martin-Pintado et al. (2005) report the detection of a new intermediate-mass protostar in the HW2 region, which appears to be within the disk structure, based on interferometric observations of several lines in SO_2 and HC_3N. There is only one thermal jet seen in HW2, which is centered upon the disk structure. This jet was recently shown to have proper motions of ~500 km s^{-1} (Curiel et al. 2006).

These observations suggest that massive stars up to about B0.5 seem likely to involve processes that are similar to the formation of low-mass stars, namely, via a disk-accretion

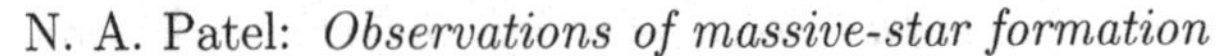

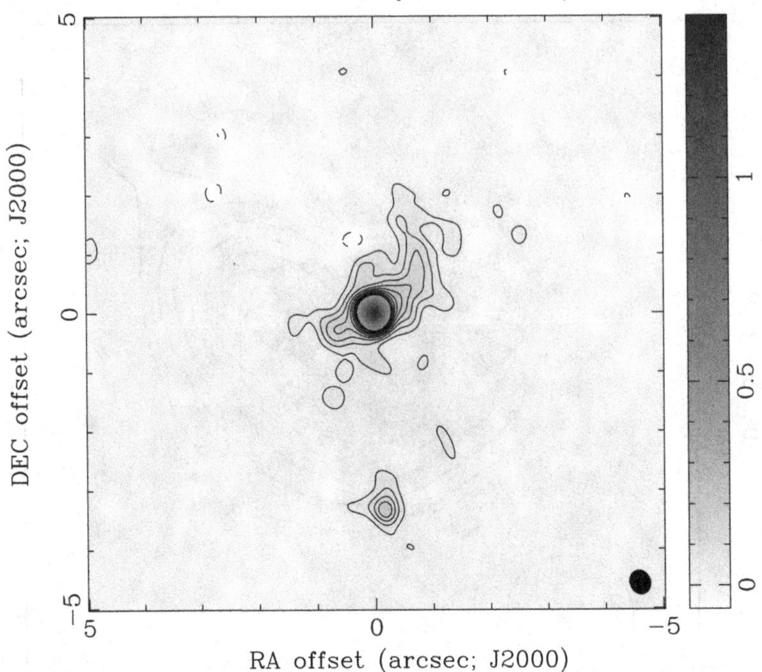

FIGURE 8. Higher angular resolution SMA observations of Cepheus−A at 0.″4 resolution, of continuum emission at 330 GHz. The previous observations shown in Figure 6, made with the *extended* array, are combined with the later observations of August 2005, taken in *very extended* configuration. Contour levels are spaced every 5σ, with intervals every 5σ, with $\sigma = 80$ mJy rms noise.

scenario. Observational evidence for more massive stars with luminosities $\geqslant 10^5\, L_\odot$ are far fewer. Van Der Tak & Menten (2005) suggest, based on their VLA observations of 43 GHz continuum emission from W3 IRS5, aFGL 2136 and AFGL 2591, that regions of $10^5\, L_\odot$ appear to produce clusters of H II regions and poorly collimated outflows, possibly due to mergers of protostars. Recent SMA observations of IRAS 16547−4247 by Franco-Hernandez et al. (2007) reveal preliminary evidence of a rotating structure with a Keplerian velocity profile that is perpendicular to the outflow mapped in radio continuum with the VLA. The SMA HC_3N (25–24) observations showing this kinematic signature might be revealing a disk associated with the most massive young star forming with the accretion process, with a luminosity of $6.2 \times 10^4\, L_\odot$.

To confirm the disk structure (to check whether the elongation is due to multiplicity of sources instead of a single flattened object), we carried out higher-angular resolution, 330 GHz continuum observations with the SMA in the *very extended* configuration with a maximum baseline length of 540 m in August 2005. The angular resolution in these observations is 0.″3. Combining these data with the previous observations, we achieve better sampling of the $u - v$ plane and higher dynamic range in the image, which is shown in Figure 7. The submillimeter continuum emission is detected only from two sources in this region: HW2 and HW3c, with HW2 as a dominant source, suggesting that the bulk of the total luminosity of 2×10^4 is most likely arising from HW2. The flattened disk structure is also confirmed, with the elongation in the same direction as seen before (Figure 6) but with more complex structure. Perhaps this "spiral arm"-like emission is indicating an instability in the disk but this requires further investigation (Patel et al. 2007). Stability criteria for disks around massive young stars has been discussed in the review by Cesaroni et al. (2006) and in theoretical modeling of fragmentation in disks (as in the poster paper by Kaitlin Kratter at this symposium). The centrally peaked, dominant source of continuum emission in HW2 shown in Figure 8 could be due to ionization during the accretion phase as previously observed in other sources in cm continuum mapped with the VLA (Keto & Wood 2006; Van Der Tak et al. 2000) and modeled by Keto (2007).

6. Summary

Recent radio observations have provided a sample of several dozen sources which are likely to be the earliest-identified regions of massive-star formation. Single-dish and interferometric radio observations of such sources have shown the presence of outflows and masers, with properties scaling-up with mass/luminosity of the source, but the exact mechanism of the generation of outflows, and the nearly isotropic and episodic ejection revealed by VLBI observations of water masers remains unclear. SMA observations of Cepheus−A HW2 have provided clear evidence of a disk associated with a B0.5-type star of about 15 $M_{\odot}$, supporting disk accretion over mergers as a mechanism for the formation of such stars. The merger scenario is still likely to be relevant for the formation of more massive stars; observational studies of such stars might have to await the commissioning of ALMA.

I am very grateful to Mario Livio and the organizers for inviting me to contribute this paper. It is a great pleasure to thank my collaborators and colleagues at the SMA in Cambridge, Hilo and Taipei. I have benefitted greatly from discussions with T. K. Sridharan, Paul Ho, Salvador Curiel, Jose Torrelles, Qizhou Zhang, Eric Keto and Jim Moran. I am very grateful to Paul Ho for his comments on this article. The SMA is a collaborative project between the Smithsonian Astrophysical Observatory and the Academia Sinica Institute of Astronomy and Astrophysics, Taiwan.

REFERENCES

Bally, J. & Zinnecker, H. 2005 *AJ* **129**, 2281.

Benjamin, R., Churchwell, E., Babler, B., et al. 2003 *PASP* **115**, 953.

Beuther, H., Hunter, T. R., Zhang, Q., Sridharan, T. K., Zhao, J.-H., Sollins, P., Ho, P. T. P., Ohashi, N., Su, Y. N., Lim, J., & Liu, S.-Y. 2004 *ApJ* **616**, L23.

Beuther, H., Schilke, P., Menten, K. M., Motte, F., Sridharan, T. K., & Wyrowski, F. 2002a *ApJ* **566**, 945.

Beuther, H., Schilke, P., Sridharan, T. K., Menten, K. M., Walmsley, C. M., & Wyrowski, F. 2002b *A&A* **383**, 892.

Beuther, H., Schilke, P., & Stanke, T. 2003 *A&A* **408**, 601.

Beuther, H., Walsh, A., Schilke, P., Sridharan, T. K., Menten, K. M., & Wyrowski, F. 2002c *A&A* **390**, 289.

Beuther, H., Zhang, Q., Reid, M., et al. 2006 *ApJ* **636**, 323.

Beuther, H., Zhang, Q., Sridharan, T. K., & Chen, Y. 2005 *ApJ* **628**, 800.

Bonnell, I. A. & Bate, M. R. 2002 *MNRAS* **336**, 659.

Carey, S. J., Clark, F. O., Egan, M. P., Price, S. D., Shipman, R. F., & Kuchar, T. A. 1998 *ApJ* **508**, 721.

Cesaroni, R., Galli, D., Lodato, G., Walmsley, C. M., & Zhang, Q. 2006 *Nature* **444**, 703.

Cesaroni, R., Neri, R., Olmi, L., Testi, L., Walmsley, C. M., & Hofner, P. 2005 *A&A* **434**, 1039.

Churchwell, E. 2000a *ARA&A* **40**, 27.

Churchwell, E. 2000b. In *Unsolved problems in Stellar Evolution* (ed. M. Livio). p. 41. Cambridge University Press.

Curiel, S., Ho, P. T. P., Patel, N. A., Torrelles, J. M., Rodrigues, L. F., Trinidad, M. A., Canto, J., Hernandez, L., Gomez, J. F., Garay, G., & Anglada, G. 2006 *ApJ* **638**, 878.

Curiel, S., Trinidad, M. A., Canto, J., Rodriguez, L. F., Torrelles, J. M., Ho, P. T. P., Patel, N. A., Greenhill, L., Gomez, J. F., Garay, G., Hernandez, L., Contreras, M. E., & Anglada, G. 2002 *ApJ* **564**, L35.

Feldt, M., Puga, E., Lenzen, R., Henning, T., Brandner, W., Stecklum, B., Lagrange, A.-M., Gendron, E., & Rousset, G. 2003 *ApJ* **599**, L91.

Franco-Hernandez, R., Moran, J. M., Rodriguez, L. F., & Garay, G. 2007 *BAAS* **39**, 145.02.

Gomez, J. F., Sargent, A., Ho, P. T. P., et al. 1999 *ApJ* **514**,287.

Greenhill, L. J., Chandler, C. J., Reid, M. J., & Humphreys, E. 2005. In *Protostars and Planets V* (eds. B. Reipurth, D. Jewitt, & K. Keil). p. 8614. University of Arizona Press.

Hirano, N., Liu S.-Y, Shang, H., Ho, P. T. P., Huang, H.,-C, Kuan, Y.,-J, McCaughrean, M. J., & Zhang, Q. 2006 *ApJ* **636**, L141.

Ho, P. T. P., Moran, J. M. & Lo, K. Y. 2004 *ApJ* **616**, L1.

Hughes, V. & Wouterloot, J. 1984 *ApJ* **276**, 204.

Keto, E. 2007 *ApJ* **666**, 976.

Keto, E., Broderick, A. E., Lada, C. J., & Narayan, R. 2006 *ApJ* **652**, 1366.

Keto, E. & Wood, K. 2006 *ApJ* **637**, 850.

Krumholtz, M., McKee, C. F., & Klein, R. I. 2005 *ApJ* **618**, L33.

Martin-Pintado, J., Jimenez-Serra, I., Rodriguez-Franco, A., Martin, S., & Thum, C. 2005 *ApJ* **628**, L61.

Massey, P. 2003 *ARA&A* **41**, 15.

Menten, K. M., Melnick, G. J., & Phillips, T. G. 1990 *ApJ* **350**, L41.

Menten, K. M., Pillai, T., & Wyrowski, F. 2005. In *Massive Star Birth: A Crossroads of Astrophysics* (eds. R. Cesaroni, M. Felli, E. Churchwell, & M. Walmsley). IAU Symposium No. 227, p. 23. Cambridge University Press.

Molinari, S., Brand, J., Cesaroni, R., & Palla, F. 1996 *A&A* **308**, 573.

Moscadelli, L. 2005. In *Massive Star Birth: A Crossroads of Astrophysics* (eds. R. Cesaroni, M. Felli, E. Churchwell, & M. Walmsley). IAU Symposium No. 227, p. 190. Cambridge University Press.

Moscadelli, L., Cesaroni, R., & Rioja, M. J. 2005 *A&A* **438**, 889.

Neufeld, D. A. & Melnick, G. J. 1991 *ApJ* **368**, 215.

Palau, A., Ho, P. T. P., Zhang, Q., Estalella, R., Hirano, N., Shang, H., Lee, C.-F, Bourke, T. L., Beuther, H., & Kuan, Y.-J. 2006 *ApJ* **636**, L137.

Patel, N. A., Curiel, S., Sridharan, T. K., Zhang, Q., Hunter, T. R., Ho, P. T. P., Torrelles, J. M., Moran, J. M., Gomez, J. F., & Anglada, G. 2005 *Nature* **437**, 109.

Patel, N. A., Curiel, S., Zhang, Q., Sridharan, T. K., Ho, P. T. P., & Torrelles, J. M. 2007 *ApJ* **658**, L55.

Patel, N. A., Goldsmith, P. F., Heyer, M. H., Snell, R. L., & Pratap, P. 1998 *ApJ* **507**, 241.

Patel, N. A., Goldsmith, P. F., Snell, R. L., Hezel, T., & Xie, T. 1995 *ApJ* **447**, 721.

Patel, N. A., Greenhill, L. J., Herrnstein, J., Zhang, Q., Moran, J., Ho, P. T. P., & Goldsmith, P. F. 2000 *ApJ* **538**, 268.

Pillai, T., Wyrowski, F., Carey, S. J., & Menten, K. M. 2006 *A&A* **450**, 569.

Ramesh, B. & Sridharan, T. K. 1997 *MNRAS* **284**, 1001.

Rathborne, J. M., Jackson, J. M., & Simon, R. 2006 *ApJ* **641**, 389.

Rodriguez, L., Ho., P. T. P., & Moran, J. 1980 *ApJ* **240**, L149.

Shang, H., Lizano, S., Glassgold, A., & Shu, F. 2004. In *Star Formation in the Interstellar Medium* (eds. D. Johnstone, F. C. Adams, D. N. C. Lin, D. A. Neufeld, & E. C. Ostriker). ASP Conf. Series 323, p. 299. ASP.

SHEPHERD, D. 2005. In *Massive Star Birth: A Crossroads of Astrophysics* (eds. R. Cesaroni, M. Felli, E. Churchwell, & M. Walmsley). IAU Symposium No. 227, p. 237. Cambridge University Press.

SHEPHERD, D. & CHURCHWELL, E. 1996 *ApJ* **457**, 267.

SHEPHERD, D., YU, K. C., BALLY, J., & TESTI, L. 2000 *ApJ* **535**, 833.

SIMON, R., JACKSON, J. M., RATHBORNE, J. M., & CHAMBERS, E. T. 2006 *ApJ* **639**, 227.

SOLLINS, P. K., HUNTER, T. R., BATTAT, J., BEUTHER, H., HO, P. T. P., LIM, J., LIU, S. Y., OHASHI, N., SRIDHARAN, T. K., SU, Y. N., ZHAO, J. H., & ZHANG, Q. 2004 *ApJ* **616**, L35.

SRIDHARAN, T. K., BEUTHER, H., SAITO, M., WYROWSKI, F., & SCHILKE, P. 2005a *ApJ* **634**, L57.

SRIDHARAN, T. K., BEUTHER, H., SCHILKE, P., MENTEN, K. M., & WYROWSKI, F. 2002 *ApJ* **566**, 931.

SRIDHARAN, T. K., WILLIAMS, S. J., & FULLER, G. 2005b *ApJ* **631**, L73.

STAHLER, S., PALLA, F., & HO, P. T. P. 2000. In *Protostars and Planets IV* (eds. V. Mannings, A. P. Boss, & S. Russell). p. 327. University of Arizona Press.

SZYMCZAK, M., PILLAI, T., & MENTEN, K. M. 2005 *A&A* **434**, 613.

TORRELLES, J. M., GOMEZ, J. F., RODRIGUEZ, L. F. , CURIEL, S., HO, P. T. P., & GARAY, G. 1996 *ApJ* **457**, 447.

TORRELLES, J. M., GOMEZ, J. F., RODRIGUEZ, L. F., HO, P. T. P., CURIEL, S., & VAZQUEZ, R. 1997 *ApJ* **489**, 744.

TORRELLES, J. M., PATEL, N. A., ANGLADA, G., GOMEZ, J. F., HO, P. T. P., LARA, L., ALBERDI, A., CANTO, J., CURIEL, S., GARAY, G., & RODRIGUEZ, L. F. 2003 *ApJ* **598**, L115.

TORRELLES, J. M., PATEL, N. A., GOMEZ, J. F., HO, P. T. P., RODRIGUEZ, L. F., ANGLADA, G., GARAY, G., GREENHILL, L., SCURIEL, S., & CANTO, J. 2001 *Nature* **411**, 277.

VAN DER TAK, F. & MENTEN, K. 2005 *A&A* **437**, 947.

VAN DER TAK, F. F. S., VAN DISHOECK, E. F., EVANS, N. J. II, & BLAKE, G. A. 2000 *ApJ* **537**, 283.

VLEMMINGS, W. H. T., DIAMOND, P. J., VAN LANGEVELDE, H. J., & TORRELLES, J. M. 2006 *A&A* **448**, 597.

YATES, J. A., FIELD, D., & GRAY, M. D. 1997 *MNRAS* **285**, 303.

YORKE, H. W. 2004. In *Star Formation at High Angular Resolution* (eds. M. Burton, R. Jayawardhana, & T. Bourke. IAU Symposium 221, p. 141. ASP.

YORKE, H. W. & SONNHALTER, C. 2002 *ApJ* **569**, 846.

WOOD, D. O. S. & CHURCHWELL, E. 1989 *A&A* **315**, 265.

WU, Y., WEI, Y., ZHAO, M., SHI, Y., YU, W., QIN, S., & HUANG, M. 2004 *A&A* **426**, 503.

ZAPATA, L., RODRIGUEZ, L. F., HO, P. T. P., ZHANG, Q., QI, C., & KURTZ, S. 2005 *ApJ* **630**, L85.

ZHANG, Q., HUNTER, T. R., BRAND, J., SRIDHARAN, T. K., MOLINARI, S., KRAMER, M., & CESARONI, R. 2001 *ApJ* **552**, L167.

ZHANG, Q., HUNTER, T., & SRIDHARAN, T. K. 1998 *ApJ* **505**, L151.

ZHANG, Q., HUNTER, T., SRIDHARAN, T. K., CESARONI, R., MOLINARI, S., WANG, J. & KRAMER, M. 2005 *ApJ* **625**, 864.

Massive-star formation in the Galactic center

By DON F. FIGER

Rochester Institute of Technology, Rochester, NY, USA

The Galactic center is a hotbed of star-formation activity, containing the most massive-star-formation site and three of the most massive young star clusters in the Galaxy. Given such a rich environment, it contains more stars with initial masses above 100 $M_{\odot}$ than anywhere else in the Galaxy. This review concerns the young stellar population in the Galactic center as it relates to massive-star formation in the region. The sample includes stars in the three massive stellar clusters, the population of younger stars in the present sites of star formation, the stars surrounding the central black hole, and the bulk of the stars in the field population. The fossil record in the Galactic center suggests that the recently formed massive stars there are present-day examples of similar populations that must have been formed through star-formation episodes stretching back to the time period when the Galaxy was forming.

---

1. Introduction

The Galactic center (GC) is an exceptional region for testing massive-star formation and evolution models. It contains 10% of the present star-formation activity in the Galaxy, yet fills only a tiny fraction of a percent of the volume in the Galactic disk.† The initial conditions for star formation in the GC are unique in the Galaxy. The molecular clouds in the region are extraordinarily dense, are under high thermal pressure, and are subject to a strong gravitational tidal field. Morris (1993) argue that these conditions may favor the preferential formation of high-mass stars. Being the closest galactic nucleus, the GC gives us an opportunity to observe processes that potentially have wide applicability in other galaxies, both in their centers and in the interaction regions of merging galaxies. Finally, the GC may be the richest site of certain exotic processes and objects in the Galaxy, i.e., runaway stellar mergers leading to intermediate-mass black holes and stellar rejuvenation through atmospheric stripping, to name a few.

This review is primarily concerned with massive-star formation in the region. For thorough reviews on a variety of topics concerning the Galactic center, see Genzel & Townes (1987), Genzel et al. (1994), Morris & Serabyn (1996), and Eckart et al. (2005).

2. The Galactic center environment and star formation

The star-formation efficiency in the GC appears to be high. Plotting the surface star-formation rate ($\Sigma_{\rm SFR} \sim 5\,M_{\odot}\,{\rm yr}^{-1}\,{\rm pc}^{-2}$) versus surface gas density ($\Sigma_{H_2} \sim 400\,M_{\odot}\,{\rm pc}^{-2}$) in a "Schmidt plot" suggests an efficiency of nearly 100%, comparable to that of the most intense infrared circumnuclear starbursts in other galaxies and a factor of twenty higher than in typical galaxies (see Figure 7 in Kennicutt 1998). It is also higher than that elsewhere in the Galaxy; commensurately, stars in the GC emit about 5–10% of the Galaxy's ionizing radiation and infrared luminosity.

† For the purposes of this review, the GC refers to a cylindrical volume with radius of ≈500 pc and thickness of ≈60 pc that is centered on the Galactic nucleus and is coincident with a region of increased dust and gas density, often referred to as the "Central Molecular Zone" (Serabyn & Morris 1996).

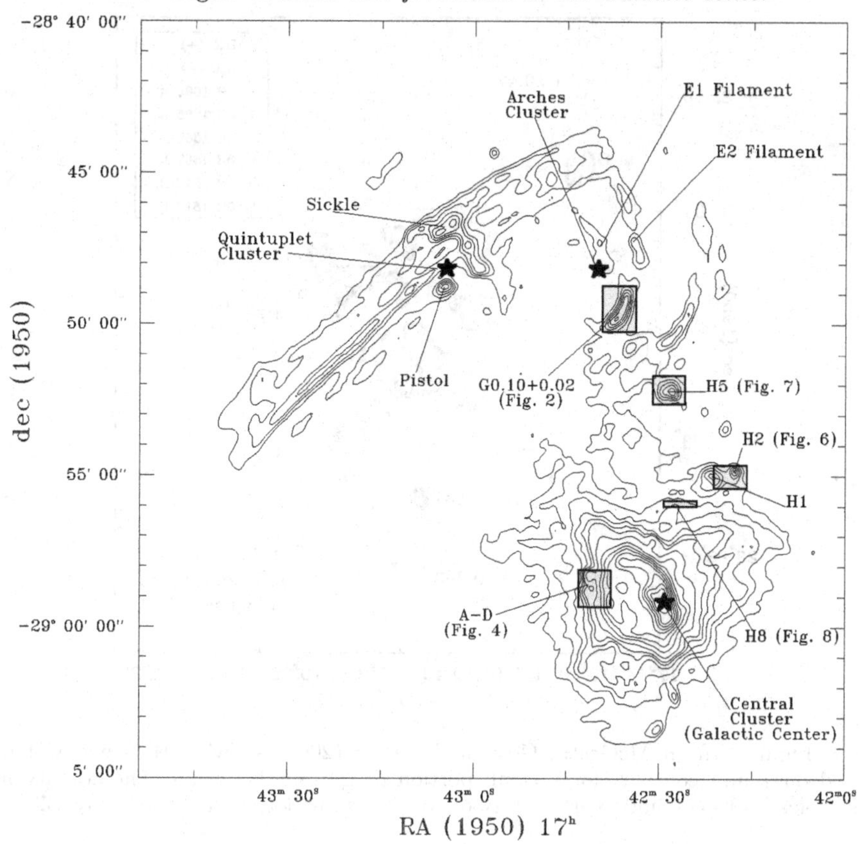

FIGURE 1. Radio emission from the GC region at 6 cm, adapted in Figure 1 by Cotera et al. (1999) from Yusef-Zadeh & Morris (1987). The star symbols represent the three massive clusters. Hot stars in the Quintuplet and Arches clusters ionize gas on the surfaces of nearby molecular clouds to produce the radio emission in the "Sickle" and "G0.10+0.02/E1/E2 Filaments," respectively. The radio emission near the Galactic center is due to a combination of thermal and non-thermal emission. The "H1–8" and "A–D" regions are ultra-compact H II regions surrounding recently formed stars.

Morris & Serabyn (1996) review the content and conditions of the interstellar medium in the "Central Molecular Zone" (CMZ), noting that the molecular clouds in the region are extraordinarily dense ($n > 10^4$ cm^{-3}) and warm ($T \sim 70$ K) with respect to those found in the disk of the Galaxy. Stark et al. (1989) argue that the density and internal velocities of clouds in the GC are a direct result of the strong tidal fields in the region, i.e., only the dense survive. Serabyn & Morris (1996) argue that the inexorable inflow of molecular material from further out in the Galaxy powers continuous and robust star-formation activity in the region.

It is still unclear how magnetic-field strength affects star formation. If it does matter, then the GC might be expected to reveal such effects. The strength of the magnetic field in the GC has been estimated through far-infrared polarized light from aligned dust grains (Hildebrand et al. 1993; Chuss et al. 2003) and Zeeman splitting of the OH molecule (Plante, Lo, & Crutcher 1995). In both cases, the field is inferred to be of milliGauss strength. However, Uchida & Guesten (1995) argue strongly that these strengths are localized to bundles that delineate the extraordinary non-thermal filaments in the region (Yusef-Zadeh & Morris 1987), and are not representative of the field strength that is

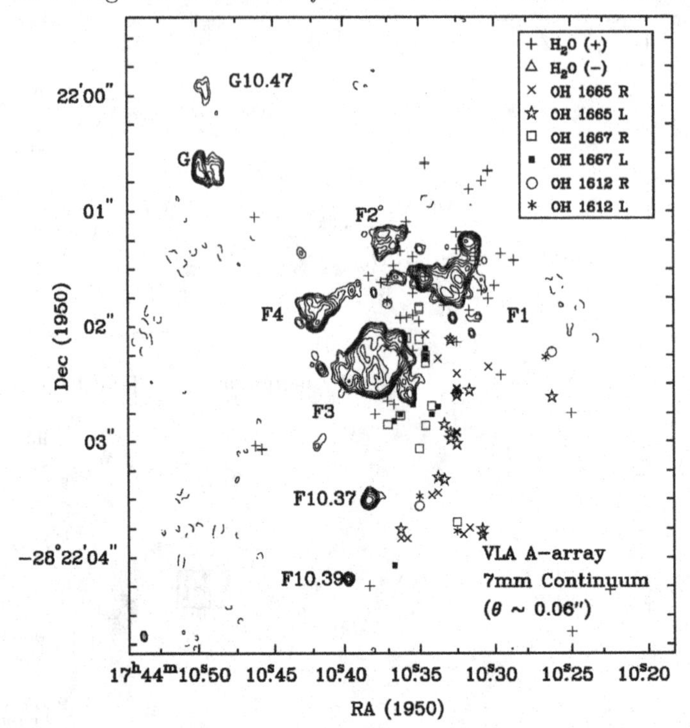

FIGURE 2. Figure 3 from McGrath, Goss, & De Pree (2004) showing H_2O and OH masers overplotted on 7 mm contours for a small portion of the Sgr B2 cloud. The activity in this region is typical of that found near the fifty or so ultra-compact H II regions in Sgr B2.

pervasive in the region. If this is correct, then the fields inside GC molecular clouds may not be so strong versus those inside disk clouds ($B \sim 3$ μG).

Metals in molecular clouds can provide cooling that aids protostellar collapse, but they also create opacity to the UV flux, winds, and bipolar outflows that emanate from newly formed stars. Measurements of metallicity in the Galactic center span a range of solar, observed in stars (Ramírez et al. 2000; Carr, Sellgren, & Balachandran 2000; Najarro et al. 2004), to twice solar, observed in the gas phase (Shields & Ferland 1994), to four times solar, observed through x-ray emission near the very center (Maeda et al. 2002). The errors from the stellar measurements are the smallest and suggest that stars in the GC are formed from material with roughly solar abundances.

3. Present-day star formation in the GC

Present-day star formation in the GC is somewhat subdued compared to the episodes that produced the massive clusters we now see. A dozen or so ultra-compact H II regions are distributed throughout the central 50 pc, each containing one or a few O stars still embedded in their natal environs. Yusef-Zadeh & Morris (1987) identify most of these sources in radio continuum observations (see Figure 1). Zhao et al. (1993) and Goss et al. (1985) infer lyman-continuum fluxes that are comparable to that expected from a single O7V star in each of the H1–H5 and A–D UCHII regions. Cotera et al. (1999) find that several of the recently formed stars in these regions have broken out of their dust shroud, revealing spectra of young massive stars; see also Figer et al. (1994) and Muno et al. (2006) for additional examples.

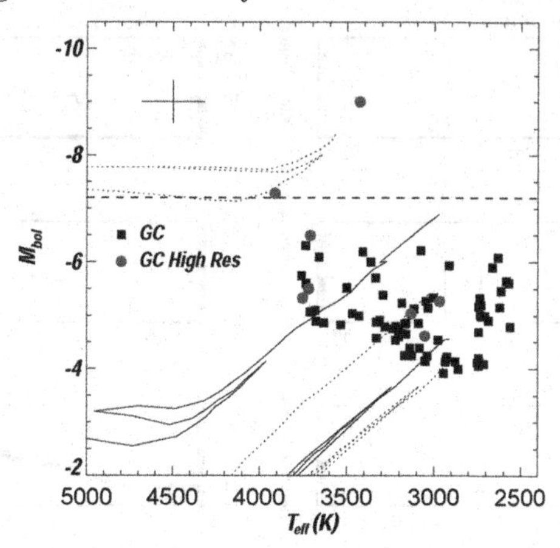

FIGURE 3. Estimates of absolute magnitude versus temperature for stars in the GC from Blum et al. (2003). The lines correspond to model isochrones having ages of 10 Myr, 100 Myr, 1 Gyr, 5 Gyr, and 12 Gyr. The supergiants (above the horizontal line) are descendant from stars having $M \approx 15\text{–}25\,M_\odot$, whereas fainter stars are descendant from lower mass main-sequence stars having a few to $15\,M_\odot$. The presence of these stars in the GC demonstrates intermediate-age star formation of massive stars.

A bit further from the GC, the Sgr B2 molecular cloud harbors a massive-star cluster in the making and is home to the most intense present-day star-formation site in the Galaxy (Gaume et al. 1995; de Pree et al. 1995; McGrath, Goss, & De Pree 2004; Takagi, Murakami, & Koyama 2002; de Vicente et al. 2000; Liu & Snyder 1999; Garay & Lizano 1999; de Pree et al. 1996). Within the next few Myr, this activity should produce a star cluster that is comparable in mass to the Arches cluster (see Figure 2). Sato et al. (2000) note evidence in support of a cloud–cloud collision as the origin for the intense star formation in Sgr B2; these include velocity gradients, magnetic-field morphology, shock-enhanced molecular emission, shock-induced molecular evaporation from dust grains, and distinctly different densities of certain molecular species throughout the cloud.

4. Continuous star formation in the GC

There is ample evidence for persistent star formation in the GC in the form of upper-tip asymptotic giant-branch stars distributed throughout the region (Lebofsky & Rieke 1987; Narayanan, Gould, & Depoy 1996; Frogel, Tiede, & Kuchinski 1999; Sjouwerman et al. 1999). Figure 3 shows a plot for some of these stars, based on spectroscopic data, overlaid with intermediate-age model isochrones (Blum et al. 2003). Note that the giants and supergiants in this plot require ages that span a few Myr to a few Gyr.

One comes to similar conclusions by analyzing photometry of the field population in the GC. Figer et al. (2004) use observed luminosity functions to determine that the star-formation rate has been roughly constant for the lifetime of the Galaxy in the GC, similar to the suggestion in Serabyn & Morris (1996) based on the sharp increase in unresolved infrared light towards the center and a mass-budget argument. Figure 4 shows model and observed luminosity functions (right) for various star-formation scenarios (left) over the lifetime of the Galaxy, assuming a Salpeter IMF (Salpeter 1955) for masses above $10\,M_\odot$, and a flat slope below this mass. The observations were obtained with *HST*/NICMOS and have been corrected for incompleteness. The "burst" models (panels 1, 2, 4, and 5) produce unrealistic ratios of bright-to-faint stars in the luminosity functions, especially

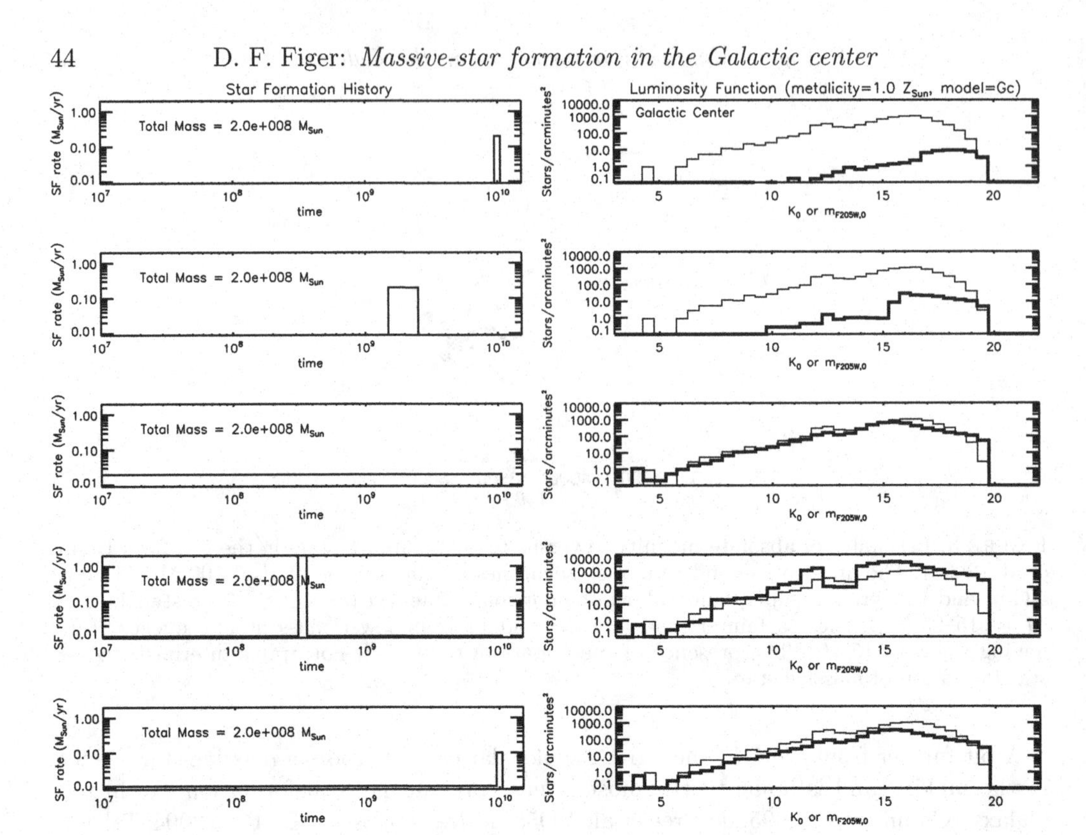

FIGURE 4. A figure adapted from Figer et al. (2004) showing various star-formation scenarios (left), and resultant model luminosity functions (right, thick) compared to observed luminosity functions (right, thin) in the GC. The models assume a Salpeter IMF slope, an elevated lower-mass turnover of $10\,M_\odot$, and are additionally constrained to produce $2 \times 10^8\,M_\odot$ in stars within the region. The observations have been corrected for incompleteness. The third panels from the top, i.e., continuous star formation, best fit the data. The observed turn-down at the faint end appears to be real and is only well fit only by assuming a very high lower-mass turnover.

for the red clump near a dereddened K-band magnitude of 12. The continuous star-formation model (panel 3) best fits the data.

5. Properties of the three massive clusters

The majority of recent star-formation activity in the GC over the past 10 Myr produced three massive clusters: the Central cluster, the Arches cluster, and the Quintuplet cluster. The following sections describe the stellar content in the clusters and the resultant implications for star formation in the region. They closely follow recent reviews (Figer et al. 1999a; Figer 2003, 2004), with updates, as summarized in Table 1.

The three clusters are similar in many respects, as they are all young and contain $\gtrsim 10^4\,M_\odot$ in stars. They have very high central stellar mass densities, up to nearly $10^6\,M_\odot$ pc^{-3}, exceeding central densities in most globular clusters. They have luminosities of $10^{7-8}\,L_\odot$, and are responsible for heating nearby molecular clouds. They also generate 10^{50-51} ionizing photons per second, enough to account for nearby giant H II regions. The primary difference between the clusters is likely to be age, where the Quintuplet and Central clusters are about twice the age of the Arches cluster. In addition, the Central cluster is unique for its population of evolved massive stars that have broad and strong helium emission lines (Krabbe et al. 1991, and references therein). While the

Cluster	Log(M1) $M_\odot$	Log(M2) $M_\odot$	Radius pc	Log(ρ1) $M_\odot$ pc^{-3}	Log(ρ2) $M_\odot$ pc^{-3}	Age Myr	Log(L) $L_\odot$	Log(Q) s^{-1}
Quintuplet	3.0	3.8	1.0	2.4	3.2	3–6	7.5	50.9
Arches†	4.1	4.1	0.19	5.6	5.6	2–3	8.0	51.0
Center‡	3.0	4.0	0.23	4.6	5.6	3–7	7.3	50.5

"M1" is the total cluster mass in observed stars. "M2" is the total cluster mass in all stars extrapolated down to a lower-mass cutoff of 1 $M_\odot$, assuming a Salpeter IMF slope and an upper-mass cutoff of 120 $M_\odot$ (unless otherwise noted). "Radius" gives the average projected separation from the centroid position. "ρ1" is M1 divided by the volume. "ρ2" is M2 divided by the volume. In either case, this is probably closer to the central density than the average density because the mass is for the whole cluster while the radius is the average projected radius. "Age" is the assumed age for the cluster. "L" gives the total measured luminosity for observed stars. "Q" is the estimated Lyman continuum flux emitted by the cluster.

† Mass estimates have been made based upon the number of stars having $M_{\rm initial} > 20\ M_\odot$ given in Figer et al. (1999b) and the mass function slope in Stolte et al. (2003). The age, luminosity and ionizing flux are from Figer et al. (2002).
‡ Krabbe et al. (1995). The mass "M2" has been estimated by assuming that a total $10^{3.5}$ stars have been formed. The age spans a range covering an initial starburst, followed by an exponential decay in the star-formation rate.

TABLE 1. Properties of massive clusters in the Galactic center

	Age (Myr)	O	LBV	WN	WC	RSG	References
Quintuplet	4	100	2	6	11	1	Figer et al. (1999a); Geballe, Najarro, & Figer (2000); Homeier et al. (2003)
Arches	2	160	0	$\gtrsim$6	0	0	Figer et al. (2002)
Center	4–7	100	$\gtrsim$1	$\gtrsim$18	$\gtrsim$12	3	Paumard et al. (2006)
Total		360	$\gtrsim$3	$\gtrsim$29	$\gtrsim$23	4	

TABLE 2. Massive stars in the Galactic center clusters

Quintuplet cluster has a few similar stars (Geballe et al. 1994; Figer et al. 1999a), the Central cluster has far more as a fraction of its total young stellar population (Paumard et al. 2006).

Table 2 summarizes the massive stellar content of the clusters.

5.1. *Central cluster*

The Central cluster contains many massive stars that have recently formed in the past 10 Myr (Becklin et al. 1978; Rieke, Telesco, & Harper 1978; Lebofsky, Rieke, & Tokunaga 1982; Forrest et al. 1987; Allen, Hyland, & Hillier 1990; Krabbe et al. 1991; Najarro et al. 1994; Krabbe et al. 1995; Najarro 1995; Libonate et al. 1995; Blum, Depoy, & Sellgren 1995a; Blum, Sellgren, & Depoy 1995b; Genzel et al. 1996; Tamblyn et al. 1996; Najarro et al. 1997). In all, there are now known to be at least 80 massive stars in the Central cluster (Eisenhauer et al. 2005; Paumard et al. 2006), including $\approx$50 OB stars on the main sequence and 30 more evolved massive stars (see Figure 5). These young stars appear to be confined to two disks (Genzel et al. 2003; Levin & Beloborodov 2003; Paumard et al. 2006; Tanner et al. 2006; Beloborodov et al. 2006). There is also a tight collection of a dozen or so B stars (the "s" stars) in the central arcsecond, highlighted in the small box in the figure. The formation of so many massive stars in the central parsec remains as

FIGURE 5. K-band image of the Central cluster obtained with NAOS/CONICA from Schödel et al. (2007). The 100 or so brightest stars in the image are evolved descendants from main-sequence O stars. The central box highlights the "s" stars that are presumably young and massive ($M_{\rm initial} \approx 20\, M_\odot$).

much a mystery now as it was at the time of the first infrared observations of the region. Most recently, this topic has largely been supplanted by the even more improbable notion that star formation can occur within a few thousand AU of the supermassive black hole, an idea that will be addressed in Section 7. See Alexander (2005) for a thorough review of the "s" stars and Paumard et al. (2006) for a review of the young population in the Central cluster.

5.2. *Arches cluster*

The Arches cluster is unique in the Galaxy for its combination of extraordinarily high mass, M $\approx 10^4\, M_\odot$, and relatively young age, $\tau = 2$ Myr (Figer et al. 2002). Being so young and massive, it contains the richest collection of O stars and WNL stars in any cluster in the Galaxy (Cotera et al. 1996; Serabyn, Shupe, & Figer 1998; Figer et al. 1999b; Blum et al. 2001; Figer et al. 2002). It is ideally suited for testing theories that predict the shape of the IMF up to the highest stellar masses formed (see Section 6).

The cluster is prominent in a broad range of observations. Figure 6 shows an *HST*/NICMOS image of the cluster—the majority of the bright stars in the image have masses greater than $20\, M_\odot$. The most massive dozen or so members of the cluster have strong emission lines at infrared wavelengths (Harris et al. 1994; Nagata et al. 1995; Cotera 1995; Figer 1995; Cotera et al. 1996; Figer et al. 1999b; Blum et al. 2001; Figer et al. 2002). These lines are produced in strong stellar winds that are also detected at radio wavelengths (Lang, Goss, & Rodríguez 2001; Yusef-Zadeh et al. 2003; Lang et al. 2005; Figer et al. 2002), and x-ray wavelengths (Yusef-Zadeh et al. 2002; Rockefeller et al. 2005; Wang, Dong, & Lang 2006).

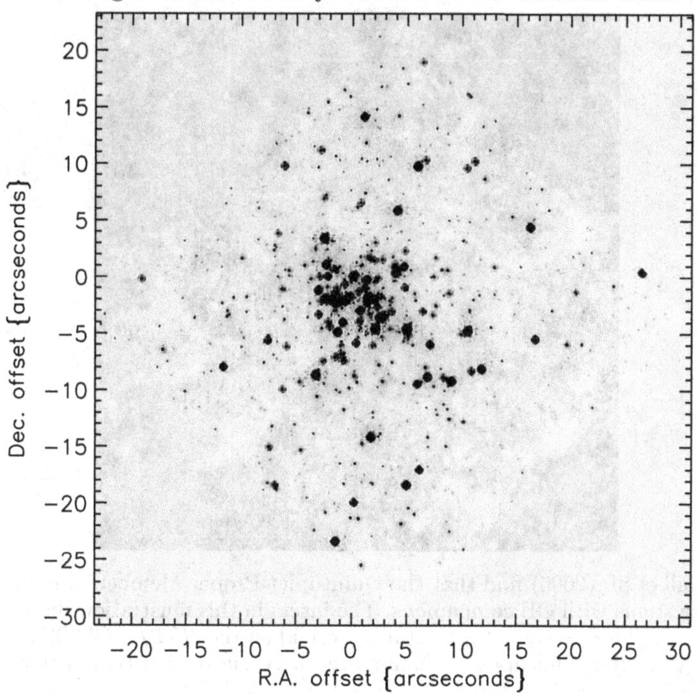

FIGURE 6. F205W image of the Arches cluster obtained by Figer et al. (2002) using *HST*/NICMOS. The brightest dozen or so stars in the cluster have $M_{\rm initial} \gtrsim 100\,M_\odot$, and there are ≈160 O stars in the cluster. The diameter is ≈1 lyr, making the cluster the densest in the Galaxy with $\rho > 10^5\,M_\odot$ pc^{-3}.

5.3. *Quintuplet cluster*

The Quintuplet cluster was originally noted for its five very bright stars, the Quintuplet Proper Members (QPMs) (Glass, Moneti, & Moorwood 1990; Okuda et al. 1990; Nagata et al. 1990). Subsequently, a number of groups identified over 30 stars evolved from massive main-sequence stars (Geballe et al. 1994; Figer, McLean, & Morris 1995; Timmermann et al. 1996; Figer et al. 1999a). Given the spectral types of the massive stars identified in the cluster, it appears that the Quintuplet cluster is ≈4 Myr old and had an initial mass of $>10^4\,M_\odot$ (Figer et al. 1999a). An accounting of the ionizing flux produced by the massive stars in the cluster conclusively demonstrates that the cluster heats and ionizes the nearby "Sickle" H II region (see Figure 1). The Quintuplet is most similar to Westerlund 1 in mass, age, and spectral content (Clark et al. 2005; Negueruela & Clark 2005; Skinner et al. 2006; Groh et al. 2006; Crowther et al. 2006).

Of particular interest in the cluster, the QPMs are very bright at infrared wavelengths, $m_K \approx 6$ to 9, and have color temperatures between ≈600 to 1,000 K. They are luminous, $L \approx 10^5\,L_\odot$, yet spectroscopically featureless, making their spectral classification ambiguous. Figer, Morris, & McLean (1996), Figer et al. (1999a), and Moneti et al. (2001) argue that these objects are not protostars, OH/IR stars, or protostellar OB stars. Instead, they claim that these stars are dust-enshrouded WCL stars (DWCLs), similar to other dusty Galactic WC stars (Williams, van der Hucht, & The 1987), i.e., WR 104 (Tuthill, Monnier, & Danchi 1999) and WR 98A (Monnier, Tuthill, & Danchi 1999). Chiar et al. (2003) tentatively identify a weak spectroscopic feature at 6.2 μm that they attribute to carbon, further supporting the hypothesis that these stars are indeed DWCLs. The stars

FIGURE 7. Tuthill et al. (2006) find that the Quintuplet Proper Members are dusty Wolf-Rayet stars in binary systems with OB companions. The insets in this illustration show high-resolution infrared imaging data for two Quintuplet stars, overlaid on the *HST*/NICMOS image from Figer et al. (1999b). All of the Quintuplet WC stars are dusty, suggesting that they are binary.

have also been detected at x-ray wavelengths (Law & Yusef-Zadeh 2004), and at radio wavelengths (Lang et al. 1999, 2005).

Recently, Tuthill et al. (2006) convincingly show that the QPMs are indeed dusty WC stars. Figure 7 shows data that reveal the pinwheel nature of their infrared emission, characteristic of binary systems containing WCL plus an OB star (Tuthill, Monnier, & Danchi 1999; Monnier, Tuthill, & Danchi 1999). This identification raises intriguing questions concerning massive-star formation and evolution. With their identifications, it becomes clear that every WC star in the Quintuplet is dusty, and presumably binary. There are two possible explanations for this result. Either the binary fraction for massive stars is extremely high (Mason et al. 1998; Nelan et al. 2004), or only binary massive stars evolve through the WCL phase (van der Hucht 2001).

The Quintuplet cluster also contains two Luminous Blue Variables, the Pistol star (Harris et al. 1994; Figer et al. 1998, 1999c), and FMM362 (Figer et al. 1999a; Geballe, Najarro, & Figer 2000). Both stars are extraordinarily luminous ($L > 10^6\,L_\odot$), yet relatively cool ($T \approx 10^4$ K), placing them in the "forbidden zone" of the Hertzsprung-Russell Diagram, above the Humphreys-Davidson limit (Humphreys & Davidson 1994). The Pistol star is particularly intriguing, in that it is surrounded by one of the most massive ($10\,M_\odot$) circumstellar ejecta in the Galaxy (see Figure 8; Figer et al. 1999c; Smith 2009). Both stars are spectroscopically (Figer et al. 1999a) and photometrically variable (Glass et al. 2001). They present difficulties for stellar evolution and formation models. Their inferred initial masses are $>100\,M_\odot$, yet such stars should have already gone supernova in a cluster that is so old, as evidenced by the existence of WC stars (Figer et al. 1999a) and the red supergiant, q7 (Moneti, Glass, & Moorwood 1994; Ramírez et al. 2000). Figer & Kim (2002) and Freitag, Rasio, & Baumgardt (2006) argue that stellar mergers might explain the youthful appearance of these stars. Alternatively, these stars might be binary, although no evidence has been found to support this assertion. Note that in a similar case, LBV1806−20 is also surrounded by a relatively evolved cluster (Eikenberry et al.

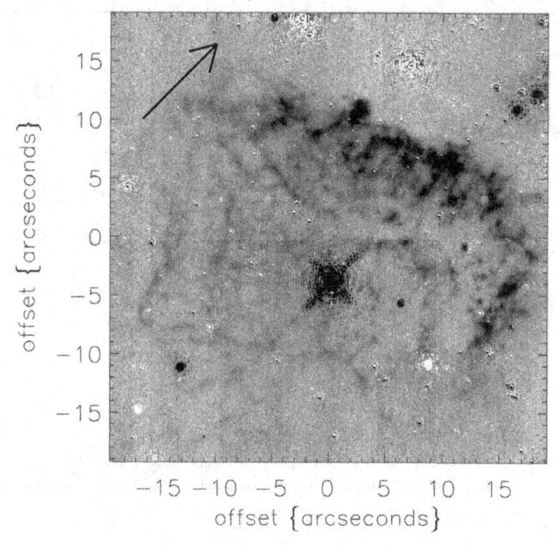

FIGURE 8. Paschen-α image of the region surrounding the Pistol star from Figer et al. (1999c). North is to the upper right, and east is to the upper left. The Pistol star ejected $\approx 10\,M_{\odot}$ of material approximately 6,000 yr ago to form what now appears to be a circumstellar nebula that is ionized by two WC stars to the north of the nebula. Moneti et al. (2001) use ISO data to show that the nebula is filled with dust that is heated by the Pistol star.

2004; Figer et al. 2005), yet it does appear to be binary (Figer, Najarro, & Kudritzki 2004).

6. The initial mass function in the Galactic center

The IMF in the Galactic center has primarily been estimated through observations of the Arches cluster (Figer et al. 1999b; Stolte et al. 2003), although there have been several attempts to extract such information through observations of the Central cluster (Genzel et al. 2003; Nayakshin & Sunyaev 2005; Paumard et al. 2006) and the background population in the region (Figer et al. 2004). These studies suggest an IMF slope that is flatter than the Salpeter value.

6.1. *The slope*

Figer et al. (1999b) and Stolte et al. (2003) estimate a relatively flat IMF slope in the Arches cluster (see Figure 9). Portegies Zwart et al. (2002) interpret the data to indicate an initial slope that is consistent with the Salpeter value, and a present-day slope that has been flattened due to dynamical evolution. Performing a similar analysis, Kim et al. (2000) arrive at the opposite conclusion—that the IMF truly was relatively flat. The primary difficulty in relating the present-day mass function to the initial mass function is the fact that n-body interactions operate on relatively short timescales to segregate the highest stellar masses toward the center of the cluster and to eject the lowest stellar masses out of the cluster. More analysis is needed to resolve this issue.

6.2. *Upper-mass cutoff*

The Arches cluster is the only cluster in the Galaxy that can be used to directly probe an upper-mass cutoff. It is massive enough to expect stars at least as massive as $400\,M_{\odot}$, young enough for its most massive members to still be visible, old enough to have broken out of its natal molecular cloud, close enough, and at a well-established distance, for us

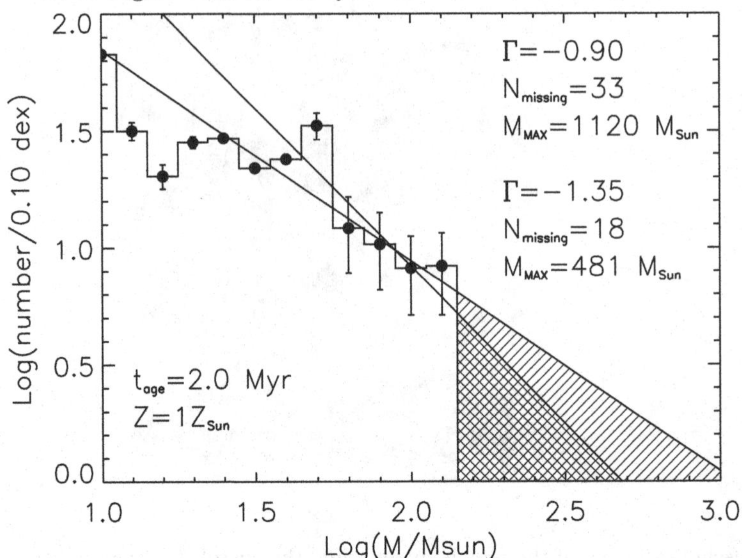

FIGURE 9. Figer (2005) find an apparent upper-mass cutoff to the IMF in the Arches cluster. Magnitudes are transformed into initial mass by assuming the Geneva models for $\tau_{\rm age} = 2$ Myr, solar metallicity, and the canonical mass-loss rates. Error bars indicate uncertainty from Poisson statistics. Two power-law mass functions are drawn through the average of the upper four mass bins, one having a slope of -0.90, as measured from the data, and another having the Salpeter slope of -1.35. Both suggest a dramatic deficit of stars with $M_{\rm initial} > 130\,M_\odot$, i.e., 33 or 18 are missing, respectively. These slopes would further suggest a single star with very large initial mass ($M_{\rm MAX}$). The analysis suggests that the probability of there not being an upper-mass cutoff is $\approx 10^{-8}$.

to discern its individual stars (Figer 2005). There appears to be an absence of stars with initial masses greater than 130 $M_\odot$ in the cluster, where the typical mass function predicts 18 (see Figure 9). Figer (2005) therefore claim a firm upper-mass limit of 150 $M_\odot$. There is additional support for such a cutoff in other environments (Weidner & Kroupa 2004; Oey & Clarke 2005; Koen 2006; Weidner & Kroupa 2006).

6.3. *Lower-mass rollover*

Morris (1993) argue for an elevated lower-mass rollover in the GC based on the environmental conditions therein, and only recently have observations been deep enough to address this claim. Stolte et al. (2005) claim observational evidence for an elevated cutoff around 6 $M_\odot$ in the Arches cluster; however, in that case, confusion and incompleteness are serious problems. In addition, even if the apparent turn-down is a real indication of the initial cluster population, the lack of low-mass stars might result from their ejection through n-body interactions (Kim et al. 2000; Portegies Zwart et al. 2002). Field observations should not suffer from such an effect, as the field should be the repository for low-mass stars ejected from massive clusters in the GC. Figure 4 reveals a turn-down in the observed luminosity function of the field in the GC at a dereddened K-band magnitude greater than 16. This appears to not be a feature of incompleteness, as the data are greater than 50% complete at these magnitudes (Figer et al. 1999b). A more convincing argument, based on this type of data, will await even deeper observations (Kim et al. 2006).

7. The "s" stars

Figure 5 shows a dense collection of about a dozen stars within 1 arcsecond (0.04 pc) of Sgr A* (Genzel et al. 1997; Ghez et al. 1998, 2000; Eckart et al. 2002; Schödel et al. 2002; Ghez et al. 2003; Schödel et al. 2003; Ghez et al. 2005). This cluster stands out for its high stellar density, even compared to the already dense field population in the GC. Schödel et al. (2003) and Ghez et al. (2005) (and references therein) have tracked the proper motions of the "s" stars, finding that they are consistent with closed orbits surrounding a massive, and dark, object having $M \approx$ 2–4 $\times 10^6 M_\odot$, consistent with previous claims based on other methods (Lynden-Bell & Rees 1971; Lacy et al. 1980; Serabyn & Lacy 1985; Genzel & Townes 1987; Sellgren et al. 1987; Rieke & Rieke 1988; McGinn et al. 1989; Lacy, Achtermann, & Serabyn 1991; Lindqvist, Habing, & Winnberg 1992; Haller et al. 1996). The orbital parameters for these stars are well determined, as seen in Figure 10 (top left), and they require the existence of a supermassive black hole in the Galactic center. While these stars are useful as gravitational test particles, they are also interesting in their own right, as they have inferred luminosities and temperatures that are similar to those of young and massive stars (Genzel et al. 1997; Eckart et al. 1999; Figer et al. 2000; Ghez et al. 2003; Eisenhauer et al. 2003, 2005; Paumard et al. 2006). Figure 10 (bottom) shows the absorption lines that suggest relatively high temperatures.

Oddly, the increased density of the young stars in the central arcsecond is not matched by the density distribution of old stars. Indeed, there is a curious absence of late-type stars in the central few arcseconds, as evidenced by a lack of stars with strong CO absorption in their K-band spectra (Lacy, Townes, & Hollenbach 1982; Phinney 1989; Sellgren et al. 1990; Haller et al. 1996; Genzel et al. 1996, 2003). This dearth of old stars represents a true "hole" in three-dimensional space, and not just a projection effect. Even the late-type stars that are projected on to the central parsec generally have relatively low velocities, suggesting dynamical evidence that the region nearest to the black hole lacks old stars (Figer et al. 2003).

The existence of such massive and young stars in the central arcseconds is puzzling, although it is perhaps only an extension of the original problem in understanding the origins of the young stars identified in the central parsec over 20 years ago. Table 3 gives a list of recent papers regarding the origin of the "s" stars. While there are over 30 papers listed in this table, they can be reduced to a few basic ideas. One class of ideas considers the "s" stars as truly young. In this case, the "origin" of the "s" stars is often reduced to the case of massive-star formation in the Galactic center region and transportation of the products to the central arcsecond. The other class regards the "s" stars as old stars that only appear to be young, i.e., via atmospheric stripping, merging, or heating. Both classes require new mechanisms that would be unique to the GC, and they both have considerable weaknesses. For example, Figer et al. (2000) argue that stripped red giants would not be as bright as the "s" stars (see Dray, King, & Davies 2006, for detailed confirmation). See Alexander (2005) for a more thorough discussion of the strengths and weaknesses of these ideas.

If the "s" stars are truly young, then that would require massive clumps to form OB stars ($M_{\rm initial} \gtrsim 20\,M_\odot$). In addition, the clumps would have to form from very high density material in order for them to be stable against tidal disruption. Assuming that the stars formed as far away from the supermassive black hole as possible, while still permitting dynamical friction to transport them into the central arcsecond during their lifetimes, then the required densities must be $>10^{11}$ cm^{-3} (Figer et al. 2000).

The average molecular cloud density in the GC is about five orders of magnitude less, so highly compressive events might be required to achieve the necessary densities.

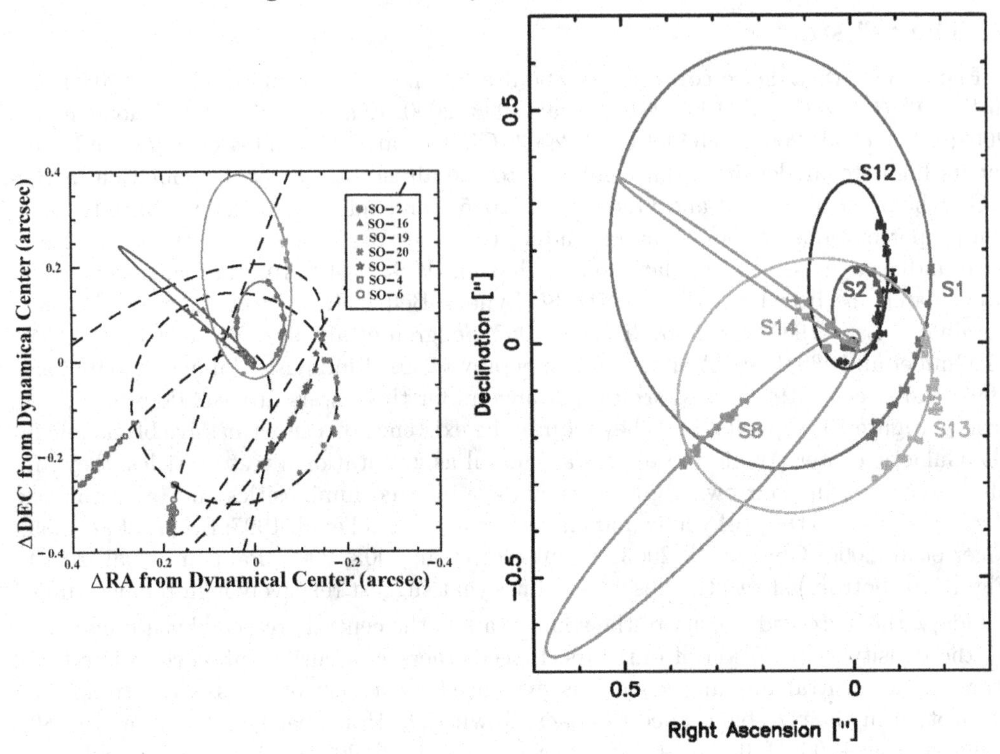

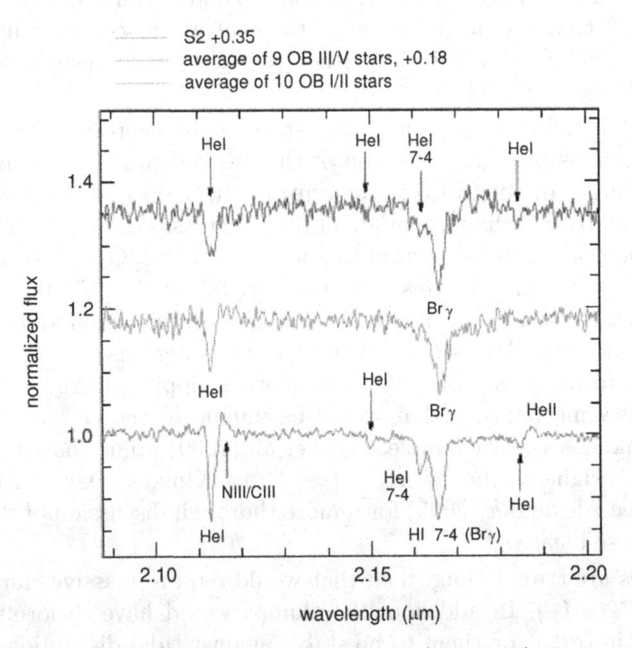

FIGURE 10. (top left) Figure 2 in Ghez et al. (2005) and (top right) Figure 6 in Schödel et al. (2003), stretched to the same scale. Both figures fit similar model orbits through separate proper motion data sets for the "s" stars. (bottom) Paumard et al. (2006) find that one of the "s" stars, S2, has a K-band spectrum that is similar to those of OB stars in the central parsec (see also Ghez et al. 2003).

Reference	*Description*
Lacy, Townes, & Hollenbach (1982)	tidal disruption of red giants
Morris (1993)	compact objects surrounded by material from red-giant envelopes ("Thorne-Zytkow objects")
Davies et al. (1998)	red-giant envelope stripping through n-body interactions
Alexander (1999)	red-giant envelope stripping through dwarf-giant interactions
Bailey & Davies (1999)	colliding red giants
Morris, Ghez, & Becklin (1999)	duty cycle of formation from infalling CND clouds and evaporation of gas reservoir by accretion and star-formation light
Gerhard (2001)	decaying massive cluster
Alexander & Morris (2003)	tidal heating of stellar envelopes to form "Squeezars"
Genzel et al. (2003)	stellar rejuvenation through red-giant mergers
Gould & Quillen (2003)	"exchange reaction" between massive-star binary and massive black hole
Hansen & Milosavljević (2003)	stars captured by inspiraling intermediate-mass black hole
Levin & Beloborodov (2003)	formation in nearby gas disk
McMillan & Portegies Zwart (2003); Kim, Figer, & Morris (2004)	inward migration of young cluster with IMBH
Portegies Zwart, McMillan, & Gerhard (2003); Kim & Morris (2003)	inward migration of young cluster
Alexander & Livio (2004)	orbital capture of young stars by MBH-SBH binaries
Milosavljević & Loeb (2004)	formation in molecular disk
Davies & King (2005)	tidal stripping of red-giant (AGB) stars (but see critique in Goodman & Paczynski 2005)
Gürkan & Rasio (2005)	decaying cluster with formation of an IMBH
Haislip & Youdin (2005)	formation in disk and orbital relaxation (but see critique in Goodman & Paczynski 2005)
Levin, Wu, & Thommes (2005)	dynamical interactions with sinking IMBH
Nayakshin & Sunyaev (2005)	*in-situ* formation within central parsec
Nayakshin & Cuadra (2005)	formation in a fragmenting star disk
Subr & Karas (2005)	formation in disk and accelerated orbital relaxation
Berukoff & Hansen (2006)	cluster inspiral and n-body interactions
Hopman & Alexander (2006)	resonant relaxation of orbits
Freitag, Amaro-Seoane, & Kalogera (2006)	mass segregation through interactions with compact remnants
Levin (2007)	star formation in fragmenting disk
Perets, Hopman, & Alexander (2006)	exchange reactions between massive-star binaries induced by efficient relaxation by massive perturbers
Dray, King, & Davies (2006)	tidal stripping of red-giant (AGB) stars (but see critique in Figer et al. 2000)

TABLE 3. Chronologically sorted list of references that explore hypotheses on the origin of the "s" stars. Some of the references primarily concern the other young stars in the central parsec and are included in the table because they propose ideas that may relate to the origins of the "s" stars. Contributions to this table have been made by Tal Alexander (priv. communication).

Alternatively, the required densities can be reduced if the stars are gravitationally bound to significant mass, i.e., a surrounding stellar cluster. Indeed, Gerhard (2001), Portegies Zwart, McMillan, & Gerhard (2003), and Kim & Morris (2003) showed that particularly massive clusters could form tens of parsecs outside of the center and be delivered into the central parsec in just a few million years. The efficiency of this method is improved with the presence of an intermediate black hole in the cluster (McMillan & Portegies Zwart 2003; Kim, Figer, & Morris 2004). It is key in any of these cluster transport models that the host system have extremely high densities of $>10^6\,M_\odot\,\mathrm{pc}^{-3}$, comparable to the highest estimated central density of the Arches cluster after core collapse (Kim & Morris 2003). Detailed n-body simulations suggest that while these ideas may be relevant for the origins of the young stars in the central parsec, it is unlikely that they could explain the existence of the "s" stars in the central arcsecond.

8. Comparisons to other massive-star populations in Galaxy

There are relatively few clusters in the Galaxy with as many massive stars as in the GC clusters. NGC 3603 has about a factor of two less mass than each of the GC clusters (Moffat et al. 2002); whereas, W1 has at least a factor of two greater mass (Clark et al. 2005; Negueruela & Clark 2005; Skinner et al. 2006; Groh et al. 2006). The next nearest similarly massive cluster is R136 in the LMC (Massey & Hunter 1998). All of these clusters, and the GC clusters, appear to have IMF slopes that are consistent with the Salpeter value (or slightly flatter) and are young enough to still possess a significant massive-star population. It is remarkable to note that these massive clusters appear quite similar in stellar content, whether in the Galactic disk, the GC, or even the lower metallicity environment of the LMC. Evidently, the star-formation processes and natal environments that gave birth to these clusters must be similar enough to produce clusters that are virtually indistinguishable.

There are probably more massive clusters yet to be found in the Galaxy. The limited sample of known massive clusters is a direct result of extinction, as most star-formation sites in the Galaxy are obscured by dust at optical wavelengths. While infrared observations have been available for over 30 years, they have not provided the necessary spatial resolution, nor survey coverage, needed to probe the Galactic disk for massive clusters. Recently, a number of groups have begun identifying candidate massive-star clusters using near-infrared surveys with arcsecond resolution (Bica et al. 2003; Dutra et al. 2003; Mercer et al. 2005). Indeed, these surveys have already yielded a cluster with approximate initial mass of 20,000 to 40,000 $M_\odot$ (Figer et al. 2006), and one would expect more to be discovered from them.

The present-day sites of massive-star formation in the Galaxy have been known for some time through radio and far-infrared observations, as their hottest members ionize and heat nearby gas in molecular clouds. As one of many examples, consider W49, which is the next most massive-star–formation site in the Galaxy compared to Sgr B2 (Homeier & Alves 2005), and wherein the star formation appears to be progressing in stages over timescales that far exceed the individual collapse times for massive-star progenitors. This suggests a stimulus that triggers the star formation, perhaps provided by a "daisy chain" effect in which newly formed stars trigger collapse in nearby parts of the cloud. Similar suggestions are proposed in the 30 Dor region surrounding R136 (Walborn, Maíz-Apellániz, & Barbá 2002). While there is no evidence for an age-dispersed population in Sgr B2, Sato et al. (2000) suggest that the cloud was triggered to form stars through a cloud–cloud interaction.

9. Conclusions

Massive-star formation in the GC has produced an extraordinary sample of stars populating the initial mass function up to a cutoff of approximately 150 $M_\odot$. The ranges of inferred masses and observed spectral types are as expected from stellar-evolution models, and the extraordinary distribution of stars in the region is a direct consequence of the large amount of mass that has fed star formation in the GC. The origin of the massive stars in the central parsec, and especially the central arcsecond, remains unresolved.

I thank the following individuals for discussions related to this work: Mark Morris, Bob Blum, Reinhard Genzel, Paco Najarro, Sungsoo Kim, and Peter Tuthill. Tal Alexander made substantial contributions to Table 3. The material in this paper is based upon work supported by NASA under award No. NNG05-GC37G, through the Long-Term Space Astrophysics program.

REFERENCES

Alexander, T. 1999 *ApJ* **527**, 835.
Alexander, T. 2005, *Physics Reports* **419**, 65.
Alexander, T. & Livio, M. 2004 *ApJ* **606**, L21.
Alexander, T. & Morris, M. 2003 *ApJ* **590**, L25.
Allen, D. A., Hyland, A. R., & Hillier, D. J. 1990 *MNRAS* **244**, 706.
Bailey, V. C. & Davies, M. B. 1999 *MNRAS* **308**, 257.
Becklin, E. E., Matthews, K., Neugebauer, G., & Willner, S. P. 1978 *ApJ* **219**, 121.
Beloborodov, A. M., Levin, Y., Eisenhauer, F., Genzel, R., Paumard, T., Gillessen, S., & Ott, T. 2006 *ApJ* **648**, 405.
Berukoff, S. J. & Hansen, B. M. S. 2006 *ApJ* **650**, 901.
Bica, E., Dutra, C. M., Soares, J., & Barbuy, B. 2003 *A&A* **404**, 223.
Blum, R. D., DePoy, D. L., & Sellgren, K. 1995a *ApJ* **441**, 603.
Blum, R. D., Ramírez, S. V., Sellgren, K., & Olsen, K. 2003 *ApJ* **597**, 323.
Blum, R. D., Schaerer, D., Pasquali, A., Heydari-Malayeri, M., Conti, P. S., & Schmutz, W. 2001 *AJ* **122**, 1875.
Blum, R. D., Sellgren, K., & DePoy, D. L. 1995b *ApJ* **440**, L17.
Carr, J. S., Sellgren, K., & Balachandran, S. C. 2000 *ApJ* **530**, 307.
Chiar, J. E., Adamson, A. J., Whittet, D. C. B., & Pendleton, Y. J. 2003 *Astronomische Nachrichten Supplement* **324**, 109.
Chuss, D. T., Davidson, J. A., Dotson, J. L., Dowell, C. D., Hildebrand, R. H., Novak, G., & Vaillancourt, J. E. 2003 *ApJ* **599**, 1116.
Clark, J. S., Negueruela, I., Crowther, P. A., & Goodwin, S. P. 2005 *A&A* **434**, 949.
Cotera, A. S. 1995 Ph.D. Thesis.
Cotera, A. S., Erickson, E. F., Colgan, S. W. J., Simpson, J. P., Allen, D. A., & Burton, M. G. 1996 *ApJ* **461**, 750.
Cotera, A. S., Simpson, J. P., Erickson, E. F., Colgan, S. W. J., Burton, M. G., & Allen, D. A. 1999 *ApJ* **510**, 747.
Crowther, P. A., Hadfield, L. J., Clark, J. S., Negueruela, I., & Vacca, W. D. 2006 *MNRAS* **372**, 1407.
Davies, M. B., Blackwell, R., Bailey, V. C., & Sigurdsson, S. 1998 *MNRAS* **301**, 745.
Davies, M. B. & King, A. 2005 *ApJ* **624**, L25.
de Pree, C. G., Gaume, R. A., Goss, W. M., & Claussen, M. J. 1995 *ApJ* **451**, 284.
de Pree, C. G., Gaume, R. A., Goss, W. M., & Claussen, M. J. 1996 *ApJ* **464**, 788.
de Vicente, P., Martín-Pintado, J., Neri, R., & Colom, P. 2000 *A&A* **361**, 1058.
Dray, L. M., King, A. R., & Davies, M. B. 2006 *MNRAS* **371**, 31.
Dutra, C. M., Bica, E., Soares, J., & Barbuy, B. 2003 *A&A* **400**, 533.
Eckart, A., Genzel, R., Ott, T., & Schödel, R. 2002 *MNRAS* **331**, 917.
Eckart, A., Ott, T., & Genzel, R. 1999 *A&A* **352**, L22.
Eckart, A., Schödel, R., & Straubmeier, C. 2005. *The Black Hole at the Center of the Milky Way.* Imperial College Press.
Eikenberry, S. S., et al. 2004 *ApJ* **616**, 506.

EISENHAUER, F., ET AL. 2005 *ApJ* **628**, 246.

EISENHAUER, F., SCHÖDEL, R., GENZEL, R., OTT, T., TECZA, M., ABUTER, R., ECKART, A., & ALEXANDER, T. 2003 *ApJ* **597**, L121.

FIGER, D. F. 1995 Ph.D. Thesis.

FIGER, D. F. 2003. In *Massive Star Odyssey, from Main Sequence to Supernova* (eds. K. A. van der Hucht, A. Herrero, & C. Esteban). Proc. IAU 212, p. 487. ASP.

FIGER, D. F. 2004. In *The Formation and Evolution of Massive Young Star Clusters* (eds. H. J. G. L. M. Lamers, L. J. Smith, & A. Nota). ASP Conf. Ser. 322, p. 49. ASP.

FIGER, D. F. 2005 *Nature* **434**, 192.

FIGER, D. F., ET AL. 2000 *ApJ* **533**, L49.

FIGER, D. F., ET AL. 2002 *ApJ* **581**, 258.

FIGER, D. F., ET AL. 2003 *ApJ* **599**, 1139.

FIGER, D. F., BECKLIN, E. E., MCLEAN, I. S., & MORRIS, M. 1994. In *Astronomy with Arrays, The Next Generation* (ed. I. S. McLean). ASSL Vol. 190, p. 545. ASSL.

FIGER, D. F. & KIM, S. S. 2002. In *Stellar Collisions, Mergers and their Consequences* (ed. M. M. Shara). ASP Conf. Ser. 263, p. 287. ASP.

FIGER, D. F., KIM, S. S., MORRIS, M., SERABYN, E., RICH, R. M., & MCLEAN, I. S. 1999b *ApJ* **525**, 750.

FIGER, D. F., MACKENTY, J. W., ROBBERTO, M., SMITH, K., NAJARRO, F., KUDRITZKI, R. P., & HERRERO, A. 2006 *ApJ* **643**, 1166.

FIGER, D. F., MCLEAN, I. S., & MORRIS, M. 1995 *ApJ* **447**, L29.

FIGER, D. F., MCLEAN, I. S., & MORRIS, M. 1999a *ApJ* **514**, 202.

FIGER, D. F., MORRIS, M., GEBALLE, T. R., RICH, R. M., SERABYN, E., MCLEAN, I. S., PUETTER, R. C., & YAHIL, A. 1999c *ApJ* **525**, 759.

FIGER, D. F., MORRIS, M., & MCLEAN, I. S. 1996. In *The Galactic Center* (ed. R. Gredel). ASP Conf. Ser. 102, p. 263. ASP.

FIGER, D. F., NAJARRO, F., GEBALLE, T. R., BLUM, R. D., & KUDRITZKI, R. P. 2005 *ApJ* **622**, L49.

FIGER, D. F., NAJARRO, F., & KUDRITZKI, R. P. 2004 *ApJ* **610**, L109.

FIGER, D. F., NAJARRO, F., MORRIS, M., MCLEAN, I. S., GEBALLE, T. R., GHEZ, A. M., & LANGER, N. 1998 *ApJ* **506**, 384.

FIGER, D. F., RICH, R. M., KIM, S. S., MORRIS, M., & SERABYN, E. 2004 *ApJ* **601**, 319.

FORREST, W. J., SHURE, M. A., PIPHER, J. L., & WOODWARD, C. E. 1987. In *The Galactic Center* (ed. D. C. Backer). AIP Conf. Proc. 155, p. 153. AIP.

FREITAG, M., AMARO-SEOANE, P., & KALOGERA, V. 2006 *ApJ* **649**, 91.

FREITAG, M., RASIO, F. A., & BAUMGARDT, H. 2006 *MNRAS* **368**, 121.

FROGEL, J. A., TIEDE, G. P., & KUCHINSKI, L. E. 1999 *AJ* **117**, 2296.

GARAY, G. & LIZANO, S. 1999 *PASP* **111**, 1049.

GAUME, R. A., CLAUSSEN, M. J., DE PREE, C. G., GOSS, W. M., & MEHRINGER, D. M. 1995 *ApJ* **449**, 663.

GEBALLE, T. R., GENZEL, R., KRABBE, A., KRENZ, T., & LUTZ, D. 1994. In *Astronomy with Arrays, The Next Generation* (ed. I. S. McLean). ASSL Vol. 190, p. 73. ASSL.

GEBALLE, T. R., NAJARRO, F., & FIGER, D. F. 2000 *ApJ* **530**, L97.

GENZEL, R., ET AL. 2003 *ApJ* **594**, 812.

GENZEL, R., ECKART, A., OTT, T., & EISENHAUER, F. 1997 *MNRAS* **291**, 219.

GENZEL, R., HOLLENBACH, D., & TOWNES, C. H. 1994 *Reports of Progress in Physics*, **57**, 417.

GENZEL, R., THATTE, N., KRABBE, A., KROKER, H., & TACCONI-GARMAN, L. E. 1996 *ApJ* **472**, 153.

GENZEL, R. & TOWNES, C. H. 1987 *ARA&A* **25**, 377.

GERHARD, O. 2001 *ApJ* **546**, L39.

GHEZ, A. M., ET AL. 2003 *ApJ* **586**, L127.

GHEZ, A. M., KLEIN, B. L., MORRIS, M., & BECKLIN, E. E. 1998 *ApJ* **509**, 678.

GHEZ, A. M., MORRIS, M., BECKLIN, E. E., TANNER, A., & KREMENEK, T. 2000 *Nature* **407**, 349.

GHEZ, A. M., SALIM, S., HORNSTEIN, S. D., TANNER, A., LU, J. R., MORRIS, M., BECKLIN, E. E., & DUCHÊNE, G. 2005 *ApJ* **620**, 744.

GLASS, I. S., MATSUMOTO, S., CARTER, B. S., & SEKIGUCHI, K. 2001 *MNRAS* **321**, 77.

GLASS, I. S., MONETI, A., & MOORWOOD, A. F. M. 1990 *MNRAS* **242**, 55P.

GOODMAN, J. & PACZYNSKI, B. 2005; astro-ph/0504079.

GOSS, W. M., SCHWARZ, U. J., VAN GORKOM, J. H., & EKERS, R. D. 1985 *MNRAS* **215**, 69P.

GOULD, A. & QUILLEN, A. C. 2003 *ApJ* **592**, 935.

GROH, J. H., DAMINELI, A., TEODORO, M., & BARBOSA, C. L. 2006 *A&A* **457**, 591.

GÜRKAN, M. A., & RASIO, F. A. 2005 *ApJ* **628**, 236.

HAISLIP, G., & YOUDIN, A. N. 2005 *American Astronomical Society Meeting Abstracts* **207**, #181.14.

HALLER, J. W., RIEKE, M. J., RIEKE, G. H., TAMBLYN, P., CLOSE, L., & MELIA, F. 1996 *ApJ* **456**, 194.

HANSEN, B. M. S. & MILOSAVLJEVIĆ, M. 2003 *ApJ* **593**, L77.

HARRIS, A. I., KRENZ, T., GENZEL, R., KRABBE, A., LUTZ, D., POLITSCH, A., TOWNES, C. H., & GEBALLE, T. R. 1994. In *The Nuclei of Normal Galaxies: Lessons from the Galactic Center* (eds. R. Genzel & A. I. Harris). NATO ASI Ser. C Proc. 445, p. 223. ASI.

HILDEBRAND, R. H., DAVIDSON, J. A., DOTSON, J., FIGER, D. F., NOVAK, G., PLATT, S. R., & TAO, L. 1993 *ApJ* **417**, 565.

HOMEIER, N. L. & ALVES, J. 2005 *A&A* **430**, 481.

HOMEIER, N. L., BLUM, R. D., PASQUALI, A., CONTI, P. S., & DAMINELI, A. 2003 *A&A* **408**, 153.

HOPMAN, C. & ALEXANDER, T. 2006 *Journal of Physics: Conference Series* **54**, 321.

HUMPHREYS, R. M. & DAVIDSON, K. 1994 *PASP* **106**, 1025.

KENNICUTT, R. C., JR. 1998 *ARA&A* **36**, 189.

KIM, S. S., FIGER, D. F., LEE, H. M., & MORRIS, M. 2000 *ApJ* **545**, 301.

KIM, S. S., FIGER, D. F., & MORRIS, M. 2004 *ApJ* **607**, L123.

KIM, S. S., FIGER, D. F., KUDRITZKI, R. P., & NAJARRO, F. N. 2006 *ApJ* **653**, L113.

KIM, S. S. & MORRIS, M. 2003 *ApJ* **597**, 312.

KOEN, C. 2006 *MNRAS* **365**, 590.

KRABBE, A., ET AL. 1995 *ApJ* **447**, L95.

KRABBE, A., GENZEL, R., DRAPATZ, S., & ROTACIUC, V. 1991 *ApJ* **382**, L19.

LACY, J. H., ACHTERMANN, J. M., & SERABYN, E. 1991 *ApJ* **380**, L71.

LACY, J. H., TOWNES, C. H., GEBALLE, T. R., & HOLLENBACH, D. J. 1980 *ApJ* **241**, 132.

LACY, J. H., TOWNES, C. H., & HOLLENBACH, D. J. 1982 *ApJ* **262**, 120.

LANG, C. C., FIGER, D. F., GOSS, W. M., & MORRIS, M. 1999 *AJ* **118**, 2327.

LANG, C. C., GOSS, W. M., & RODRÍGUEZ, L. F. 2001 *ApJ* **551**, L143.

LANG, C. C., JOHNSON, K. E., GOSS, W. M., & RODRÍGUEZ, L. F. 2005 *AJ* **130**, 2185.

LAW, C. & YUSEF-ZADEH, F. 2004 *ApJ* **611**, 858.

LEBOFSKY, M. J. & RIEKE, G. H. 1987. In *The Galactic Center* (ed. D. C. Backer). AIP Conf. Proc. 155, p. 79. AIP.

LEBOFSKY, M. J., RIEKE, G. H., & TOKUNAGA, A. T. 1982 *ApJ* **263**, 736.

LEVIN, Y. 2007 *MNRAS* **374**, 515.

LEVIN, Y. & BELOBORODOV, A. M. 2003 *ApJ* **590**, L33.

LEVIN, Y., WU, A., & THOMMES, E. 2005 *ApJ* **635**, 341.

LIBONATE, S., PIPHER, J. L., FORREST, W. J., & ASHBY, M. L. N. 1995 *ApJ* **439**, 202.

LINDQVIST, M., HABING, H. J., & WINNBERG, A. 1992 *A&A* **259**, 118.

LIU, S.-Y. & SNYDER, L. E. 1999 *ApJ* **523**, 683.

LYNDEN-BELL, D. & REES, M. J. 1971 *MNRAS* **152**, 461.

MAEDA, Y., ET AL. 2002 *ApJ* **570**, 671.

MASON, B. D., GIES, D. R., HARTKOPF, W. I., BAGNUOLO, W. G., JR., TEN BRUMMELAAR, T., & MCALISTER, H. A. 1998 *AJ* **115**, 821.

MASSEY, P. & HUNTER, D. A. 1998 *ApJ* **493**, 180.

MCGINN, M. T., SELLGREN, K., BECKLIN, E. E., & HALL, D. N. B. 1989 *ApJ* **338**, 824.

McGrath, E. J., Goss, W. M., & De Pree, C. G. 2004 *ApJS* **155**, 577.
McMillan, S. L. W. & Portegies Zwart, S. F. 2003 *ApJ* **596**, 314.
Mercer, E. P., et al. 2005 *ApJ* **635**, 560.
Milosavljević, M. & Loeb, A. 2004 *ApJ* **604**, L45.
Moffat, A. F. J., et al. 2002 *ApJ* **573**, 191.
Moneti, A., Glass, I. S., & Moorwood, A. F. M. 1994 *MNRAS* **268**, 194.
Moneti, A., Stolovy, S., Blommaert, J. A. D. L., Figer, D. F., & Najarro, F. 2001 *A&A* **366**, 106.
Monnier, J. D., Tuthill, P. G., & Danchi, W. C. 1999 *ApJ* **525**, L97.
Morris, M. 1993 *ApJ* **408**, 496.
Morris, M., Ghez, A. M., & Becklin, E. E. 1999 *Advances in Space Research* **23**, 959.
Morris, M. & Serabyn, E. 1996 *ARA&A* **34**, 645.
Muno, M. P., Bower, G. C., Burgasser, A. J., Baganoff, F. K., Morris, M. R., & Brandt, W. N. 2006 *ApJ* **638**, 183.
Nagata, T., Woodward, C. E., Shure, M., & Kobayashi, N. 1995 *AJ* **109**, 1676.
Nagata, T., Woodward, C. E., Shure, M., Pipher, J. L., & Okuda, H. 1990 *ApJ* **351**, 83.
Najarro, F. 1995 Ph.D. Thesis.
Najarro, F., Figer, D. F., Hillier, D. J., & Kudritzki, R. P. 2004 *ApJ* **611**, L105.
Najarro, F., Hillier, D. J., Kudritzki, R. P., Krabbe, A., Genzel, R., Lutz, D., Drapatz, S., & Geballe, T. R. 1994 *A&A* **285**, 573.
Najarro, F., Krabbe, A., Genzel, R., Lutz, D., Kudritzki, R. P., & Hillier, D. J. 1997 *A&A* **325**, 700.
Nayakshin, S. & Cuadra, J. 2005 *A&A* **437**, 437.
Narayanan, V. K., Gould, A., & Depoy, D. L. 1996 *ApJ* **472**, 183.
Nayakshin, S. & Sunyaev, R. 2005 *MNRAS* **364**, L23.
Negueruela, I. & Clark, J. S. 2005 *A&A* **436**, 541.
Nelan, E. P., Walborn, N. R., Wallace, D. J., Moffat, A. F. J., Makidon, R. B., Gies, D. R., & Panagia, N. 2004 *AJ* **128**, 323.
Oey, M. S. & Clarke, C. J. 2005 *ApJ* **620**, L43.
Okuda, H., et al. 1990 *ApJ* **351**, 89.
Paumard, T., et al. 2006 *ApJ* **643**, 1011.
Perets, H. B., Hopman, C., & Alexander, T. 2006 *ApJ* **656**, 709.
Phinney, E. S. 1989. In *The Center of the Galaxy* (ed. M. Morris). IAU Symp. 136, p. 543. Kluwer Academic Press.
Plante, R. L., Lo, K. Y., & Crutcher, R. M. 1995 *ApJ* **445**, L113.
Portegies Zwart, S. F., Makino, J., McMillan, S. L. W., & Hut, P. 2002 *ApJ* **565**, 265.
Portegies Zwart, S. F., McMillan, S. L. W., & Gerhard, O. 2003 *ApJ* **593**, 352.
Ramírez, S. V., Sellgren, K., Carr, J. S., Balachandran, S. C., Blum, R., Terndrup, D. M., & Steed, A. 2000 *ApJ* **537**, 205.
Rieke, G. H. & Rieke, M. J. 1988 *ApJ* **330**, L33.
Rieke, G. H., Telesco, C. M., & Harper, D. A. 1978 *ApJ* **220**, 556.
Rockefeller, G., Fryer, C. L., Melia, F., & Wang, Q. D. 2005 *ApJ* **623**, 171.
Salpeter, E. E. 1955 *ApJ* **121**, 161.
Sato, F., Hasegawa, T., Whiteoak, J. B., & Miyawaki, R. 2000 *ApJ* **535**, 857.
Schödel, R., et al. 2002 *Nature* **419**, 694.
Schödel, R., Ott, T., Genzel, R., Eckart, A., Mouawad, N., & Alexander, T. 2003 *ApJ* **596**, 1015.
Schödel, R., et al. 2007 *A&A* **462**, L1.
Sellgren, K., Hall, D. N. B., Kleinmann, S. G., & Scoville, N. Z. 1987 *ApJ* **317**, 881.
Sellgren, K., McGinn, M. T., Becklin, E. E., & Hall, D. N. 1990 *ApJ* **359**, 112.
Serabyn, E. & Lacy, J. H. 1985 *ApJ* **293**, 445.
Serabyn, E. & Morris, M. 1996 *Nature* **382**, 602.
Serabyn, E., Shupe, D., & Figer, D. F. 1998 *Nature* **394**, 448.
Shields, J. C., & Ferland, G. J. 1994 *ApJ* **430**, 236.

Sjouwerman, L. O., Habing, H. J., Lindqvist, M., van Langevelde, H. J., & Winnberg, A. 1999. In *The Central Parsecs of the Galaxy* (eds. H. Falcke, A. Cotera, W. J. Duschl, F. Melia, & M. J. Rieke). ASP Conf. Ser. 186, p. 379. ASP.

Skinner, S. L., Simmons, A. E., Zhekov, S. A., Teodoro, M., Damineli, A., & Palla, F. 2006 *ApJ* **639**, L35.

Smith, N. 2009; *this volume.*

Stark, A. A., Bally, J., Wilson, R. W., & Pound, M. W. 1989. In *The Center of the Galaxy* (ed. M. Morris). IAU Symp. 136, p. 129. Kluwer Academic Press.

Stolte, A., Brandner, W., Grebel, E. K., Figer, D. F., Eisenhauer, F., Lenzen, R., & Harayama, Y. 2003 *The Messenger* **111**, 9.

Stolte, A., Brandner, W., Grebel, E. K., Lenzen, R., & Lagrange, A.-M. 2005 *ApJ* **628**, L113.

Subr, L. & Karas 2005 *A&A* **433**, 405.

Takagi, S.-i., Murakami, H., & Koyama, K. 2002 *ApJ* **573**, 275.

Tamblyn, P., Rieke, G. H., Hanson, M. M., Close, L. M., McCarthy, D. W., Jr., & Rieke, M. J. 1996 *ApJ* **456**, 206.

Tanner, A., et al. 2006 *ApJ* **641**, 891.

Timmermann, R., Genzel, R., Poglitsch, A., Lutz, D., Madden, S. C., Nikola, T., Geis, N., & Townes, C. H. 1996 *ApJ* **466**, 242.

Tuthill, P. G., Monnier, J. D., & Danchi, W. C. 1999 *Nature* **398**, 487.

Tuthill, P., Monnier, J., Tanner, A., Figer, D., & Ghez, A. 2006 *Science* **313**, 935.

Uchida, K. I., & Guesten, R. 1995 *A&A* **298**, 473.

van der Hucht, K. A. 2001 *New Astronomy Review* **45**, 135.

Walborn, N. R., Maíz-Apellániz, J., & Barbá, R. H. 2002 *AJ* **124**, 1601.

Wang, Q. D., Dong, H., & Lang, C. 2006 *MNRAS* **371**, 38.

Weidner, C. & Kroupa, P. 2004 *MNRAS* **348**, 187.

Weidner, C. & Kroupa, P. 2006 *MNRAS* **365**, 1333.

Williams, P. M., van der Hucht, K. A., & The, P. S. 1987 *Quarterly Journal of the Royal Astronomical Society* **28**, 248.

Yusef-Zadeh, F., Law, C., Wardle, M., Wang, Q. D., Fruscione, A., Lang, C. C., & Cotera, A. 2002 *ApJ* **570**, 665.

Yusef-Zadeh, F. & Morris, M. 1987 *ApJ* **320**, 545.

Yusef-Zadeh, F., Nord, M., Wardle, M., Law, C., Lang, C., & Lazio, T. J. W. 2003 *ApJ* **590**, L103.

Zhao, J.-H., Desai, K., Goss, W. M., & Yusef-Zadeh, F. 1993 *ApJ* **418**, 235.

An x-ray tour of massive-star–forming regions with *Chandra*

By LEISA K. TOWNSLEY

Department of Astronomy and Astrophysics, Pennsylvania State University,
525 Davey Laboratory, University Park, PA 16802, USA

The *Chandra X-ray Observatory* is providing fascinating new views of massive-star–forming regions, revealing all stages in the life cycles of massive stars and their effects on their surroundings. I present a *Chandra* tour of some of the most famous of these regions: M17, NGC 3576, W3, Tr14 in Carina, and 30 Doradus. *Chandra* highlights the physical processes that characterize the lives of these clusters, from the ionizing sources of ultracompact H II regions (W3) to superbubbles so large that they shape our views of galaxies (30 Dor). X-ray observations usually reveal hundreds of pre-main sequence (lower-mass) stars accompanying the OB stars that power these great H II region complexes, although in one case (W3 North) this population is mysteriously absent. The most massive stars themselves are often anomalously hard x-ray emitters; this may be a new indicator of close binarity. These complexes are sometimes suffused by soft diffuse x-rays (M17, NGC 3576), signatures of multi-million-degree plasmas created by fast O-star winds. In older regions we see the x-ray remains of the deaths of massive stars that stayed close to their birthplaces (Tr14, 30 Dor), exploding as cavity supernovae within the superbubbles that these clusters created.

---

1. Revealing the life cycle of a massive stellar cluster

High-resolution x-ray images from the *Chandra X-ray Observatory* and *XMM-Newton* elucidate all stages in the life cycles of massive stars—from ultracompact H II (UCHII) regions to supernova remnants—and the effects that those massive stars have on their surroundings. x-ray studies of massive-star–forming regions (MSFRs) thus give insight into the massive stars themselves, the accompanying lower-mass cluster population, new generations of stars that may be triggered by the massive cluster, interactions of massive-star winds with themselves and with the surrounding neutral medium, and the fate of massive stars that die as cavity supernovae inside the wind-blown bubbles they created.

In the era of *Spitzer* and excellent ground-based near-infrared (IR) data, why are x-ray studies of MSFRs important? *Chandra* routinely penetrates $A_V > 100$ mag of extinction with little confusion or contamination, revealing young stellar populations in a manner that is unbiased by the presence of inner disks around these stars. These vast pre-main sequence (pre-MS) populations are easily seen by *Chandra* and the $17' \times 17'$ imaging array of its Advanced CCD Imaging Spectrometer (ACIS-I); a typical ACIS-I observation of a nearby ($D < 3$ kpc) MSFR, lasting 40–100 ksec, finds 500–1500 young stars, tracing the initial mass function (IMF) down to $\sim 1\,M_{\odot}$. These observations can increase the number of known cluster members by as much as a factor of 50 in poorly-studied regions (e.g., NGC 6357; J. Wang et al. 2007).

In the first seven months of 2006, at least 18 refereed papers on x-ray observations of MSFRs (using *Chandra* or *XMM*) were published or submitted. These include studies of 30 Doradus (Townsley et al. 2006a,b), NGC 6334 (Ezoe et al. 2006), σ Orionis (Franciosini et al. 2006), Westerlund 1 (Skinner et al. 2006; Muno et al. 2006), RCW 38 (Wolk et al. 2006), Cepheus B/OB3b (Getman et al. 2006), NGC 6357 (J. Wang et al. 2007), M17 (Broos et al. 2007), NGC 2362 (Delgado et al. 2006), NGC 2264 (Flaccomio et al. 2006; Rebull et al. 2006), Trumpler 16 in Carina (Sanchawala et al. 2006), NGC 6231 (Sana et al. 2006), the Orion Nebula Cluster (Stassun et al. 2006), the Arches and Quintuplet clusters (Q. Wang et al. 2006), and W49A (Tsujimoto et al. 2006). Early *Chandra* and *XMM* studies were summarized in Townsley et al. (2003); more recent work is described in Feigelson et al. (2007). Other MSFRs currently under study at Penn State include RCW 49 (Townsley et al. 2004), NGC 7538 (Tsujimoto et al. 2005), W51A (Townsley et al. 2004, 2005), and the Rosette Nebula (J. Wang et al. 2008).

This contribution begins with recent x-ray results on the young MSFRs M17 and NGC 3576, two regions at different evolutionary stages, but with notably similar outflows of hot, x-ray–emitting gas. Next I introduce a new ACIS mosaic of the Westerhout 3 complex, showing how triggered star formation in the same molecular cloud can result in very different stellar clusters. Then I address the rich and complicated point source and diffuse x-ray emission seen towards Trumpler 14 in Carina. Finally I provide a first look at a new *Chandra* observation of 30 Doradus obtained in early 2006.

2. M17, the Omega Nebula: An x-ray champagne flow

The Omega Nebula (M17) is the second-brightest H II region in the sky (after the Orion Nebula). Its OB cluster NGC 6618 probably has well over 5000 members, extending up to masses $M \sim 70 + 70\,M_{\odot}$ (spectral type O4V+O4V) in its central binary. It is situated at the edge of one of the Galaxy's most massive and dense molecular cloud cores, M17SW, at a distance of 1.6 kpc (Nielbock et al. 2001). It is a blister H II region viewed nearly edge-on, on the periphery of a giant molecular cloud (GMC) containing an UCHII region with UV-excited water masers and active star formation. The ionization front of the H II region encounters the GMC along two photodissociation regions, called the northern and southern bars. With an age of <1 Myr (Hanson et al. 1997) and no evolved stars, M17 is one of the few bright MSFRs that has sufficient stellar wind power to produce an x-ray outflow and yet is unlikely to have hosted any supernovae.

We observed M17's central cluster NGC 6618 with *Chandra*/ACIS-I for 40 ksec in March 2002. The extensive soft diffuse emission seen in M17 is rare; it was described by Townsley et al. (2003) and is shown in Figure 1. The 877 point sources detected in this observation are described, cataloged, and compared to IR data by Broos et al. (2007). Fewer than 10% of these point sources (in M17 and in the other Galactic MSFRs described here) are unrelated to the MSFR (foreground stars and background AGN, J. Wang et al. 2007). *Chandra* has a point spread function (PSF) that is sub-arcsecond on-axis, but degrades radially to become many arcseconds wide at the field edge. This effect is apparent in Figure 1 and in other ACIS images shown in this paper.

Figure 1 (right) shows the 5.8 μm GLIMPSE image of M17 from the *Spitzer Space Telescope*'s Infrared Array Camera (IRAC). The early *Chandra* and *Spitzer* images in Figure 1 show dramatic and complex interacting components of hot and cold material. On the western side of the cluster, the parsec-scale thermalized O-star winds shock the parental GMC, triggering new massive-star formation to the southwest. To the east, the winds escape the confinement of the blister H II region producing the brightest known x-ray champagne flow, heating and chemically enriching the ISM (Townsley et al. 2003).

Chandra's strongly-varying PSF and the ACIS-I camera's large field of view also make it difficult to convey the sub-arcsecond imaging quality in full-field smoothed images. Figure 2 (top) shows the central $\sim 25''$ of the M17 ACIS image, binned to $0.25''$ pixels. NGC 6618's central binary is separated by $\sim 2''$ and is clearly resolved. These are the brightest x-ray sources in the field (Broos et al. 2007). Figure 2 (bottom) shows their spectra and model fits. Each source was fit using *XSPEC* (Arnaud 1996) with a two-temperature *apec* thermal plasma model (R. Smith et al. 2001), including absorption. The soft plasma components in these O4 stars are of similar strength and are typical of

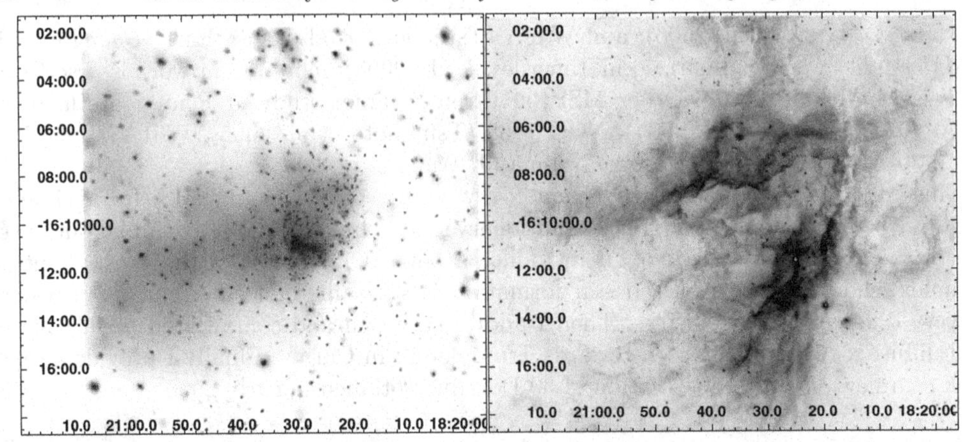

FIGURE 1. Left panel: Smoothed 0.5–2 keV ACIS-I image of M17 (∼8 pc across), highlighting the soft diffuse x-ray emission. Here and throughout this paper, coordinates are J2000 RA and Dec. Right panel: *Spitzer*/IRAC 5.8 μm image of the same M17 field, part of the GLIMPSE survey and provided by the GLIMPSE team. This pair of images highlights both the rich stellar population revealed through x-ray/IR comparison studies and the complementarity of the two Great Observatories for mapping diffuse emission in the region.

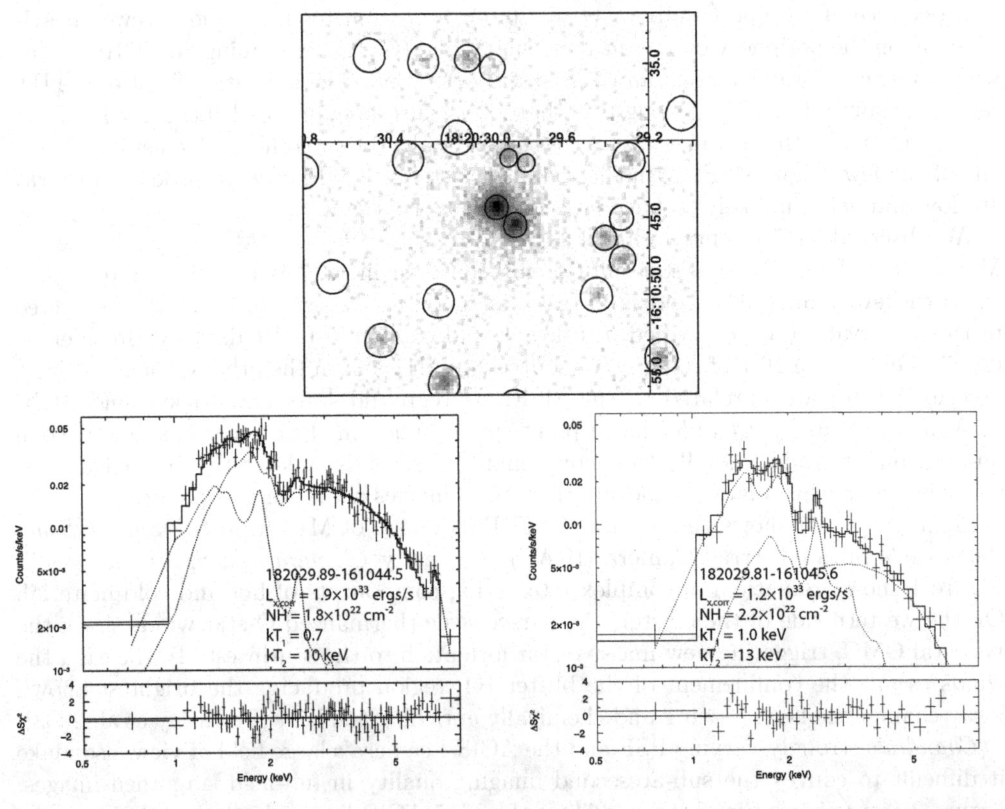

FIGURE 2. Upper panel: High-resolution ACIS image of the central region of NGC 6618 in M17, highlighting the bright O4+O4 binary. The nearly circular polygons mark the photometric and spectral extraction regions for each source. Lower panels: ACIS spectra of the two O4 stars. For each source, the top panel shows the event data as a histogram, each component of the model (convolved with the instrumental response) as dotted lines, and the composite model as a solid line. The lowest panels show the fit residuals.

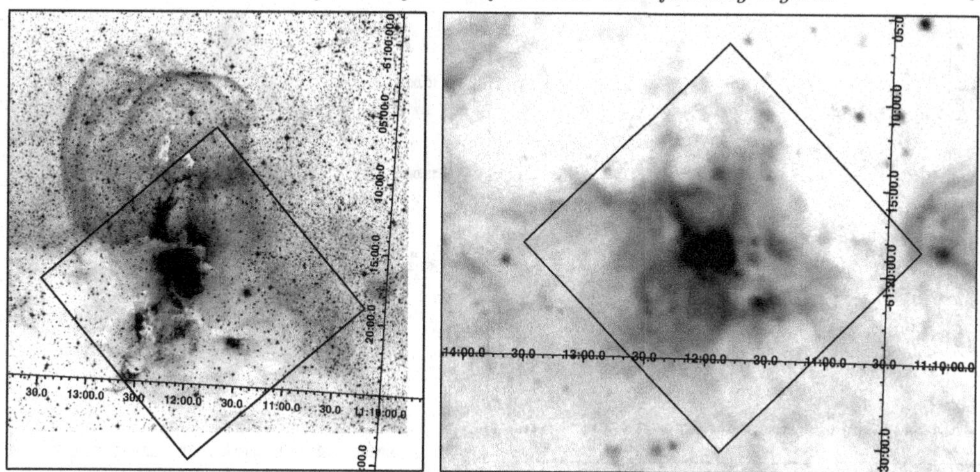

FIGURE 3. NGC 3576 images. Left panel: Digital Sky Survey red band. Right panel: *MSX* 8 μm. The black squares show the size and approximate roll angle of the ACIS-I array.

O stars; they are thought to be due to microshocks in the powerful stellar winds. Both of these sources, however, show very hard thermal plasmas as well; the brighter O4 star's spectrum is dominated by this hard component. A reasonable explanation for this hard emission is that it comes from colliding winds in an as-yet-unrecognized close binary system. Since both O4 stars exhibit this hard emission, it is possible that the O4+O4 binary actually consists of no fewer than four massive components.

3. The giant H II region NGC 3576: An M17 analog?

NGC 3576 is a Galactic giant H II region, located at a distance of 2.8 kpc (de Pree et al. 1999). It contains several known O and early B stars, but these are not sufficient to account for the Lyα ionizing photons inferred from radio data (Figuerêdo et al. 2002). An embedded massive IR star cluster is located at the edge of its GMC (Persi et al. 1994). It contains at least 51 stars earlier than A0, most with large IR excesses. It likely includes as yet unrecognized O stars that would account for the number of Lyα photons.

A radio study of NGC 3576 (de Pree et al. 1999) revealed a large north-south velocity gradient in the ionized gas of the nebula, indicating a large-scale ionized outflow. This flow may contribute to the large loops and filaments seen in visual data (Figure 3, left). The 8μm *MSX* image (Figure 3, right) is complementary to the visual data, revealing heavy obscuration in the southern half of the field and what appears to be a bipolar bubble likely blown by the massive stars in NGC 3576's central cluster. This bubble appears closed to the south, where it encounters the dense GMC, but open to the north, in the same direction as the large visual loops.

In July 2005, we obtained a 60-ksec ACIS observation of NGC 3576 to see if a soft x-ray bubble suggested by the *ROSAT* data was indicating that NGC 3576 possessed an x-ray outflow similar to that seen in M17. The ACIS aimpoint was directed at the strong infrared source IRS 1, at the core of the embedded young stellar cluster, to search for embedded protostars and the stellar sources responsible for ionizing NGC 3576, as well as to resolve the brightest flaring pre-MS stars. Our ACIS field here is similar to that for M17: only part of the bubble seen by *ROSAT* is captured. It was important to keep the aimpoint on IRS 1, though, to study the embedded population and the IR cluster, which hold the key to understanding the source of the *ROSAT* bubble.

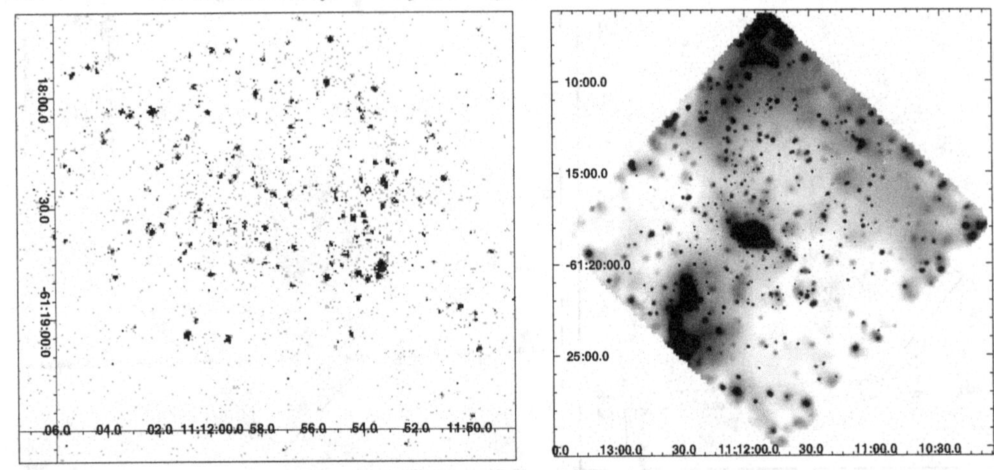

FIGURE 4. NGC 3576 ACIS-I data. Left panel: The central cluster. The approximate position of IRS 1 is marked with a small black circle. Right panel: Full-band (0.5–8 keV) smoothed image.

Figure 4 (left) shows that this observation indeed reveals the massive stellar engine powering NGC 3576, shredding the molecular cloud from which it formed. Many sources not known even from mid-IR studies (e.g., Maercker et al. 2006) are seen in x-rays. The brightest of these is a hard, deeply-embedded source $\sim$20″ south of IRS 1. This and other new x-ray sources may be the long-sought embedded massive stars providing the extra ionization for NGC 3576.

Much of the x-ray bubble seen by *ROSAT* is resolved by *Chandra* into a wide array of point sources distributed across the field. Diffuse x-rays remain, however, near the top of the ACIS field, and in a more concentrated patch southeast of the central cluster. Comparing the smoothed ACIS image (Figure 4, right) to the visual and mid-IR data in Figure 3, it appears that the northern visual loops outline (at least in part) soft diffuse x-ray emission in and beyond the northern, more open lobe of the bipolar bubble. Throughout the field the diffuse x-ray emission is complementary to the mid-IR emission, suggesting either that warm dust has displaced the hot x-ray gas or that this material is shadowing the x-rays, absorbing and obscuring any soft x-ray emission that may lie behind it. The southern lobe is also traced by soft diffuse x-ray gas; the lack of shadowing here implies that this part of the wind-blown bubble lies on the near side of the GMC. There are clear relationships between NGC 3576's x-ray plasma, velocity and temperature gradients measured in the radio, the impressive visual loops, and the mid-IR bipolar bubble. Spectral analysis of this diffuse x-ray emission is ongoing and should give some insight into the energetics of this important complex.

Chandra has revealed that NGC 3576 and M17 are two examples of the same phenomenon, although NGC 3576's cluster is younger and more deeply embedded in its natal cloud, leading to the more complex spatial morphology in its soft diffuse x-ray emission. It appears that hot, flowing x-ray gas from OB winds may be a common component of young blister H II regions, strongly affecting their morphology and dynamics.

4. W3: Ionizing sources revealed and a missing cluster

W3 is a large, nearby ($D = 2.0$ kpc, Hachisuka et al. 2006) MSFR in the Perseus Arm of the Milky Way. The W3 complex contains one of the most massive molecular clouds in the outer Galaxy (Heyer & Terebey 1998), massive embedded protostars (Megeath

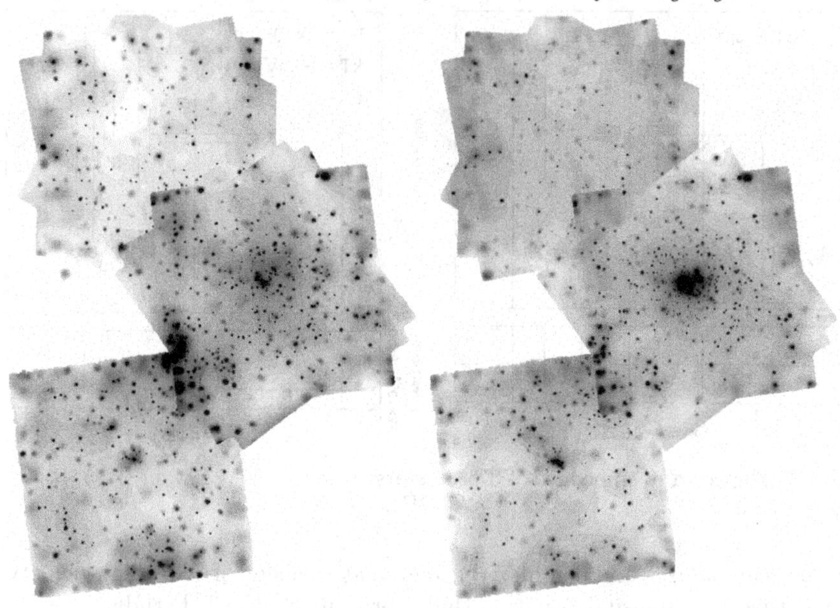

FIGURE 5. Left panel: Smoothed soft-band (0.5–2 keV) ACIS-I mosaic of W3 North (top), W3 Main (middle), and W3(OH) (bottom). The slightly older OB cluster IC 1795 is partially imaged between W3 Main and W3(OH). Right panel: Hard-band (2–8 keV) mosaic of the same W3 regions.

et al. 2005), near- and mid-IR sources (Kraemer et al. 2003), masers, and outflows. It is unique in containing all morphological classes of H II regions, from hypercompact to diffuse, 0.01 to 1 pc in diameter, with ages 10^3–10^6 yrs (Tieftrunk et al. 1997).

Many authors have argued that star formation in W3 is being induced by the expansion of the adjacent W4 superbubble, which is sweeping up molecular gas into a high-density layer, within which stars are forming. Oey et al. (2005) revised this scenario, proposing that the young (3–5 Myr) OB cluster IC 1795, triggered to form by W4, is blowing its own second-generation superbubble at the molecular cloud interface, triggering in turn the W3 complex of massive-star formation. W3 is an ideal testbed for understanding recent, ongoing, and triggered star formation; x-ray sources there are all at the same distance, yet exist in a range of dynamical settings, providing a series of controlled environments in which to study the violent interactions of massive stars with their natal cloud.

With many observations throughout 2005, ACIS-I was used to image the three major star-forming complexes, W3 North, W3 Main, and W3(OH), amassing ~80 ksec on each field. A mosaic of these observations is shown in Figure 5. ACIS resolves the OB stars powering the H II regions, embedded massive stars and protostellar objects, and the pre-MS population, yielding >1300 x-ray sources in the three ~80 ksec pointings.

W3 North is a well-developed parsec-scale H II region powered by the O6V star GSC 04050−02567 that may be isolated, with no lower-mass accompanying population (Carpenter et al. 2000). The H II region is so bright in the near-IR due to nebular emission that it is difficult to perform an IR search for an underlying cluster. *Chandra* detects the O6 star; it is a modest, soft x-ray emitter, typical of single O stars. However Figure 5 clearly reveals that W3 North is not an Orion-like field; no cluster of young x-ray–emitting pre-MS stars is seen. This may indicate that the region has an anomalous IMF, that this O6 star is a young example of a massive star that formed in isolation (de Wit et al. 2005), or that the O6 star is a runaway from either IC 1795 or W3 Main that has somehow

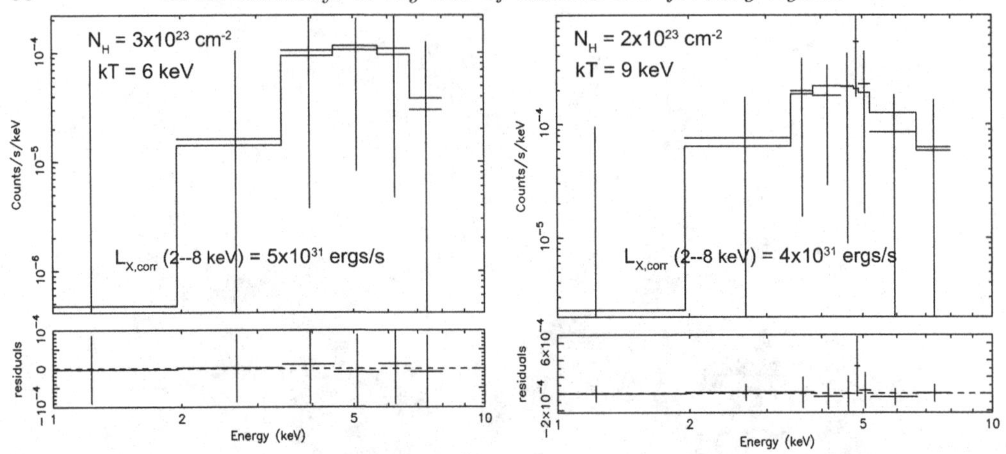

FIGURE 6. ACIS spectra of embedded W3 protostars. Left panel: W3 Main IRS5. Right panel: W3(OH).

managed to create a large and complex H II region at the edge of W3's high-density layer. In any case, this is an unusual situation that merits more detailed study.

W3 Main harbors the strongest CO peak in W3 and IR clusters with over 200 stars (Carpenter et al. 2000). An early 40-ksec ACIS-I observation showed possible extended emission spatially coincident with the H II regions; over 100 of the 236 ACIS sources detected in this original dataset are near the W3 core (Hofner et al. 2002). We added another 40 ksec to this original observation; in the composite image, it is clear that W3 Main is the dominant star-forming center of W3, with an extensive population of young stars nearly filling the $17' \times 17'$ ACIS-I field of view. The full extent of W3 Main was not clear from earlier IR studies. Of the >1300 sources in the W3 ACIS mosaic, over half of them are captured in the W3 Main pointing.

W3(OH) contains the B0.5 star IRAS 02232+6138 surrounded by a cluster of more than 200 stars (Carpenter et al. 2000), an UCHII region ($2''$ in diameter), OH and H_2O masers, outflows, IR sources, and strong CO indicating massive embedded protostars. ACIS detects a point source associated with the W3(OH) UCHII region and resolves the string of known IR stellar clusters northeast of W3(OH) that may indicate a broader region of ongoing star formation (Tieftrunk et al. 1997).

Figure 6 shows the ACIS spectra of two interesting embedded sources in the W3 complex: IRS5 in W3 Main and the x-ray source in W3(OH), presumably the source of ionization for the UCHII region. IRS5 is thought to be a multiple system of perhaps five proto-OB stars (Megeath et al. 2005). We don't resolve the components with ACIS, but we do find a very hard spectrum for the x-ray source coincident with IRS5, again consistent with colliding-wind binary emission, although this hard emission could be coming from a cluster wind generated by the combined effects of the winds from all of these massive protostars (Cantó et al. 2000). The W3(OH) source is also extremely hard. Due to the high obscuration toward these sources, we have no information on any soft x-ray emission that they might be generating. The luminosities that we note in Figure 6 are corrected for this absorption, but are only given for the hard (2–8 keV) x-ray band.

5. Trumpler 14: Swarms of stars and misplaced hot gas

The Carina complex is part of the Sagittarius-Carina spiral arm, at a distance of ~2.3 kpc (N. Smith 2006b). It is a remarkably rich star-forming region, containing 8 open

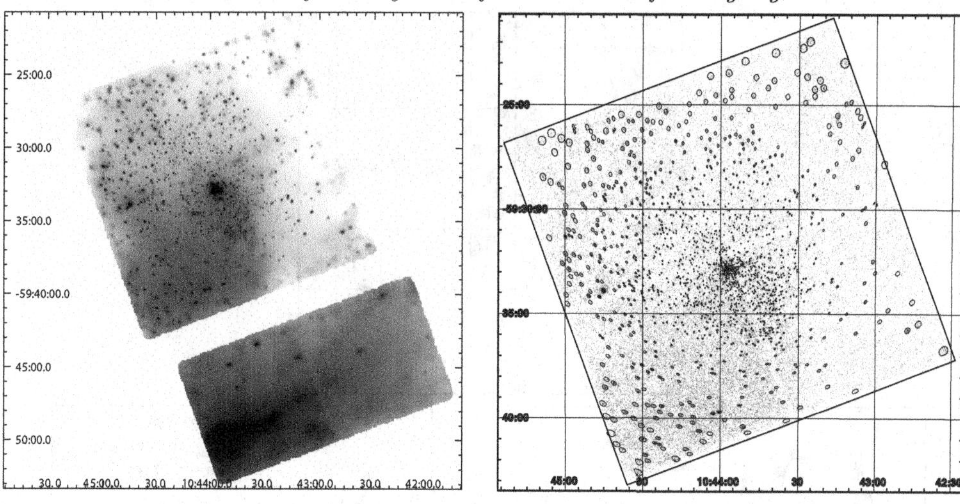

FIGURE 7. ACIS images of Tr14 in Carina. Left panel: Full-field, full-band smoothed image. Right panel: The ACIS-I image (binned by 2″), showing ∼1600 point source extraction regions.

clusters with at least 66 O stars, several Wolf-Rayet (WR) stars, and the luminous blue variable Eta Carinae (N. Smith 2006a). The combined Carina OB clusters Tr16, Tr14, and Cr228 contain the nearest rich concentration of early O stars; their ionizing flux and winds may be fueling a young bipolar superbubble (N. Smith et al. 2000). High ISM velocities throughout the complex (Walborn et al. 2002b) and the presence of evolved stars may imply that a supernova might already have occurred in this region, although no well-defined remnant has ever been seen.

Tr14 is an extremely rich, young (∼1 My), compact OB cluster near the center of the Carina complex, containing at least 30 O and early B stars (Vázquez et al. 1996). The radio H II region Carina I is situated just west of Tr14, at the edge of the GMC and near a strong CO peak. It is ionized by Tr14 and is carving out a cavity in the molecular cloud which now contains *IRAS* sources, high-mass protostars, their associated UCHII regions, and ionization fronts (Brooks et al. 2001). Although star formation has ceased in the Keyhole region of Carina due to the harsh environment created by the massive stars there, the proximity of Tr14 to the GMC has triggered a new generation of star formation in Carina I (Rathborne et al. 2002).

In September 2004 we obtained a 57-ksec ACIS-I observation of Tr14 (Figure 7). The aimpoint of this observation is the central star in the cluster, HD 93129AB, a very early-type (O2I–O3.5V) binary (Walborn et al. 2002a), with the two components separated by ∼3″. Our ACIS-I observation reveals ∼1600 x-ray point sources in the Tr14 region, suffused by bright, soft diffuse emission. Since the Tr16 and Tr14 clusters overlap, some sources toward the east and southeast are most likely Tr16 members; a few sources in the northeast corner of our field may be members of the nearby Tr15 cluster. The soft diffuse emission pervading the eastern half of our field is sharply cut off to the west by the GMC. The density of x-ray point sources also falls sharply to the west, due in part to reduced sensitivity because of higher extinction and in part because star formation is just now being triggered in that region, as described above.

Figure 8 (top) shows the crowded central region of Tr14 in x-rays, with the source extraction regions drawn. The brightest source near field center is HD 93129A; it is clearly resolved from HD 93129B (seen slightly to the southeast) and from another, fainter star just to its south (west of HD 93129B). The ACIS spectra of these early O stars are shown

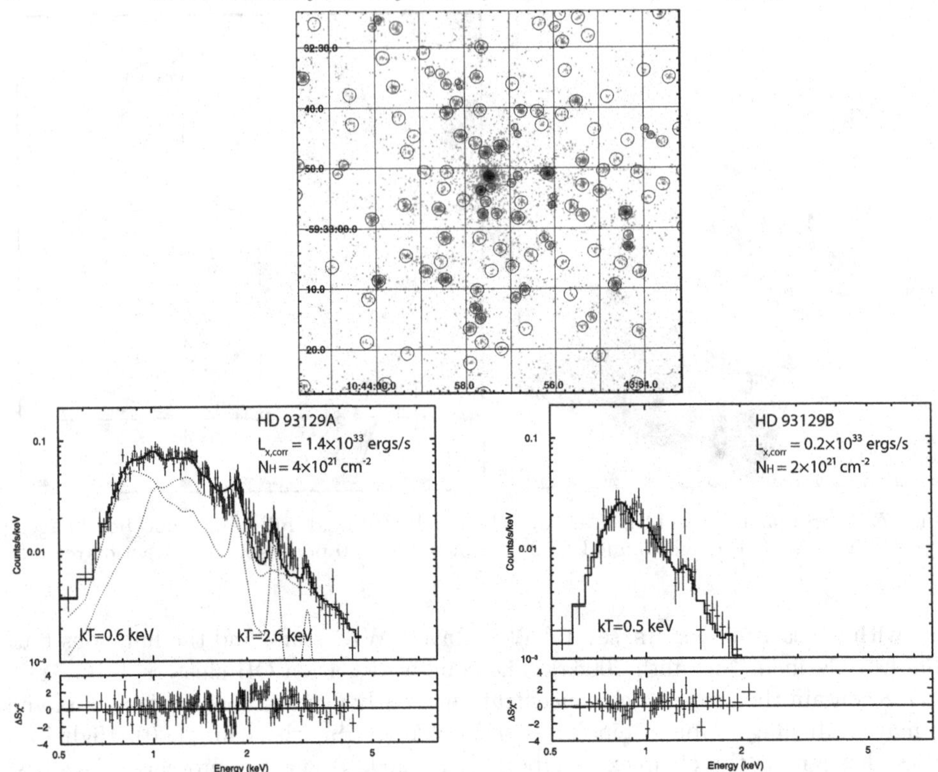

FIGURE 8. Top panel: The central part of the ACIS-I observation of Tr14 with extraction regions marked. Lower left: ACIS spectrum of the resolved massive binary at the aimpoint of the Tr14 observation, HD 93129A (O2If) and lower right panel: HD 93129B (O3V).

in Figure 8 (lower-right panel). Although these two stars are of similar early spectral type, they show remarkably different x-ray spectra. While both exhibit the expected soft thermal plasma with $kT \sim 0.5$ keV, HD 93129A also requires a second, harder thermal component and is seven times brighter in x-rays than HD 93129B. We find similar results for other O stars in the field: while the O3V star HD 93128 is soft ($kT = 0.3$ keV) and faint (absorption-corrected luminosity $L_{\mathrm{x,corr}} = 0.1 \times 10^{33}$ ergs s^{-1}), the O3V star HD 93250 is the brightest source in the ACIS field, with $L_{\mathrm{x,corr}} = 2.1 \times 10^{33}$ ergs s^{-1} and requiring both soft ($kT = 0.6$ keV) and hard ($kT = 3.3$ keV) thermal plasma components. As for other sources described above, this leads us to speculate that HD 93129A and HD 93250 could be colliding-wind binaries; alternatively the hard components of their spectra are soft enough that they could be magnetically active early O stars, similar to θ^1 Ori C (Gagné et al. 2005). In contrast, HD 93129B and HD 93128 are likely more "normal" O stars without these phenomena.

The soft diffuse emission seen in this ACIS observation of Tr14 (Figure 9, top) is also remarkable. On the ACIS-I array, this diffuse emission is brightest in the southeast quadrant, in the region between Tr14 and Tr16. Surprisingly, it is not centered on the Tr14 cluster. The apparent surface brightness of this emission may be affected by gradients in the absorbing column across the field. We serendipitously imaged very bright diffuse emission in the off-axis ACIS-S CCDs that were also operational for this observation. This emission is far from the known massive stars in the Carina complex. Although its apparent surface brightness is much higher than the diffuse emission seen on the ACIS-I

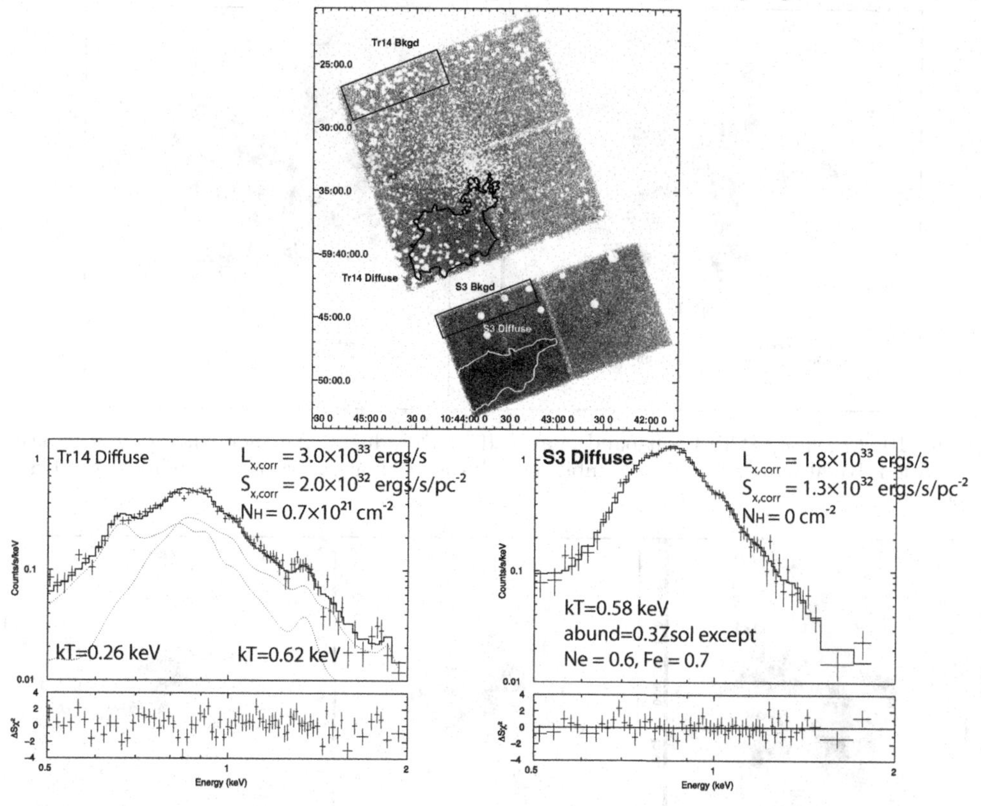

FIGURE 9. Diffuse emission in Tr14. Top: Image showing point sources removed, with diffuse extraction regions marked. Lower left and right: Spectra of the two diffuse regions.

array, it is intrinsically fainter; we see it as brighter because it suffers no measurable absorption. This may indicate that it lies in front of the Tr14 cluster, perhaps partially filling the lower lobe of the Carina superbubble seen in the mid-IR.

While both regions of diffuse emission exhibit soft spectra (Figure 9), they require quite different model fits. The emission in the southeast quadrant of the I array requires a two-temperature fit ($kT = 0.3$ keV and $kT = 0.6$ keV), some absorption, and typical low abundances of $Z = 0.3\,Z_{\odot}$. The S-array emission requires just a single thermal component with $kT = 0.6$ keV, but no absorbing column is seen and a good fit is only obtained by increasing the abundances of Ne and Fe, possibly indicating a supernova origin for this emission. Earlier x-ray data suggested that the Carina complex is pervaded by diffuse x-ray emission not concentrated on the star clusters. This detailed view from *Chandra* now shows that the diffuse emission is spatially and spectrally complex and, while it may be due in part to massive-star winds, one or more cavity supernovae probably also contribute to its complexity. More high-resolution observations of the Carina MSFR are necessary to untangle this complicated mix of pointlike and diffuse x-ray sources.

6. 30 Doradus: Live fast, die young, blow some bubbles

30 Doradus in the Large Magellanic Cloud is the most luminous Giant Extragalactic H II Region and "starburst cluster" in the Local Group. At the center of 30 Dor is the "super star cluster" R136, with dozens of 1–2 Myr-old $>50\,M_{\odot}$ O and Wolf-Rayet stars

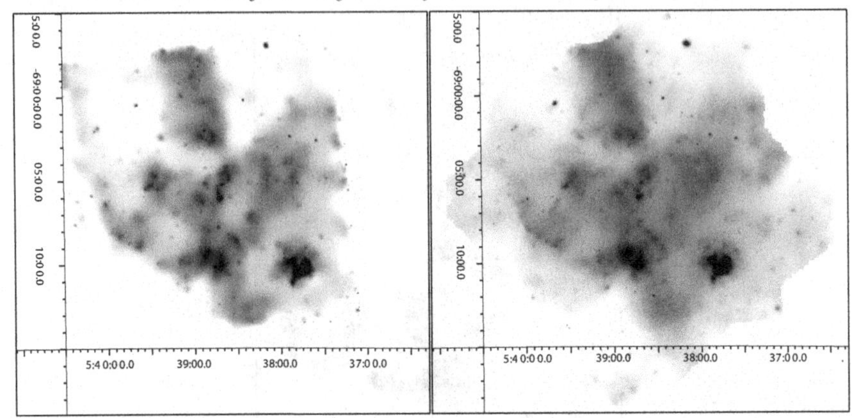

FIGURE 10. *Chandra*/ACIS-I soft-band images of 30 Doradus. Left panel: The original 23 ksec observation. Right panel: New data combined with the original observation, making a ∼110 ksec image.

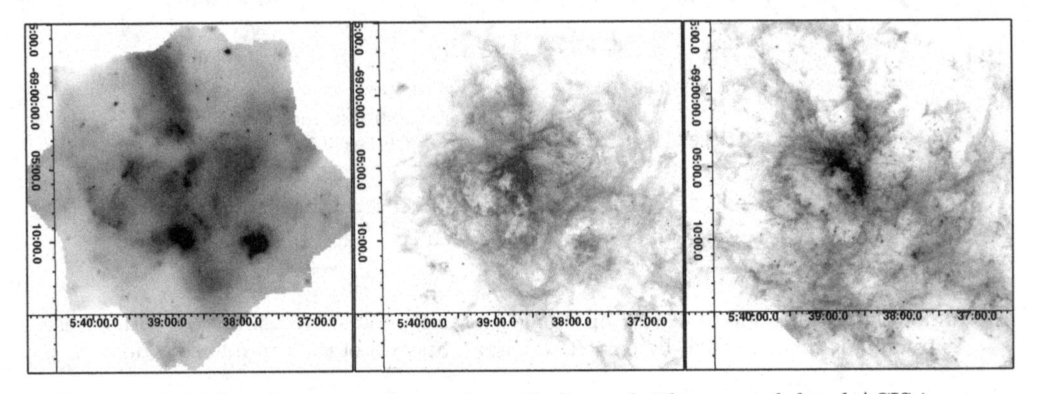

FIGURE 11. 30 Doradus across the spectrum. Left panel: The new soft-band ACIS image. Center: An Hα image from MCELS. Right panel: The *Spitzer* 8 μm image.

(Massey & Hunter 1998). 30 Dor's superbubbles are well-known bright x-ray sources, where multiple cavity supernovae from past OB stars produce soft x-rays filling bubbles 50–100 pc in size with $L_{\mathrm{x}} \sim 10^{35-36}$ ergs s^{-1} (Chu & Mac Low 1990).

Figure 10 (left) shows the smoothed ACIS-I image from our original 23-ksec observation of 30 Dor, described in Townsley et al. (2006a,b). Using super-resolution techniques, we find ∼100 point sources in R136, but clearly the dominant x-ray structures in 30 Dor remain the plasma-filled superbubbles. We constructed maps of the diffuse x-ray emission in those superbubbles, showing variations in plasma temperature ($T = 3$–9 million degrees), absorption ($N_H = 1$–6×10^{21} cm^{-2}), and absorption-corrected x-ray surface brightness ($S_{\mathrm{x,corr}} = 3$–126×10^{31} ergs s^{-1} pc^{-2}). Some new x-ray concentrations $\simeq 30''$ (7 pc) in extent are spatially associated with high-velocity optical emission line clouds (Chu & Kennicutt 1994). Figure 10 (right) shows the original short ACIS-I observation combined with a new 90-ksec dataset obtained in January 2006. As expected, more point sources are resolved in the longer observation. Smaller-scale diffuse structures are also emerging, likely tracing more subtle shock features across the complex. No obvious hard diffuse emission from a cluster wind in R136 is seen, even in this longer observation.

Figure 11 shows the deeper ACIS image in context with visual and IR data. The center panel shows the recently released Hα image from the Magellanic Cloud Emission Line Survey (MCELS, C. Smith et al. 2000), while the right panel shows the 8 μm

Spitzer/IRAC image (*Spitzer* press release by B. Brandl, 2004). The MCELS and *Spitzer* data gives new insight into the diffuse x-ray structures: the hot superbubbles fill the interiors of ionization fronts outlined by Hα emission, which in turn are outlined by shells of warm dust; the x-ray confinement and morphology are fully appreciated only when anchored by these multiwavelength data.

Our spectral analysis of the superbubbles seen in the early ACIS data revealed a range of absorptions, plasma temperatures, and abundance variations. This short observation limits our ability to study diffuse structures at small scales; in order to get enough counts to do good spectral fitting, we must average over many tens of parsecs. We suspect that we are thus averaging over many distinct plasma components, with different pressures, temperatures, absorbing columns, and possibly even different abundances. The new observation will allow us to refine our spectral fitting and to make higher-resolution maps of the physical parameters that govern 30 Dor's diffuse x-ray emission. We will also be able to study more completely the massive stars in R136 and throughout the complex.

7. Summary

High-resolution x-ray studies of MSFRs reveal magnetically active pre-MS stars, massive-star microshocks, and colliding-wind binaries. *Chandra* provides a disk-unbiased stellar sample for IR study. Diffuse x-ray emission is also seen, caused by O-star winds interacting with themselves and with the surrounding media. Supernovae occurring in the cavities blown by OB clusters produce brighter soft x-rays, dominating the x-ray emission in large multi-cluster and multi-generation star-forming complexes.

Several *Chandra* discoveries are noted in this short overview. For all regions, early O stars often show anomalously hard spectra, perhaps indicating close binarity. For individual regions, discoveries include:

- M17: First clear detection of an x-ray outflow.
- NGC 3576: Bipolar bubble seen in *MSX* data is likely the wind-blown bubble from an embedded OB cluster; this field shows the second clear example of x-ray outflowing gas; hard x-rays reveal cluster members not seen in the IR—these are likely the missing ionizing massive stars.
- W3: W3 Main is huge compared to other W3 young clusters; the W3 Main IRS5 and W3(OH) massive protostars are very hard x-ray sources; W3 North is missing its cluster.
- Tr14: Some early O stars are likely binary or magnetically active, some are not; soft diffuse emission appears between Tr14 and Tr16, while separate, unabsorbed soft diffuse emission located far from massive stars may indicate a cavity supernova remnant inside the Carina complex's young superbubble.
- 30 Dor: More spatial structure and point sources emerge in the longer observation; no clear sign of hard x-rays from a cluster wind in R136 are seen.

I am most indebted to my colleagues Patrick Broos for data analysis and figure production, Eric Feigelson for ideas and interpretation, and Gordon Garmire, the ACIS PI, for using part of his *Chandra* guaranteed time on the MSFR project. I thank Barbara Whitney and Ed Churchwell for supplying the GLIMPSE image of M17, You-Hua Chu for the MCELS image of 30 Dor, and Bernhard Brandl for the IRAC image of 30 Dor. Support for this work was provided by NASA through contract NAS8-38252 and *Chandra* Awards GO4-5006X, GO4-5007X, GO5-6080X, GO5-6143X, and SV4-74018 issued by the Chandra X-ray Observatory Center, operated by the Smithsonian Astrophysical Observatory for and on behalf of NASA under contract NAS8-03060.

REFERENCES

Arnaud, K. A. 1996. In *Astronomical Data Analysis Software and Systems V* (eds. G. Jacoby & J. Barnes), ASP Conf. Ser. 101, p. 17. Astronomical Society of the Pacific.

Brooks, K. J., Storey, J. W. V., & Whiteoak, J. B. 2001 *MNRAS* **327**, 46.

Broos, P. S., Getman, K. V., Townsley, L. K., Feigelson, E. D., Wang, J., Garmire, G. P., Jiang, Z., & Tsuboi, Y. 2007 *ApJS* **169**, 353.

Cantó, J., Raga, A. C., & Rodríguez, L. F. 2000 *ApJ* **536**, 896.

Carpenter, J. M., Heyer, M. H., & Snell, R. L. 2000 *ApJS* **130**, 381.

Chu, Y. & Mac Low, M. 1990 *ApJ* **365**, 510.

Chu, Y.-H. & Kennicutt, R. C., Jr. 1994 *ApJ* **425**, 720.

de Pree, C. G., Nysewander, M. C., & Goss, W. M. 1999 *AJ* **117**, 2902.

de Wit, W. J., Testi, L., Palla, F., & Zinnecker, H. 2005 *A&A* **437**, 247.

Delgado, A. J., Gonzalez-Martin, O., Alfaro, E. J., & Yun, J. L. 2006 *ApJ* **646**, 269.

Ezoe, Y., Kokubun, M., Makishima, K., Sekimoto, Y., & Matsuzaki, K. 2006 *ApJ* **638**, 860.

Feigelson, E., Townsley, L., Gudel, M., & Stassun, K. 2007. In *Protostars and Planets V* (eds. B. Reipurth, D. Jewitt, & K. Keil). p. 313. University of Arizona Press.

Figuerêdo, E., Blum, R. D., Damineli, A., & Conti, P. S. 2002 *AJ* **124**, 2739.

Flaccomio, E., Micela, G., & Sciortino, S. 2006 *A&A* **455**, 903.

Franciosini, E., Pallavicini, R., & Sanz-Forcada, J. 2006 *A&A* **446**, 501.

Gagné, M., Oksala, M. E., Cohen, D. H., Tonnesen, S. K., ud-Doula, A., Owocki, S. P., Townsend, R. H. D., & MacFarlane, J. J. 2005 *ApJ* **628**, 986.

Getman, K. V., Feigelson, E. D., Townsley, L., Broos, P., Garmire, G., & Tsujimoto, M. 2006 *ApJS* **163**, 306.

Hachisuka, K., et al. 2006 *ApJ* **645**, 337.

Hanson, M. M., Howarth, I. D., & Conti, P. S. 1997 *ApJ* **489**, 698.

Heyer, M. H. & Terebey, S. 1998 *ApJ* **502**, 265.

Hofner, P., Delgado, H., Whitney, B., Churchwell, E., & Linz, H. 2002 *ApJ* **579**, L95.

Kraemer, K. E., Shipman, R. F., Price, S. D., Mizuno, D. R., Kuchar, T., & Carey, S. J. 2003 *AJ* **126**, 1423.

Maercker, M., Burton, M. G., & Wright, C. M. 2006 *A&A* **450**, 253.

Massey, P. & Hunter, D. A. 1998 *ApJ* **493**, 180.

Megeath, S. T., Wilson, T. L., & Corbin, M. R. 2005 *ApJ* **622**, L141.

Muno, M. P., Law, C., Clark, J. S., Dougherty, S. M., de Grijs, R., Portegies Zwart, S., & Yusef-Zadeh, F. 2006 *ApJ* **650**, 203.

Nielbock, M., Chini, R., Jütte, M., & Manthey, E. 2001 *A&A* **377**, 273.

Oey, M. S., Watson, A. M., Kern, K., & Walth, G. L. 2005 *AJ* **129**, 393.

Persi, P., Roth, M., Tapia, M., Ferrari-Toniolo, M., & Marenzi, A. R. 1994 *A&A* **282**, 474.

Rathborne, J. M., Burton, M. G., Brooks, K. J., Cohen, M., Ashley, M. C. B., & Storey, J. W. V. 2002 *MNRAS* **331**, 85.

Rebull, L. M., Stauffer, J. R., Ramirez, S. V., Flaccomio, E., Sciortino, S., Micela, G., Strom, S. E., & Wolff, S. C. 2006 *AJ* **131**, 2934.

Sana, H., Gosset, E., Rauw, G., Sung, H., & Vreux, J.-M. 2006 *A&A* **454**, 1047.

Sanchawala, K., Chen, W.-P., Lee, H.-T., Nakajima, Y., Tamura, M., Baba, D., Sato, S., & Chu, Y.-H. 2006; arXiv:astro-ph/0603043.

Skinner, S. L., Simmons, A. E., Zhekov, S. A., Teodoro, M., Damineli, A., & Palla, F. 2006 *ApJ* **639**, L35.

Smith, C., Leiton, R., & Pizarro, S. 2000. In *Stars, Gas and Dust in Galaxies: Exploring the Links* (eds. D. Alloin, K. Olsen, & G. Galaz), ASP Conf. Ser. 221, p. 83. Astronomical Society of the Pacific.

Smith, N. 2006a *MNRAS* **367**, 763.

Smith, N. 2006b *ApJ* **644**, 1151.

Smith, N., Egan, M. P., Carey, S., Price, S. D., Morse, J. A., & Price, P. A. 2000 *ApJ* **532**, L145.

Smith, R. K., Brickhouse, N. S., Liedahl, D. A., & Raymond, J. C. 2001 *ApJ* **556**, L91.

Stassun, K. G., van den Berg, M., Feigelson, E., & Flaccomio, E. 2006 *ApJ* **649**, 914.

Tieftrunk, A. R., Gaume, R. A., Claussen, M. J., Wilson, T. L., & Johnston, K. J. 1997 *A&A* **318**, 931.

Townsley, L. K., Broos, P. S., Feigelson, E. D., Brandl, B. R., Chu, Y.-H., Garmire, G. P., & Pavlov, G. G. 2006a *AJ* **131**, 2140.

Townsley, L. K., Broos, P. S., Feigelson, E. D., & Garmire, G. P. 2005. In *Massive Star Birth: A Crossroads of Astrophysics* (eds. R. Cesaroni, E. Churchwell, M. Felli, & C. M. Walmsley). IAU Symp. 227, p. 297. Cambridge University Press.

Townsley, L. K., Broos, P. S., Feigelson, E. D., Garmire, G. P., & Getman, K. V. 2006b *AJ* **131**, 2164.

Townsley, L. K., Feigelson, E. D., Montmerle, T., Broos, P. S., Chu, Y., & Garmire, G. P. 2003 *ApJ* **593**, 874.

Townsley, L. K., Feigelson, E. D., Montmerle, T., Broos, P., Chu, Y.-H., Garmire, G., & Getman, K. 2004. In *X-ray and Radio Connections* (eds. L. O. Sjouwerman & K. K. Dyer). Published electronically at http://www.aoc.nrao.edu/events/xraydio/meetingcont/3.4_townsley.pdf/.

Tsujimoto, M., Hosokawa, T., Feigelson, E. D., Getman, K. V., & Broos, P. 2006 *ApJ* **653**, 409.

Tsujimoto, M., Townsley, L., Feigelson, E. D., Broos, P., Getman, K. V., & Garmire, G. 2005. In *Protostars and Planets V* (eds. B. Reipurth, D. Jewitt, & K. Keil). p. 8307. University of Arizona Press.

Vázquez, R. A., Baume, G., Feinstein, A., & Prado, P. 1996 *A&AS* **116**, 75.

Walborn, N. R., et al. 2002a *AJ* **123**, 2754.

Walborn, N. R., Danks, A. C., Vieira, G., & Landsman, W. B. 2002b *ApJS* **140**, 407.

Wang, J., Townsley, L. K., Feigelson, E. D., Broos, P. S., Getman, K. V., Román-Zúñiga, C., & Lada, E. 2008 *ApJ* **675**, 464.

Wang, J., Townsley, L. K., Feigelson, E. D., Getman, K. V., Broos, P. S., Garmire, G. P., & Tsujimoto, M. 2007 *ApJS* **168**, 100.

Wang, Q. D., Dong, H., & Lang, C. 2006 *MNRAS* **371**, 38.

Wolk, S. J., Spitzbart, B. D., Bourke, T. L., & Alves, J. 2006 *ApJ* **132**, 1100.

Massive stars: Feedback effects in the local universe

By M. S. OEY[1] AND C. J. CLARKE[2]

[1]Department of Astronomy, 830 Dennison Building, University of Michigan, Ann Arbor, MI 48109-1042, USA

[2]Institute of Astronomy, University of Cambridge, Madingley Road, Cambridge CB3 0HA, UK

Massive stars as a population are the source of various feedback effects that critically impact the evolution of their host galaxies. We examine parameterizations of the high-mass stellar population and self-consistent parameterizations of the resulting feedback effects, including mechanical feedback, radiative feedback, and chemical feedback, as we understand them in the local universe. To date, it appears that the massive-star population follows a simple power-law clustering law that extends down to individual field massive stars, and the robust stellar IMF appears to have a constant upper-mass limit. These properties result in specific patterns in the H II-region luminosity function, and the ionization of the diffuse, warm, ionized medium. The resulting supernovae generate a population of superbubbles whose distributions in size and expansion velocity are also described by simple power laws, and from which a galaxy's porosity parameter is easily derived. A critical star-formation threshold can then be estimated, above which the escape of Lyman-continuum photons, hot gas, and nucleosynthetic products is predicted. A first comparison with a large sample of Hα observations of galaxies is broadly consistent with this prediction, and suggests that ionizing photons are likely to escape from starburst galaxies. The superbubble size distribution also offers a basis for a Simple Inhomogeneous Model for galactic chemical evolution, which is especially applicable to metal-poor systems and instantaneous metallicity distributions. This model offers an alternative interpretation of the Galactic halo metallicity distribution and emphasizes the relative importance of star-formation intensity, in addition to age, in a system's evolution. The fraction of zero-metallicity, Population III stars is easily predicted for any such model. We emphasize that all these phenomena can be modeled in a simple, analytic framework over an extreme range in scale, offering powerful tools for understanding the role of massive stars in the cosmos.

---

1. Introduction

Massive stars are of great interest because of their profound feedback effects that alter the surrounding environment on local, global, and cosmic scales. Their radiative feedback causes ionization of neutral gas; their supernova (SN) explosions drive mechanical feedback that shock-heats gas to $\gtrsim 10^6$ K; and the nucleosynthesis processes within these stars and their SNe produce most of the elements that are tracers of past stellar populations. In short, massive stars are one of the principal drivers of galactic and cosmic evolution.

2. The massive-star population

By "massive stars," we consider those stars having masses above, say, $10\,M_\odot$. If we are to understand the global feedback effects from massive stars, then it is important to understand their properties as a *population*. This population of stars is characterized by its distribution in (*a*) mass, and (*b*) space.

2.1. *The IMF and upper-mass limit*

The stellar mass distribution is parameterized by the familiar stellar initial mass function (IMF). The massive-star IMF has been evaluated many times for OB associations and clusters in the Galaxy and Magellanic Clouds (e.g., Massey 2003), and it appears to be fairly robustly consistent with the Salpeter (1955) slope:

$$n(m)\, dm \propto m^{-2.35}\, dm \ , \tag{2.1}$$

where $n(m)$ is the number of stars in the mass range m to $m+dm$. Cruder evaluations of extragalactic massive-star populations using integrated colors and properties of galaxies support this result (e.g., Elmegreen 2006; Fernandes *et al.* 2004; Bell & de Jong 2001; Baldry & Glazebrook 2003). An important possible exception may be the field massive-star IMF (see below). Beware, however, that it is difficult to disentangle effects of the IMF slope and upper-mass cutoff.

While the slope of the upper IMF is fairly well determined, the upper-mass limit is less so. There are theoretical considerations supporting the existence of a stellar upper-mass limit based on physical instability arguments (e.g., Ledoux 1941; Schwarzschild & Härm 1959; Stothers 1992), as well as limitations related to the high rate and short timescales of accretion that are needed to overcome the protostar's own radiation pressure (e.g., Larson & Starrfield 1971; Kahn 1974; Elmegreen & Lada 1977; Wolfire & Cassinelli 1987; Bonnell, Bate, & Zinnecker 1998). These issues are also discussed elsewhere in these proceedings (e.g., review by Krumholz). While the physical processes remain to be fully understood, we can, in the meantime, empirically evaluate the upper-mass limit, or lack thereof.

The IMF dictates that the highest-mass stars are the rarest, and so the obvious place to search for such stars is in the richest clusters that are young enough ($\lesssim$3 Myr) to preclude any having exploded as SNe. These rich, extremely young clusters are likewise rare (see below), but we are fortunate that local examples do exist. The R136a cluster in the 30 Doradus star-forming complex in the Large Magellanic Cloud (LMC; Figure 1) is one such example. Selman et al. (1999b) examined this region using extensive ground-based observations and suggested that R136a exhibits an upper-mass cutoff around 150 $M_\odot$, based on fairly qualitative arguments. Massey & Hunter (1998) and Hunter et al. (1997) evaluated zero-age main sequence (ZAMS) masses of hundreds of the most massive stellar candidates in R136a, based on photometry and spectroscopic classifications from *HST*. Weidner & Kroupa (2004) and Oey & Clarke (2005) both examined the Massey & Hunter statistics of these reported data for R136a and independently concluded that this region quantitatively demonstrates an upper-mass limit around 150–200 $M_\odot$. Another example of an extremely young and extremely rich cluster is the Arches Cluster in the Galactic center environment. It, too, exhibits an upper-mass limit around 150–200 $M_\odot$ (Figer 2005).

But R136a and the Arches Cluster are only two specific regions. Since star formation presumably is a stochastic process, it may be that we were simply extremely unlucky to have picked two clusters that both happened to have rendered unusually low maximum stellar masses, even though the physical mass limit may be much higher. Furthermore, only one or two more examples of rich clusters suitable for deterministic evaluation of an upper-mass limit may be accessible for such detailed, empirical, stellar-mass analyses. However, *if* the IMF indeed behaves like a universal probability density function (PDF), then we can also use a combined ensemble of the stellar contents of the numerous, ordinary OB associations to evaluate an upper-mass limit. Note that such treatment of the IMF as a PDF is indeed the conventional way in which it is usually considered.

FIGURE 1. Three-color image of the R136a cluster in 30 Doradus, imaged with the NTT. Medium, light, and dark shades of gray correspond to V, B, and U, respectively (Selman et al. 1999a).

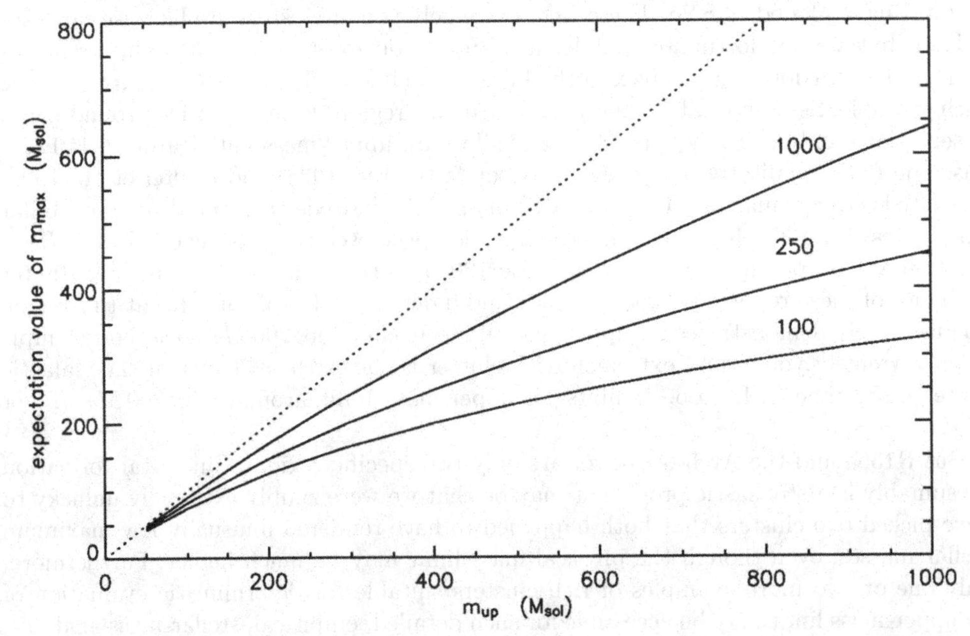

FIGURE 2. Expectation value of the maximum-mass star in a cluster, given N stars having $m \geqslant 10\,M_{\odot}$, as a function of physical upper-mass limit $m_{\rm up}$ (Oey & Clarke 2005).

We can then assemble a large number of the youngest massive stars and examine the form of the upper IMF. For a Salpeter IMF, we can write the expectation value of the maximum stellar mass $m_{\rm max}$ for an ensemble of N_* stars having masses $m \geqslant 10\ M_\odot$, given a physical upper-mass limit $m_{\rm up}$:

$$\langle m_{\rm max} \rangle = m_{\rm up} - \int_0^{m_{\rm up}} \left[\int_0^M \phi(m)\ dm \right]^N\ dM \quad , \tag{2.2}$$

where $\phi(m)$ is the IMF. Oey & Clarke (2005) considered data for eight OB associations in the Galaxy and LMC in which the upper IMF was fully evaluated by Massey et al. (1995), based on spectroscopic classifications, and which they found to be $\leqslant$3 Myr old. Among these associations, there is a total of 263 stars having $m \geqslant 10\, M_\odot$. For a physical $m_{\rm up} = 1000\, M_\odot$, equation (2.2) predicts an expected $m_{\rm max} > 450\, M_\odot$ (Figure 2), whereas the observed maximum mass is $\lesssim 150\, M_\odot$. Since R136a was studied by the same group, we can include it in this uniform sample, raising the total number to 913 stars having $m \geqslant 10\, M_\odot$. For the same parent IMF assumptions, Figure 2 shows that the predicted $m_{\rm max}$ is now $\sim 600\, M_\odot$. However, the observed maximum stellar mass in the entire ensemble is still only around $150\, M_\odot$, and so Figure 2 implies a similar physical upper-mass cutoff.

Once again, we could be exceedingly unlucky in considering an extraordinary sample of nine associations, all of which happened to render an unusually low $m_{\rm max}$. Oey & Clarke (2005) quantified the likelihood of this occurrence as well, for physical $m_{\rm up}$ ranging from ∞ to $120\, M_\odot$. Only $m_{\rm up} \sim 120$–$150\, M_\odot$ yielded significant probabilities for the observed $m_{\rm max}$ in the individual clusters simultaneously, therefore clearly demonstrating an upper-mass cutoff around those values. Koen (2006) used an alternative statistical analysis with these same data that confirms this result.

Clearly, this upper-mass limit is only demonstrated for this particular sample of objects, and assuming that they are all pre-SN. It seems significant, however, that the *same* $m_{\rm up}$ is found across grossly varying environments encompassing the extreme conditions near the Galactic Center for the Arches Cluster; the highly active, yet much less extreme conditions for R136a; and the relatively unremarkable conditions for the OB associations. These findings suggest a universal upper-mass limit around $150\, M_\odot$ in the local universe. Elmegreen (2000, 2006) makes similar arguments in considering aggregate stellar populations of galaxies.

2.2. *The clustering law*

With a fairly well-defined parameterization of the stellar mass distribution for the upper IMF, we now turn to the spatial distribution of the massive-star population. We can parameterize the space distribution by defining a *clustering law*, $N(N_*)$, which describes the distribution in N_*, the number of massive stars per cluster. Over the last decade, it has become apparent that the clustering law for young, massive clusters is robustly consistent with a power law, similar to the stellar IMF:

$$N(N_*)\ dN_* \propto N_*^{-2}\ dN_* \quad . \tag{2.3}$$

This is equivalent to the initial cluster mass function and is seen in a variety of environments, including starbursts and populations of super star clusters (e.g., Meurer et al. 1995; Zhang & Fall 1999) and extrapolated from the globular cluster present-day mass function (e.g., Elmegreen & Efremov 1997; Hunter et al. 2003).

The upper-mass cutoff to the clustering law appears to vary in different systems (see below), but no physical maximum has yet been suggested. In the opposite, low-mass

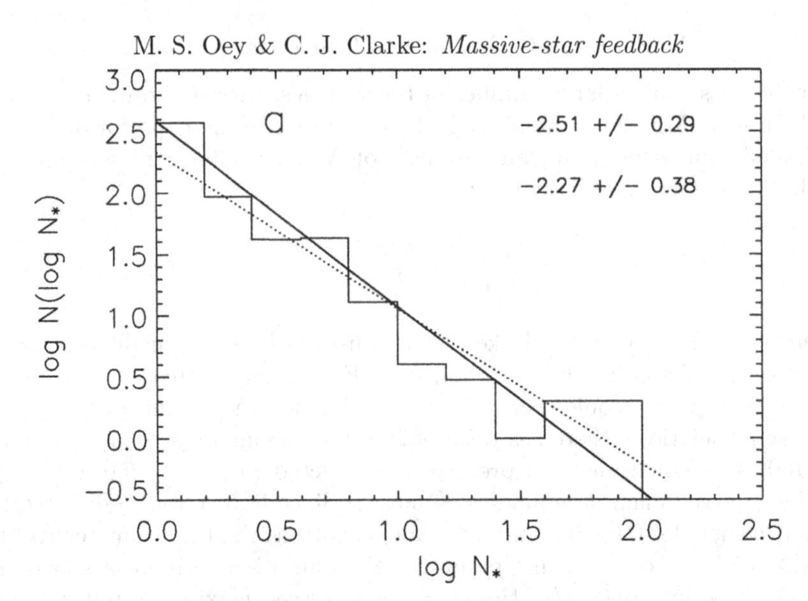

FIGURE 3. SMC clustering law for OB star candidates, from Oey et al. (2004). The solid line shows a fit to the entire dataset, while the dotted line shows a fit omitting the field stars (first bin, at $\log N_* = 0$). These fitted slope values are shown, respectively.

extreme, the "clusters" reduce to single, individual-field massive stars. With the definition of "massive" stars as above, "field" massive stars include both those that are genuinely isolated from any other stars, if they exist, and those that are the "tip of the iceberg" for a small group whose remaining members all have m less than would qualify for our definition of a "massive" star.

What is the relationship between the field massive stars and those in clusters? Oey et al. (2004) studied a uniformly selected sample of massive-star candidates from *UBVR* photometry (Massey et al. 2002) and recalibrated FUV *B*5 (Parker et al. 1998) photometry across most of the Small Magellanic Cloud (SMC) and determined the clustering law using a friends-of-friends algorithm. For that galaxy, they found that the massive-star clustering follows a smooth power law all the way down to $N_* = 1$, and that the power-law exponent is consistent with equation (2.3) (Figure 3).

At face value, this suggests the co-existence of a universal IMF and universal clustering law, of the respective forms in equations (2.1) and (2.3). However, we caution that the field star IMF has been suggested to be significantly steeper than in clusters, based on both observations (Massey 2002) and theoretical arguments (Kroupa & Weidner 2003). If this is indeed the case, then a flattening should be observed in the clustering law near $N_* = 1$, since this implies fewer field stars. Because this flattening is not observed in Figure 3, there must be a corresponding steepening in the clustering law in this regime, in order to recover the smooth power law that we see in the SMC data. Thus reality may be somewhat subtle and more complex than is seen at face value; Oey et al. (2004) discuss this issue in more detail.

Regardless of any underlying complexities, the resulting observed clustering law in the SMC does show a smooth power law described by equation (2.3). If this applies generally, then we can directly estimate the fraction of field massive stars as (Oey et al. 2004):

$$f_{\text{field}} = (\ln N_{*,\max} + 0.5772)^{-1} \quad , \tag{2.4}$$

where $N_{*,\max}$ is the number of massive stars in the richest cluster of the ensemble. Because of this dependence on $N_{*,\max}$, we see that the field star fraction has a modest

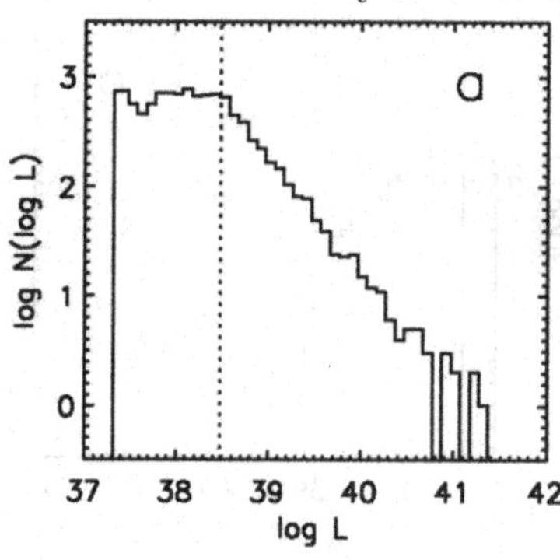

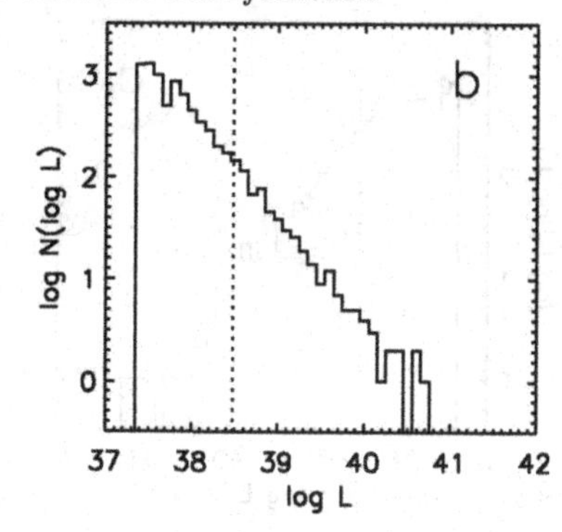

FIGURE 4. Monte Carlo models of the zero-age H II-region luminosity function (panel a) and the same distribution evolved to 7 Myr (panel b). The dotted lines show the Hα luminosity associated with the most massive star in the IMF (Oey & Clarke 1998a).

inverse dependence on the total star-formation rate of the system. For typical star-forming galaxies, $f_{\rm field} \sim$ 20–25%.

The preceding thus describes a well-defined parameterization for the massive-star population in star-forming galaxies, given by equations (2.1) and (2.3), and adopting a stellar upper-mass cutoff $m_{\rm up}$= 150 $M_{\odot}$.

3. Radiative feedback

Now turning to the feedback effects from this massive-star population, we begin by considering the radiative feedback, which refers to the photoionization caused by these stars. There are two principal effects: the generation of ordinary H II regions, and the generation of the diffuse, nebular component of the interstellar medium (ISM). The latter is usually referred to as the warm ionized medium (WIM), or alternatively, diffuse ionized gas (DIG).

3.1. *The H II-region luminosity function*

The Hα luminosity is a direct probe of the cumulative massive-star population in an H II region, and the clustering law naturally results in a corresponding power-law luminosity function for the classical H II regions, which has been empirically examined in many nearby galaxies. However, the H II-region luminosity function (H IILF) is often seen to deviate from a smooth N_*^{-2} power law in the following ways: (1) a two-slope form is often seen, with the lower-luminosity population showing a shallower power-law slope; (2) inter-arm nebular populations in grand-design spirals often show shallower slopes than the arm populations; and (3) there is a correlation with galaxy type, such that the early-type galaxies show much steeper slopes than late types. Oey & Clarke (1998a) demonstrated, using Monte Carlo models, that these variations are all fully consistent with, and indeed expected, from the properties of the massive-star population described above. One of the most important effects is a flattening that is seen in the H IILF at low luminosities that results from stochastic sampling of the stellar IMF in this "unsaturated" regime (Figure 4). For luminous nebulae generated by rich, "saturated" clusters that fully sample the IMF through the upper-mass limit $m_{\rm up}$, the H II-region luminosity is directly proportional to N_*; whereas for sparse, "unsaturated" clusters, the resulting H II-region luminosity is subject to the specific stellar population. The stochastic flattening in the low-luminosity end of the H IILF quantitatively explains the observed effects (1) and (2) described above. The second of these, the observed steepening seen for inter-arm H II

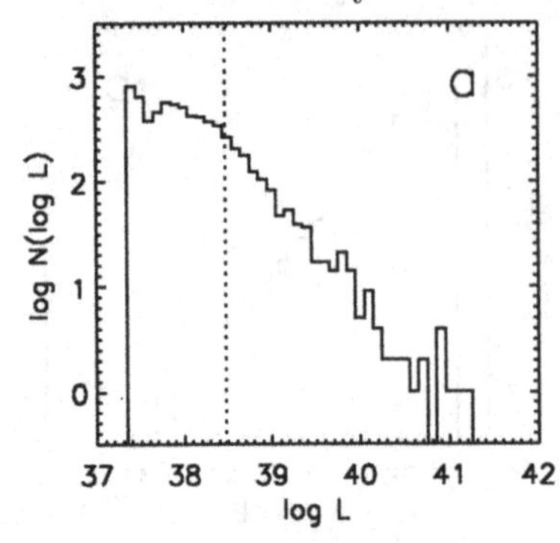

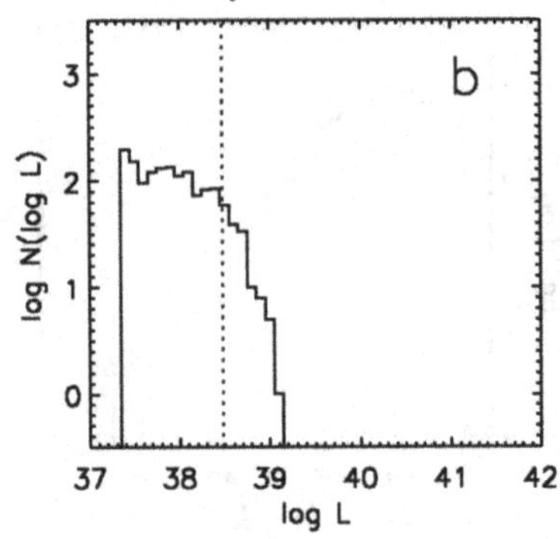

FIGURE 5. Monte Carlo models of the H II-region luminosity function for continuous star formation. Panel a shows the distribution for no upper-mass limit to the clusters, and panel b shows the same, but imposing a limit of $N_* = 10$ (Oey & Clarke 1998a).

regions, can be explained if these show an evolved slope as in Figure 4b, while the zero-age population in the arms corresponds to the model in Figure 4a. As a single-age population evolves, the entire distribution fades, and the low-luminosity regime becomes dominated by evolved saturated objects. Oey & Clarke (1998a) demonstrate this effect in nearby grand-design spirals, where the inter-arm regions are presumably a more evolved population left behind in the wake of the spiral density waves.

The third effect, the trend with galaxy Hubble type, can be explained by a varying upper cutoff in N_*, while preserving the -2 power-law slope. Figure 5a shows a Monte Carlo model with no upper-mass limit to the clusters, while Figure 5b shows the same model H IILF, but imposing a limit of $N_* \leqslant 10$. The former are qualitatively and quantitatively similar to the observed H IILF in late-type galaxies (e.g., Kennicutt et al. 1989; Rand 1992; Banfi et al. 1993; Rozas et al. 1996) while the former agree with the observations for Sa galaxies (Caldwell et al. 1991). Thus, all three of the observed patterns in the H IILF are fully consistent with, and indeed expected, for the universal clustering law. Oey & Clarke (1998a) describe these phenomena in detail.

3.2. *The diffuse, warm ionized medium*

Classical H II regions account for only about half of the total Hα emission from star-forming galaxies (e.g., Walterbos 1998; Ferguson et al. 1996). The remaining half originates from the widespread, diffuse WIM. While the ionization of the WIM remains to be fully understood, it is generally thought also to originate from massive stars (e.g., Reynolds 1984). Direct comparisons of the stellar populations in LMC OB associations with their associated nebular luminosities suggests that up to 50%, and in some cases, more, of the ionizing radiation could escape from H II regions (Oey & Kennicutt 1997; Voges et al. 2005). Hoopes & Walterbos (2000) came to a similar conclusion based on FUV observations of M33 from *UIT*.

The other half of WIM ionization can be accounted for by field stars, assuming that the universal clustering law indeed extends to individual massive stars representing the "tip of the iceberg" on sparse clusters and groups, as found above for the SMC. Equation (2.4) predicts that typically about 25% of the massive-star population resides in the field as defined in this way, a fraction confirmed empirically for the SMC (Oey et al. 2004). Thus, field stars can account for one-quarter of a galaxy's total Hα luminosity. For the WIM constituting half of that total, then the field stars can ionize about half again of the WIM.

There are a few caveats; for example, other ionization processes are implicated by apparent detailed ionization states observed in the WIM (e.g., Reynolds et al. 1999; Rand 2000). Also, the most recent hot star atmosphere models (e.g., Martins et al. 2005; Repolust et al. 2005; Smith et al. 2002) are suggesting softer ionizing fluxes which may reduce the role of massive stars. However, it seems clear that this stellar population dominates production of the WIM.

Radiative feedback to the IGM is also a topic of vital current interest, and is discussed further below.

4. Mechanical feedback

We now turn to the mechanical energy produced by the core-collapse supernovae of the massive-star population. Given the short ($\leqslant$40 Myr) lifetimes of these stars, the overwhelming majority remain in the OB associations where they were born, and so the subsequent SNe are spatially clustered. Our universal N_*^{-2} clustering law translates directly into a mechanical luminosity function, that parameterizes the kinetic energy for the ensemble of clusters. Assuming that all SNe yield the same kinetic energy, we can write the mechanical luminosity function as

$$N(L)\, dL \propto L^{-2}\, dL \quad , \tag{4.1}$$

where L is the "mechanical luminosity" or SN power expected from a given cluster. This makes the rough approximation that the discrete SNe represent a continuous energy input over the 40 Myr timescale (e.g., McCray & Kafatos 1987).

4.1. *Superbubbles in the ISM*

The evolution of multi-SN superbubbles is given by simple, Sedov-like relations (e.g., Weaver et al. 1977):

$$R \propto (L/n)^{1/5}\, t^{2/5} \tag{4.2}$$

$$P_i \propto L^{2/5}\, n^{3/5}\, t^{-2/5} \quad , \tag{4.3}$$

where R and P_i are, respectively, the superbubble radius and interior pressure, and n and t are the ambient ISM density and object age. These relations assume adiabatic evolution, namely, that the shells are pressure-driven by the shock-heated interior gas with no thermal losses.

For these simple analytic relations given by equations (4.1)–(4.3), Oey & Clarke (1997) derived global parameterizations of superbubble populations, assuming continuous or burst creation scenarios and that the shells stop growing when they are pressure-confined by the ambient ISM. For example, it is straightforward to derive that the steady-state size distribution for the mechanical luminosity function in equation (4.1) and a continuous creation rate is,

$$N(R) \propto R^{-3}\, dR \quad . \tag{4.4}$$

At present, the only galaxies for which this prediction can be reliably tested are the LMC and SMC, both of which have deep H I surveys and shell catalogs. The top panel of Figure 6 shows the size distribution for SMC H I shell catalog (Staveley-Smith et al. 1997), which is in excellent agreement with the general prediction of equation (4.4) (Oey & Clarke 1997). The distribution in shell expansion velocities v, can be similarly derived for the same population parameterizations (Oey & Clarke 1998b):

$$N(v)\, dv \propto v^{-7/2}\, dv \quad . \tag{4.5}$$

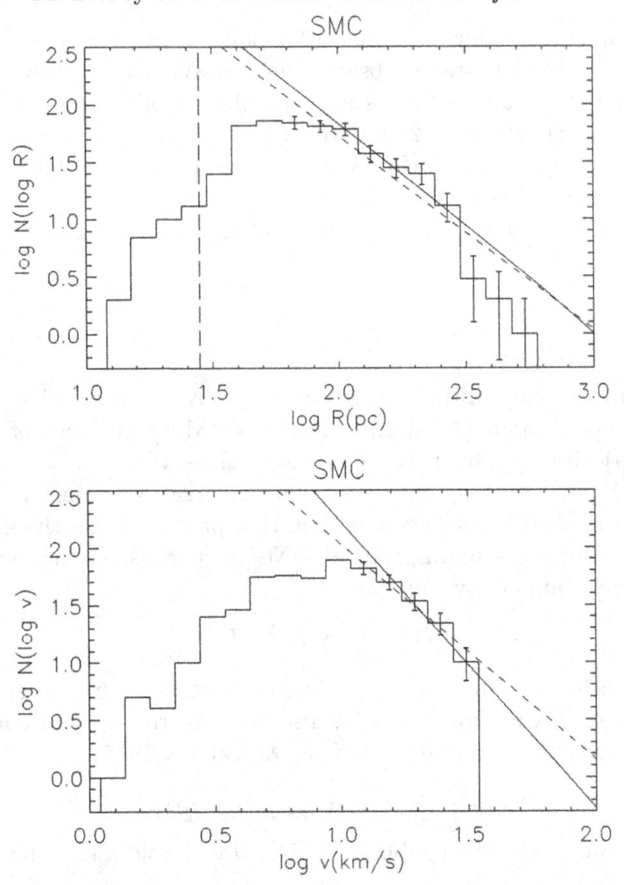

FIGURE 6. SMC H I shell size distribution (top) and expansion velocity distribution (bottom) from the survey by Staveley-Smith et al. (1997). The overplotted lines in the top panel show a power-law fit of 2.7 ± 0.6 to the data (dashed), and a slope of 2.8 ± 0.4 (solid) predicted from the observed H IILF for this galaxy. The spatial resolution of the H I survey is shown by the vertical long-dashed line. In the bottom panel, the solid line shows the predicted slope of -3.5, and the dashed line is a fit to the data, of -2.9 ± 1.4. Note that only the high-velocity tail of the distribution corresponds to the expanding shells; the remainder are near the sound speed, and in pressure equilibrium with the ambient ISM (Oey & Clarke 1998b).

The bottom panel of Figure 6 again shows agreement with this prediction for the SMC shells, for those objects that are still expanding (Oey & Clarke 1998b). On the other hand, the LMC shell population (Kim et al. 1999) is different in nature. Whereas the SMC catalog has >500 distinct H I shells (Staveley-Smith et al. 1997), the LMC catalog has only 126 coherent objects, in a survey with similar instrumental sensitivities. Since the LMC is larger and has a substantially higher star-formation rate than the SMC, the smaller shell population is at first sight counter intuitive. However, as discussed below, the LMC's star-formation rate appears to be high enough that the shells frequently interact and merge, thereby losing their individual entities. Our predictions for the shell size distribution and other global parameters cannot apply in such circumstances.

4.2. *The threshold SFR for feedback to the IGM*

Indeed, we can derive a threshold star-formation rate (SFR) above which we expect this condition of shell interactions and ISM shredding. The porosity parameter Q is a

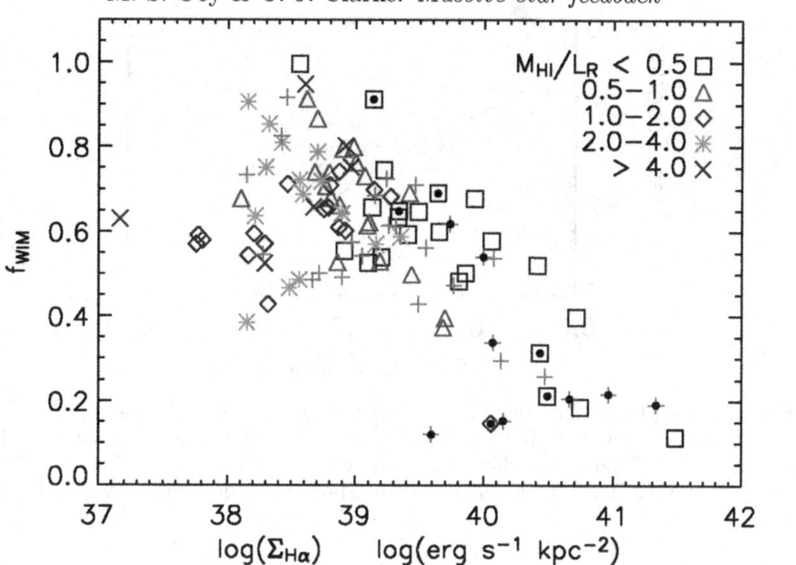

FIGURE 7. Fraction $f_{\rm WIM}$ of diffuse emission relative to total Hα luminosity for 109 galaxies from the SINGG survey as a function of Hα surface brightness $\Sigma_{\rm H\alpha}$, computed within the Hα half-light radius. Symbols are assigned by H I gas fraction $M_{\rm HI}/L_R$ as shown, and black dots indicate galaxies dominated by nuclear star formation (Oey et al. 2007).

conventional way to parameterize the hot (10^6 K) ionized medium (HIM) in galaxies. Q is essentially the filling factor of this hot gas, and, since it is thought to originate from shock-heating by SN explosions, Q can be estimated as the total volume of superbubbles relative to the relevant galaxy volume. For $Q > 1$, the galaxy is generating more hot gas than it can contain, and an outflow is predicted. This also implies that the neutral ISM is shredded, thereby allowing the escape of ionizing photons from the massive-star population. Oey & Clarke (1997) derived specific relations for Q, for two- and three-dimensional situations, and Clarke & Oey (2002) derived the critical star-formation rate $\rm SFR_{crit}$ in general terms:

$$\mathrm{SFR_{crit}} = 0.15 \left(\frac{M_{\rm ISM,10}\tilde{v}_{10}^2}{f_d} \right) M_\odot \ \mathrm{yr}^{-1} \quad , \tag{4.6}$$

where $M_{\rm ISM,10}$ is the ISM mass in units of $10^{10}\,M_\odot$, $\tilde{v}_{10}$ is the ISM thermal velocity in units of 10 $\rm km\,s^{-1}$, and f_d is a geometric correction factor for disk galaxies.

For the Milky Way, Clarke & Oey (2002) found that $\mathrm{SFR_{crit}} \sim 1\,M_\odot\ \mathrm{yr}^{-1}$, similar to our Galaxy's estimated star-formation rate, and implying that it is close to this threshold. They also found that most local starburst galaxies might be expected to exceed this criterion, since they often have smaller $\rm SFR_{crit}$ but larger star-formation rates in comparison to our Galaxy. Lyman-break galaxies should also show escaping ionizing radiation. Note that the $\rm SFR_{crit}$ criterion is not based on an escape velocity, but rather an over-pressure.

A large sample of nearby galaxies that can be used to test this model is the Survey for Ionization in Neutral-Gas Galaxies (SINGG; Meurer et al. 2006), which is an Hα survey of an optically-blind, H I-selected galaxy sample. For the first dataset of 109 galaxies, we used the HIIphot software of Thilker et al. (2000) to define the boundaries of the classical H II regions, assigning all remaining Hα emission to the WIM. Figure 7 shows the WIM fraction $f_{\rm WIM}$ of the Hα emission vs. the Hα surface brightness $\Sigma_{\rm H\alpha}$

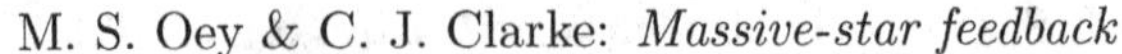

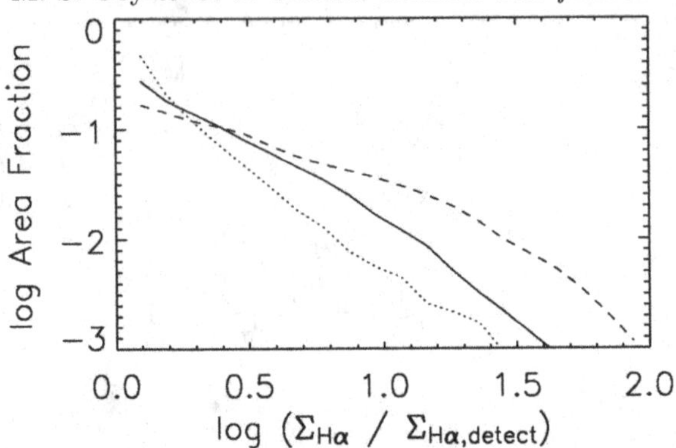

FIGURE 8. Co-added Hα surface brightness distributions for the SINGG galaxies. The dashed line shows "starburst" galaxies having $\log \Sigma_{\mathrm{H}\alpha} > 39.7$, the solid line shows galaxies with $38.7 < \log \Sigma_{\mathrm{H}\alpha} \leqslant 39.7$, and the dotted line shows galaxies with $\log \Sigma_{\mathrm{H}\alpha} \leqslant 38.7$; units of $\Sigma_{\mathrm{H}\alpha}$ are erg s^{-1} kpc^{-2}. Only pixels with a signal above a 3σ detection level are included (Oey et al. 2007).

for the sample. Galaxies with the highest $\Sigma_{\mathrm{H}\alpha}$ within the star-forming disk show the lowest f_{WIM}. We refer to galaxies having $\Sigma_{\mathrm{H}\alpha} > 2.5 \times 10^{39}$ erg s^{-1} kpc^{-2} as "starburst" galaxies here, although Heckman (2005) defines starbursts as much more intense systems having $\Sigma_{\mathrm{H}\alpha} > 10^{41}$ erg s^{-1} kpc^{-2}. Figure 8 shows co-added Hα surface brightness distributions of galaxies in the SINGG sample, in three bins of $\Sigma_{\mathrm{H}\alpha}$. It is again apparent that our starburst galaxies have the flattest slope for the lowest surface brightnesses, also demonstrating that they have the lowest f_{WIM}.

A possible cause for the lower Hα diffuse fraction in starbursts could be a lower fraction of ionizing field stars, as implied by equation (2.4). If this is the dominant effect, then we should similarly see it reflected in a relation between f_{WIM} and the total SFR as measured, for example, by the total Hα luminosity. However, Oey et al. (2007) show that such an effect is not seen, and thus, the lower fraction of field stars is not the dominant cause of the trend in Figure 7.

We do note that local starbursts are expected to exceed the SFR$_{\mathrm{crit}}$ threshold criterion for the escape of ionizing radiation [equation (4.6)]. Figure 7 shows that the galaxies with the lowest H I gas fractions, as measured by M_{HI}/L_R, are those that most strongly exhibit the anti-correlation in f_{WIM} vs. $\Sigma_{\mathrm{H}\alpha}$, with M_{HI} corresponding to the H I gas mass, and L_R to the R-band luminosity. The low H I gas fractions are consistent with a model in which there is not enough neutral gas for the total ionizing luminosity, and some of the radiation is lost from these galaxies. Note that ionizing photons could be lost through the shredded geometry of the ISM, as suggested in the Clarke & Oey (2002) model, or the ISM could simply be fully ionized and density-bounded. Since neutral gas is detected in all of the galaxies, we favor the former model, but it may also be possible that the star-forming disk is fully ionized and that the unresolved H I detections result from the outer regions of the galaxies. Oey et al. (2007) present a more complete discussion, in which they show that a substantial fraction of the SINGG galaxies exceed the SFR$_{\mathrm{crit}}$ threshold for the escape of ionizing radiation, including all of the starburst galaxies. At the same time, they also show that the relation between f_{WIM} and $\Sigma_{\mathrm{H}\alpha}$ in Figure 7 is consistent with the simplest expectations for density bounding. These results are strongly suggestive that ionizing radiation is escaping from starburst galaxies through at least one of these mechanisms. Nevertheless, we caution that several searches for Lyman continuum emission from galaxies have yielded negative results (e.g., Heckman et al. 2001;

Leitherer et al. 1995); whereas more recently, the blue compact dwarf galaxy Haro 11 does show a detection (Bergvall et al. 2006), and at least two Lyman-break galaxies also show unambiguous Lyman continuum emission (Shapley et al. 2006). However, the positive detections correspond to low escape fractions ($\lesssim$5%) of the total ionizing radiation. Further studies are necessary to resolve these important and tantalizing issues regarding the Hα diffuse fraction and implied consequences.

5. Chemical feedback

The third feedback process is the nucleosynthesis by massive stars and their core-collapse supernovae. The element enrichment of the ISM in galaxies and their environments drives the chemical evolution of galaxies and the cosmos. As another massive-star feedback process, chemical evolution can again be modeled with the same parameterizations for the massive-star population and mechanical feedback processes. The supernova activity heats the coronal gas within superbubbles, which is the immediate medium into which the nucleosynthetic products are injected. The elements can mix and disperse efficiently within this hot gas (Oey 2003; Tenorio-Tagle 1996), but their dispersal into cooler environments is a complex and poorly understood process. Pioneering simulations of the mixing and dispersal process are only recently emerging (e.g., Balsara & Kim 2005; de Avillez & Mac Low 2002; see also Scalo & Elmegreen 2004).

Oey (2000, 2003) introduced a rudimentary analytic model for galactic chemical evolution that is based on the simple parameterizations above. The model assumes that the enrichment volume for an OB association scales directly with the SN-driven superbubble volume, with the former being the volume into which the SN products are uniformly diluted. The superbubble size distribution given by equation (4.4) can therefore be used to derive the relative metallicity distribution in the ISM for the ensemble of massive-star clusters given by equation (2.3). Note that the largest volumes dilute the products to the lowest metallicities, and so Oey (2000) obtains:

$$N(Z) \propto Z^{-2}\, dZ \quad , \quad Z_{\rm min} < Z < Z_{\rm max} \quad . \tag{5.1}$$

where Z is the ISM metallicity, uniformly distributed within each enrichment volume. We impose the condition that no further dispersal of the elements occurs beyond these individual volumes. Thus, this represents an extreme inhomogeneous, no-mixing model, which can be contrasted to the opposite extreme of the pure, homogeneous Simple Model (Schmidt 1963; Pagel & Patchett 1975) that is used as the standard reference for most studies of galactic chemical evolution.

Assuming that subsequent generations of massive stars generate the same metallicity distribution [equation (5.1)], we can model the enrichment process by progressively summing the metallicities as the volumes overlap. Thus, after n generations of star formation, the instantaneous metallicity distribution function (MDF) is given by

$$N_{\rm inst}(Z) = \sum_{j=1}^{n} P_j\, N_j(Z) \quad , \tag{5.2}$$

where $N_j(Z)$ is the MDF for the ensemble of j overlapping volumes, which can be generated from the parent MDF given by equation (5.1). Note that any other form of the parent MDF can also be used in place of equation (5.1), and also that the Central Limit Theorem implies that $N_j(Z)$ approximates a Gaussian distribution in the limit of large j. P_j is the probability of obtaining j overlapping regions, and is given by the binomial

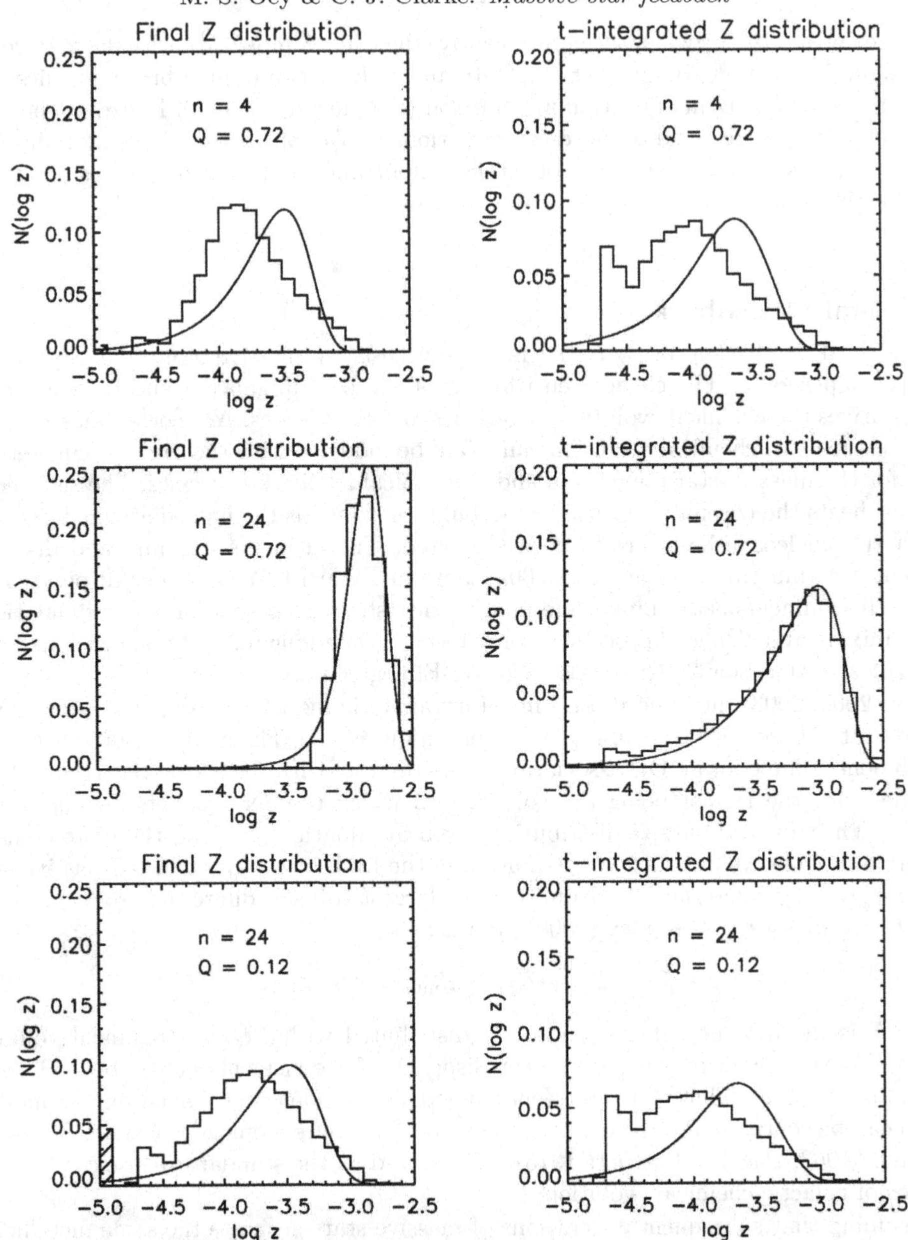

FIGURE 9. SIM models for different combinations of n and Q as shown, which correspond roughly to age and star-formation intensity, respectively. Final, instantaneous MDFs are shown on the left side, and the corresponding time-integrated MDFs are on the right. The SIM models are shown in the histograms, and analytic approximations, valid for large nQ, are shown by the curves.

distribution:

$$P_j = \binom{n}{j} \; Q^j \; (1-Q)^{n-j} \;, \quad 1 \leqslant j \leqslant n \quad , \tag{5.3}$$

where Q is the volume filling factor, that is again simply scaled from the porosity parameter considered above.

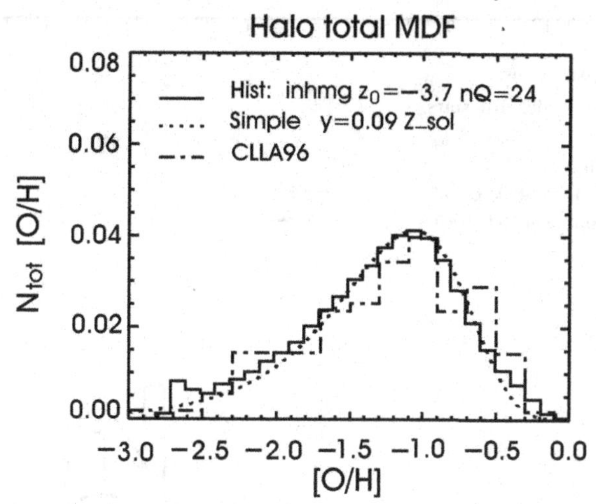

FIGURE 10. Halo metallicity distribution function converted to [O/H] from data by Carney et al. (1996, dot-dashed line). A Simple Inhomogeneous Model is overplotted with the solid line, and a homogeneous Simple Model is overplotted with the dotted line. The two models imply very different evolutionary states (based on Oey 2003).

Equation (5.2) therefore offers a Simple Inhomogeneous Model (SIM), which is in the spirit of the homogeneous Simple Model, but which can be compared to observed *instantaneous* MDFs. The SIM is especially relevant to the most metal-poor systems, where stochastic effects dominate the evolution. It also emphasizes that the star-formation intensity, not merely the simple age of the system, is a major driver of the chemical evolution. The parameter that describes the evolutionary state in the SIM models is the product nQ, which is also the mean of the binomial distribution. Figure 9 shows a series of SIM models for different combinations of n and Q, with instantaneous MDFs shown on the left. The right column shows corresponding models for time-integrated MDFs (see below). For the top and bottom models, the product nQ is the same, although the individual values of n and Q differ; we see that the resulting MDFs are qualitatively and quantitatively similar. In contrast, the middle panels show a model where the individual values of n and Q are the same as values in the top or bottom, yet the product nQ is much larger. For this model, the evolutionary state is much more evolved. Thus, an old system with a low star-formation intensity will have a similar MDF to a young system with a high star-formation intensity. This can straightforwardly explain the co-existence of old, metal-rich systems like the Galactic bulge, and extremely metal-poor systems like I Zw 18 that also show old stellar populations. These two systems may have similar ages, but extremely different time-integrated star-formation intensities, and they therefore show very different metallicities and evolutionary state, perhaps analogous to the middle and bottom models in Figure 9.

Oey (2000, 2003) also derived a cumulative, time-integrated MDF for all objects ever created out of the ISM whose enrichment proceeds in this way, namely, long-lived stars:

$$N_{\rm tot} = \frac{1}{n} \sum_{j=1}^{n} \sum_{k=j}^{n} D_{k-1}\, P_{j,k}\, N_j(Z) \quad . \tag{5.4}$$

We now include a depletion factor D_k, which is the fraction of gas remaining after k generations of star formation. Figure 10 compares the time-integrated SIM model (solid histogram) to the Galactic halo MDF from the data of Carney et al. (1996; dot-dashed

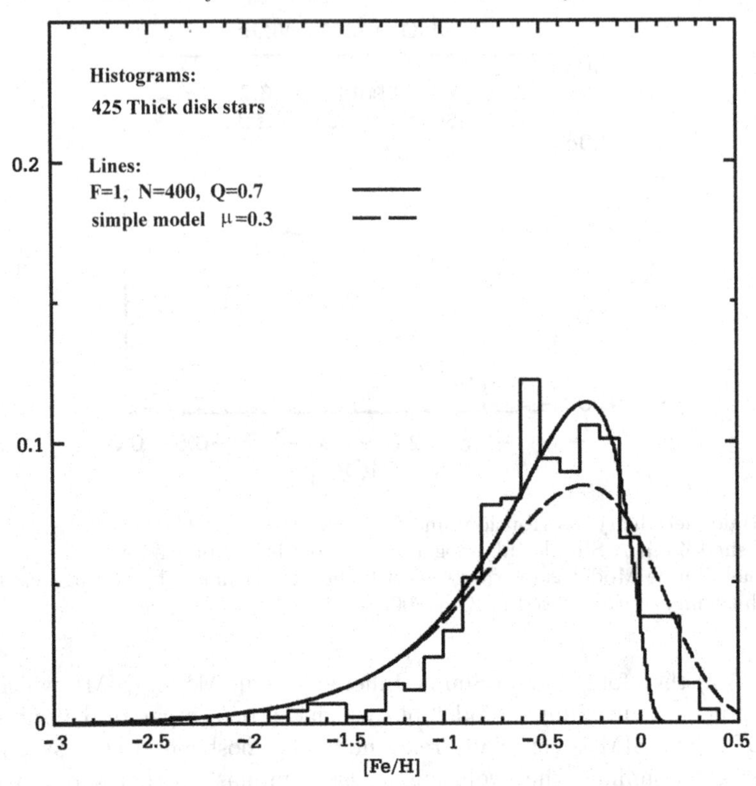

FIGURE 11. Metallicity distribution function for the Galactic thick disk from data by Nordström et al. (2004) and selection criteria of Bensby et al. (2003, 2005; solid histogram). The overplotted curves show a Simple Inhomogeneous Model (solid) and a Simple Model (dashed; Bensby & Oey 2006, in preparation).

line). A homogeneous Simple Model is also overplotted (dotted line). We see that both models agree well with the data. However, Oey (2003) shows that the interpretations of these two models are extremely different: the Simple Model implies that the halo is a highly evolved system, because the decrease in high-metallicity stars is caused by a lack of remaining gas to form such stars. In contrast, the SIM implies that the halo is a relatively unevolved system, in which the decreasing high-metallicity tail is still dominated by the form of the parent MDF which is given by the Z^{-2} distribution [equation (5.1)]. Thus we see that further empirical constraints are necessary to distinguish between these dramatically different possibilities.

Real systems should follow evolution that is intermediate between these extremes described by the homogeneous Simple Model and inhomogeneous SIM. Figure 11 shows a preliminary MDF for the Galactic thick disk. The data correspond to F and G dwarfs from the sample of Nordström et al. (2004), selected according to kinematic criteria of Bensby et al. (2003, 2005). We overplot both a Simple Model (dashed line) and SIM model (solid line). We see that the thick-disk data do, in fact, lie between the two models. The only exception is in the low-metallicity tail, where the so-called G-dwarf Problem is seen: there is a lack of the lowest-metallicity stars, in comparison to both models, although the discrepancy is not as extreme as for the Galactic thin disk.

Another useful feature of the SIM model is that it offers a straightforward prediction for the fraction of zero-metallicity, Population III stars. For any given SIM model, it is simply the fraction of stars corresponding to $j = 1$, which do not overlap any preceding generations of contamination. Thus,

$$F_{\rm III} = \sum_{k=1}^{n} D_{k-1} P_{1,k} \Bigg/ \sum_{j=1}^{n} \sum_{k=j}^{n} D_{k-1} P_{j,k} \quad . \tag{5.5}$$

For the Galactic halo SIM model shown in Figure 10, $F_{\rm III} = 2 \times 10^{-2}$. As discussed by Oey (2003), this is two orders of magnitude below the empirical upper limit of 2×10^{-4}, demonstrating that the halo also shows a G-dwarf Problem, as also found by others (e.g., Prantzos 2003). Thus, we see the power of these simple analytic models in raising fundamental issues regarding the formation and evolution of our Galaxy.

6. Summary

In summary, we see that empirical evidence thus far supports simple parameterizations of the massive-star population in terms of their spatial distribution, given by the N_*^{-2} clustering law [equation (2.3)], and mass distribution, given by a Salpeter (or similar) power-law IMF. The evidence is also suggestive, as described above, for a stellar upper-mass limit around $150\,M_\odot$, at least locally. The clustering law implies that about 10–30% of high-mass stars are field stars [equation (2.4)], defined as those having no other massive-star siblings in the host cluster. These simple parameterizations of massive stars as a population have powerful applications for analytically describing their feedback effects.

The ionizing radiation from these stars drives radiative feedback. The clustering law directly results in the H II-region luminosity function, which shows a similar -2 power-law exponent, provided that the H II regions have enough ionizing stars to fully sample the stellar IMF. There is evidence that the upper limits to the H IILF vary across the Hubble Sequence, although the slope of the H IILF appears to be constant. Star-forming galaxies possess an ISM component at 10^4 K, which also appears to be ionized by the high-mass stellar population. Typically, about half of the total Hα emission from such galaxies is observed to originate from ordinary H II regions, while the remaining half originates from the diffuse, warm ionized medium. The diffuse WIM is likely ionized by both field stars and radiation escaping from the H II regions, in roughly equal parts. However, starburst galaxies show lower WIM fractions. The origin of this trend is unclear: data from the SINGG survey are consistent with predictions for escaping radiation based on shredding of the ISM by mechanical feedback; on the other hand, direct searches for Lyman continuum radiation from galaxies have shown extremely low ($\lesssim$5%) escape fractions.

The global mechanical feedback from the massive-star population is dominated by their core-collapse SNe; stellar winds only play a significant role for the youngest populations, in which the highest-mass stars remain unevolved and contribute the strongest stellar winds. In a steady-state scenario with constant global star-formation rate, the clustering law implies a resulting superbubble size distribution with a dependence of R^{-3} [equation (4.4)] and expansion velocity distribution dependence of $v^{-7/2}$ [equation (4.5)]. These relations agree well for the H I shell catalog for the SMC. From the superbubble size distribution, we can derive a galaxy's porosity parameter or volume filling factor for the superbubbles. This is a standard convention for estimating the magnitude of the hot (10^6 K) component of the ISM, which is thought to originate within the SN-heated interiors of the superbubbles. As the star-formation rate increases, the shells merge, shred the cooler ISM, and generate more hot gas than the galaxy can contain, thus driving a galactic outflow or superwind. It is possible to define a critical star-formation rate SFR_{crit}, based on our previous parameterizations and ISM properties [equation (4.6)]; we find that it is on the order of a few $M_\odot$ yr^{-1} for an L^* galaxy. We note that the LMC shows only 1/4 the total number of coherent H I shells compared to the SMC, despite its larger size and star-formation rate. This is consistent with the prediction that it is near SFR_{crit}, so that the shells are merging and interacting. Thus, SFR_{crit} represents a threshold condition for the outflow of ionizing photons through the shredded ISM, as alluded to above, in addition to the outflow of hot gas and the newly produced SN products. We note that this represents a pressure-driven model, independent of a galaxy's escape velocity.

The clustering law and resulting superbubble size distribution also offer a convenient framework for understanding the stochastic, inhomogeneous progression of galactic chemical evolution. Whereas the standard Simple Model is purely homogeneous at all times, the Simple Inhomogeneous Model can make predictions for the *instantaneous* metallicity distribution function at any snapshot in time [equation (5.2)]. It can also predict cumulative MDFs [equation (5.4)], and so it can also be applied to samples of long-lived stars. The SIM is applicable, in particular, to the most metal-poor conditions, and it agrees well with the Galactic halo metallicity distribution. This offers an alternative interpretation of the halo as a relatively unevolved population, in contrast to the highly evolved status implied by the Simple Model. The SIM also offers an opposite extreme to the homogeneous Simple Model, within which observations may be bracketed. In addition, it emphasizes that a system's evolutionary status depends as much on the star-formation intensity, as on mere age. Finally, the SIM provides simple, straightforward predictions for the fraction of zero-metallicity Population III stars for any given model [equation (5.5)]. Those for the Galactic halo confirm a discrepancy of at least two orders of magnitude in the observed lack of zero-metallicity stars compared to predicted fraction.

These diverse, massive-star feedback effects are all unified by the same set of analytic parameterizations for this energetic stellar population. The observations are largely, and remarkably, consistent across varying physical phenomena over an extreme range in scale and age. Thus, this simple, self-consistent framework offers powerful tools and insight on the role of the massive-star population in the cosmos.

MSO thanks the Symposium organizers for their hospitality and travel support. Some of this work was supported by the National Science Foundation, grants AST-0448893 and AST-0448900; and the NASA Astrophysics Data Program, grant NAG5-10768.

REFERENCES

Baldry, I. K. & Glazebrook, K. 2003 *ApJ* **593**, 258.
Banfi, M., Rampazzo, R., Chincarini, G., & Henry, R. B. C. 1993 *A&A* **280**, 373.
Balsara, D. S. & Kim, J. 2005 *ApJ* **634**, 390.
Bell, E. F. & de Jong, R. S. 2001 *ApJ* **550**, 212.
Bensby, T., Feltzing, S., & Lundström, I. 2003 *A&A* **410**, 527.
Bensby, T., Feltzing, S., Lundström, I., & Ilyin, I. 2005 *A&A* **433**, 185.
Bergvall, N., Zackrisson, E., Andersson, B.-G., Arnberg, D., Masegosa, J., & Östlin, G. 2006 *A&A* **448**, 513.
Bonnell, I. A., Bate, M. R., & Zinnecker, H. 1998, *MNRAS* **298**, 93.
Caldwell, N., Kennicutt, R., Phillips, A. C., & Schommer, R. A. 1991 *ApJ* **370**, 526.
Carney, B. W., Laird, J. B., Latham, D. W., & Aguilar, L. A. 1996 *AJ* **112**, 668.
Clarke, C. J. & Oey, M. S. 2002, *MNRAS* **337**, 1299.
de Avillez, M. & Mac Low, M.-M. 2002 *ApJ* **581**, 1047.

Elmegreen, B. G. 2000 *ApJ* **539**, 342.

Elmegreen, B. G. 2006 *ApJ* **648**, 572.

Elmegreen, B. G. & Efremov, Y. N. 1997 *ApJ* **480**, 235.

Elmegreen, B. G. & Lada, C. 1977 *ApJ* **214**, 725.

Fernandes, I. F., de Carvalho, R., Contini, T., & Gal, R. R. 2004 *MNRAS* **355**, 728.

Ferguson, A. M. N., Wyse, R. F. G., Gallagher, J. S., & Hunter, D. A. 1996 *AJ* **111**, 2265.

Figer, D. F. 2005 *Nature* **434**, 192.

Heckman, T. M. 2005. In *Starbursts: From 30 Doradus to Lyman Break Galaxies* (eds. R. de Grijs & R. M. González Delgado). Ap&SS Library 329, p. 3. Springer.

Heckman, T. M., Sembach, K. R., Meurer, G. R., Leitherer, C., Calzetti, D., & Martin, C. L. 2001 *ApJ* **558**, 56.

Hoopes, C. G. & Walterbos, R. A. M. 2000 *ApJ* **541**, 597.

Hunter, D. A., Elmegreen, B. G., Dupuy, T. J., & Mortonson, M. 2003 *AJ* **126**, 1836.

Hunter, D. A., Vacca, W. D., Massey, P., Lynds, R., & O'Neil, E. J. 1997 *AJ* **113**, 1691.

Kahn, F. D. 1974 *A&A* **37**, 149.

Kennicutt, R. C., Edgar, B. K., & Hodge, P. W. 1989 *ApJ* **337**, 761.

Kim, S., Dopita, M. A., Staveley-Smith, L., & Bessell, M. S. 1999 *AJ* **118**, 2797.

Koen, C. 2006 *MNRAS* **365**, 590.

Kroupa, P. & Weidner, C. 2003 *ApJ* **598**, 1076.

Larson, R. B. & Starrfield, S. 1971 *A&A* **13**, 190.

Ledoux, P. 1941 *ApJ* **94**, 537.

Leitherer, C., Ferguson, H. C., Heckman, T. M., & Lowenthal, J. 1995 *ApJ* **454**, L19.

Martins, F., Schaerer, D., & Hillier, D. J. 2005 *A&A* **436**, 1049.

Massey, P. 2002 *ApJS* **141**, 81.

Massey, P. 2003 *ARA&A* **41**, 15.

Massey, P. & Hunter, D. A. 1998 *ApJ* **493**, 180.

Massey, P., Johnson, K. E., & DeGioia-Eastwood, K. 1995 *ApJ* **454**, 151.

McCray, R. & Kafatos, M. 1987 *ApJ* **317**, 190.

Meurer, G. R., et al. 2006 *ApJS* **165**, 307.

Meurer, G. R., Heckman, T. M., Leitherer, C., Kinney, A., Robert, C., & Garnett, D. R. 1995 *AJ* **110**, 2665.

Nordström, B., et al. 2004 *A&A* **418**, 989.

Oey, M. S. 2000 *ApJ* **542**, L25.

Oey, M. S. 2003 *MNRAS* **339**, 849.

Oey, M. S., et al. 2007 *ApJ* **661**, 801.

Oey, M. S. & Clarke, C. J. 1997 *MNRAS* **289**, 570.

Oey, M. S. & Clarke, C. J. 1998a *AJ* **115**, 1543.

Oey, M. S. & Clarke, C. J. 1998b. In *Interstellar Turbulence* (eds. J. Franco & A. Carramiñana). p. 112. Cambridge University Press.

Oey, M. S. & Clarke, C. J. 2005 *ApJ* **620**, L43.

Oey, M. S. & Kennicutt, R. C. 1997 *MNRAS* **291**, 827.

Oey, M. S., King, N. L., & Parker, J. W. 2004 *AJ* **127**, 1632.

Pagel, B. E. J. & Patchett, B. E. 1975 *MNRAS* **172**, 13.

Parker, J. W., et al. 1998 *AJ* **116**, 180.

Prantzos, N. 2003 *A&A* **404**, 211.

Rand, R. J. 1992 *AJ* **103**, 815.

Rand, R. J. 2000 *ApJ* **537**, L13.

Repolust, T., Puls, J., & Herrero, A. 2004 *A&A* **415**, 349.

Reynolds R. J. 1984 *ApJ* **282**, 191.

Reynolds, R. J., Haffner, L. M., & Tufte, S. L. 1999 *ApJ* **525**, L21.

Rozas, M., Beckman, J. E., & Knapen, J. H. 1996 *A&A* **307**, 735.

Salpeter, E. E. 1955 *ApJ* **121**, 161.

Scalo, J. & Elmegreen, B. G. 2004 *ARAA* **42**, 275.

Scharzschild, M. & Härm, R. 1959 *ApJ* **129**, 637.
Schmidt, M. 1963 *ApJ* **137**, 758.
Selman, F., Melnick, J., Bosch, G., & Terlevich, R. 1999a *A&A* **341**, 98.
Selman, F., Melnick, J., Bosch, G., & Terlevich, R. 1999b *A&A* **347**, 532.
Shapley, A. E., Steidel, C. C., Pettini, M., Adelberger, K. L., & Erb, D. K. 2006, *ApJ* **651**, 688.
Smith, L. J., Norris, R. P. F., & Crowther, P. A. 2002 *MNRAS* **337**, 1309.
Staveley-Smith, L., Sault, R. J., Hatzidimitriou, D., Kesteven, M. J., & McConnell, D. 1997 *MNRAS* **289**, 225.
Stothers, R. B. 1992 *ApJ* **392**, 706.
Tenorio-Tagle, G. 1996 *AJ* **111**, 1641.
Thilker, D. A., Braun, R., & Walterbos, R. A. M. 2000 *AJ* **120**, 3070.
Voges, E. S., Walterbos, R. A. M., Hoopes, C. G., & Oey, M. S. 2005. In *Extra-Planar Gas* (ed. R. Braun). ASP Conf. Ser. 331, p. 225. ASP.
Walterbos, R. A. M. 1998, *PASA* **15**, 99.
Weaver, R., McCray, R., Castor, J., Shapiro, P., & Moore, R. 1977 *ApJ* **218**, 377.
Weidner, C. & Kroupa, P. 2004 *MNRAS* **348**, 187.
Wolfire, M. G. & Cassinelli, J. P. 1987 *ApJ* **319**, 850.
Zhang, Q. & Fall, S. M. 1999 *ApJ* **527**, L81.

The initial mass function in clusters

By BRUCE G. ELMEGREEN

IBM Research Division, T. J. Watson Research Center, Yorktown Hts., NY 10598, USA

The stellar initial mass function (IMF) in star clusters is reviewed. Uncertainties in the observations are emphasized. We suggest there is a distinct possibility that cluster IMFs vary systematically with density or pressure. Dense clusters could have additional formation processes for massive stars that are not present in low-density regions, making the slope of the upper-mass IMF somewhat shallower in clusters. Observations of shallow IMFs in some super star clusters and in elliptical galaxies are reviewed. We also review mass segregation and the likelihood that peculiar IMFs, as in the Arches cluster, result from segregation and stripping, rather than an intrinsically different IMF. The theory of the IMF is reviewed in some detail. Several problems introduced by the lack of a magnetic field in SPH simulations are discussed. The universality of the IMF in simulations suggests that something more fundamental than the physical details of a particular model is at work. Hierarchical fragmentation by any of a variety of processes may be the dominant cause of the power-law slope. Physical differences from region to region may make a slight difference in the slope and also appear in the low-mass turnover point.

---

1. Introduction: Uncertainties

The stellar initial mass function (IMF) is difficult to measure because of systematic uncertainties, selection effects, and statistical variance. Stars in clusters may all have the same age and distance, making their masses relatively straightforward to determine, but mass segregation, field star contamination, variable extinction, and small number statistics can be problems in determining the IMF. Nearby clusters show the low-mass stars well, but these clusters also tend to be the most common, and therefore among the lowest in mass, so they do not sample far enough in the high-mass IMF to contain massive stars. High-mass clusters contain massive stars, but these clusters are rare, and the nearest are typically too far away to reveal their lowest-mass stars and brown dwarfs. The Orion trapezium cluster is one of the few regions where an IMF can be determined throughout all stellar types, but even then the highest-mass star is only one-half or one-third the mass of the highest possible mass for a star (Hillenbrand & Carpenter 2000). No cluster has yet been observed over the entire stellar mass range. Most of what we know about the IMF in clusters is from piecing together different parts of the IMF from different clusters.

Stars in an OB association are also at about the same distance from the Sun, but they typically span a range of ages that is longer than the shortest lifetime of a massive star, making formation rate and stellar evolution corrections necessary before determining the IMF. OB associations also tend to have variable extinction, and their dispersal has to be considered to reconstruct which stars actually formed there.

The IMF in the field comes partly from stars that formed in the field (in small molecular clouds), partly from stars that drifted there out of nearby OB associations, and partly from old dissolved clusters. The advantage of IMF determinations in the field is that tens of thousands of stars can be included in a survey (e.g., Parker et al. 1998), as can a wide range of stellar masses. However, there are many uncertainties in converting what is observed—the present-day mass function—into what is desired—the initial mass function. For example, this conversion depends on the star-formation history and the rate of vertical disk heating. Stellar evolution and the mass–luminosity relation are also important.

The IMF in whole galaxies comes from the summed IMFs of all the star-forming regions, i.e., from the clusters, loose groups, associations, and even the accreted satellite stars. Average IMFs are typically derived from abundance ratios (e.g., iron comes mostly from low-mass stars and oxygen comes mostly from high-mass stars), color-magnitude diagrams, Hα-equivalent widths, etc. However, resolution limits, faintness, unknown star-formation histories, variable extinction, crowding, and many other problems can arise in the determination of a galaxy-wide IMF.

2. IMFs in clusters: Should we expect systematic variations?

Many dense clusters have an IMF with a slope at intermediate to high mass that is close to the Salpeter IMF slope, $\Gamma = 1.35$ on a plot with log-mass intervals ($\xi(M) d\log M \propto M^{-\Gamma} d\log M$), which is the same as a negative slope of $1 + \Gamma$ on a plot with linear intervals in mass. There is considerable variation around this slope (± 0.5 in Scalo 1998), but this could be from sampling statistics (Elmegreen 1999; Kroupa 2001). The 30 Dor cluster has a slope remarkably close to the Salpeter value (Massey & Hunter 1998), as do the clusters h and χ Persei (Slesnick, Hillenbrand, & Massey 2002), NGC 604 in M33 (González Delgado & Perez 2000), NGC 1960 and NGC 2194 (Sanner et al. 2000), NGC 6611 (Belikov et al. 2000), and many others.

The Sco-Cen OB association has an IMF significantly steeper than the Salpeter function (Preibisch et al. 2002); the slope is -1.7 to -1.8 instead of -1.35. The massive stars in W51 ($M \geqslant 4\ M_\odot$) also have a steep IMF slope, -1.8, but two of the four subgroups in this region have a statistically significant excess of stars at the highest mass ($\sim 60\ M_\odot$; Okumura et al. 2000). We cannot tell whether these are physical variations or statistical fluctuations. Peretto, Andre, & Belloche (2006) suggested that three cores in the massive-star-forming region NGC 2264 appear to be headed for a merger. If some massive stars form by mergers or other peculiar events, then fluctuations at the high-mass end of the IMF can be large, considering that most clusters form only a few high-mass stars anyway.

Indeed, if there are two routes to forming a high-mass star, then there have to be two IMFs, one for the regions that favor one process and another for the regions that favor the second process. Two such processes could, for example, include gas core contraction in a turbulent medium (Tan & McKee 2004; Krumholz, McKee, & Klein 2005), and gas accretion from an intercore medium (Bonnell et al. 2007). Another process could be protostar coalescence in a cluster core. The first of these processes has the stellar mass defined by core formation, rather than accretion or coalescence after core formation, and this first process may apply to a wide range of environments, including—but not limited to—dense clusters. The second and third of these processes may work best in dense clusters. If this is the case, then the cluster IMF would have more routes to the formation of massive stars than other regions, and could therefore have a systematically flatter slope. This does not mean it would have an IMF flatter than the Salpeter IMF, because the turbulent core process could have a high-mass IMF that is steeper than Salpeter. In that case, all the additional processes that work in a cluster may serve only to flatten the cluster IMF to the Salpeter slope, with only the most extreme cluster conditions flattening it more than the Salpeter slope. Evidence for independent variations at both the high- and low-mass ends of the IMF were summarized by Elmegreen (2004). The observations seem to support the view that low-density regions have slightly steeper

IMFs than the Salpeter slope. This is consistent with the existence of multiple routes to massive stars.

3. Do massive stars need the cluster environment?

Testi, Palla, & Natta (1999) suggested that Herbig Ae/Be stars have a correlation between maximum mass and the surrounding cluster density. The observation was really that more massive clusters have a more massive upper end of the IMF, but because all of their clusters have about the same radius, the cluster mass translates into a cluster density. Bonnell & Clarke (1999) showed their result could be from sampling statistics: more massive clusters sample further out in the IMF. A similar debate took place 16 years earlier (Larson 1982; Elmegreen 1983). There have been also several other attempts to correlate cluster mass with maximum stellar mass (e.g., Khersonsky 1997). The size-of-sample effect is strong, however, and it can disguise a physical link between cloud mass and stellar mass, making the physical effect difficult to demonstrate. At the moment, there are no clear correlations between maximum stellar mass and the cloud or cluster mass that are in excess of expectations from sampling statistics.

The important question is whether the cluster environment affects the final stellar-mass distribution. We mentioned how it might in the previous section (denser regions flatten the IMF), but know of no clear evidence for it one way or another. De Wit et al. (2005) turned the question around and investigated whether massive stars ever form alone in the field. They observed 43 local "field" O-type stars and looked for evidence that they escaped from a cluster where they might have formed. Most of these O stars could reasonably be placed with some nearby cluster, but a few, 4% overall, could have formed in isolation. De Wit et al. pointed out that this percentage is consistent with a cluster mass function that extends down to a single $\sim$100 $M_\odot$ star with slope of $\beta = 1.7$.

Oey, King, & Parker (2004) did a similar study in the LMC, finding the distribution function of the number of O-type stars in clusters. This distribution also went smoothly down to clusters containing a single O star. The difference between the Oey et al. result and the de Wit et al. result is that the Oey et al. clusters also contain other stars, but the local isolated O stars do not occur in clusters and are truly isolated. If these isolated O stars cannot be traced to clusters, and if they really formed alone or in a loose group, then it would appear that massive stars do not need the cluster environment. It would be very interesting to know the IMF of stars which do not form in dense clusters. The above discussion suggests that this "isolated" IMF (not to be confused with a "field" IMF, which is a blend) should be steeper than the cluster IMF.

Figure 1 shows the positions of all massive stars in the 30 Doradus region of the LMC, using data from Massey & Hunter (1998). The various symbols represent stars in different mass intervals. Clearly the high-mass stars appear all over the region, even outside the dense cluster core, which is R136. This distribution is not surprising because the peripheral gas is still dense and fragmented, and it is also compressed by the stars in the core, leading to triggering (Walborn et al. 1999). Continuing the discussion of the previous paragraph, it would be interesting to know the IMF of triggered star formation.

There are many young regions that recently formed O-type stars, but show no evidence for clusters at all. NGC 604 in the galaxy M33 is an example. Hunter et al. (1996) estimated the massive-star IMF there and derived a slope of -1.6. These regions have been called super OB associations by Maíz-Apellániz (2001), who studied other examples. One would think if the O stars formed in clusters, there would be some remnant or core of those clusters remaining during the short massive-star lifetime. The lack of such cores implies the O stars formed in relative isolation.

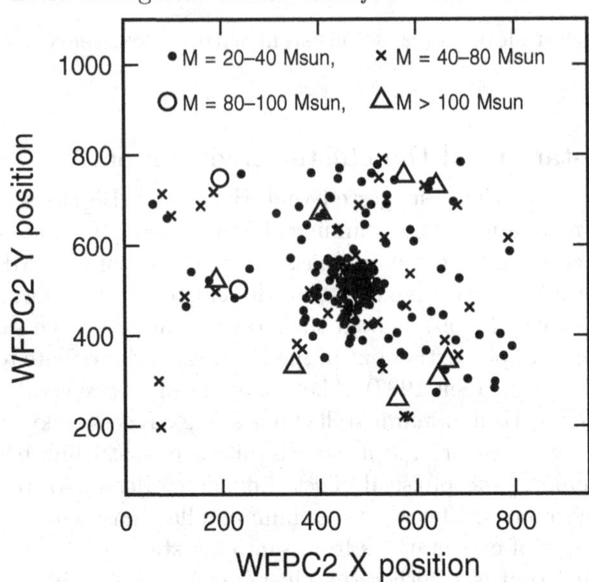

FIGURE 1. Distribution of O-type stars in the 30 Dor region, using data from Massey & Hunter (1998). High-mass stars form all over the region, not just in the dense core.

4. Applications to starbursts and young elliptical galaxies

Starbursts and mergers have a large fraction of their young stars in clusters (Larsen & Richtler 2000), and because the total number of clusters can be large, the samples can include rare supermassive clusters (Whitmore 2003). It is unknown whether star formation at a high rate in bursts produces the same range of cluster masses and the same IMF as star formation at a lower rate for a longer time, both producing the same total mass in stars. If so, then the cluster mass function and the IMF are sampling from universal functions. But this need not be the case. Starbursts could produce more massive clusters in a short time than normal galaxies in a long time because the pressure is always higher in a starburst and massive clusters are high-pressure regions. If also follows that if the IMF depends on the cluster environment, or if there are two IMFs, one of which depends more on the cluster environment than the other, then starburst regions with a significant population of massive or unusually dense clusters could produce a flatter IMF than normal galaxies.

Whole starburst regions do not appear to have IMFs noticeably flatter than the Salpeter function (see review in Elmegreen 2005), but some regions may have individual clusters with top-heavy or bottom-light IMFs. Sternberg (1998) found a high light-to-mass ratio in NGC 1705−1 that implies either an IMF slope shallower than $\Gamma = 1$ or an inner cutoff to the IMF that removes low-mass stars. Smith & Gallagher (2001) found the same for the cluster M82F: an inner cutoff around 2 to 3 $M_{\odot}$ for $\Gamma = 1.3$. Alonso-Herrero et al. (2001) derived a high L/M ratio for clusters in the starburst galaxy NGC 1614. McCrady et al. (2003) observed another cluster in M82, MGG−11, and inferred a deficit in low-mass stars. Mengel et al. (2002) had the same conclusion for clusters in NGC 4038/9.

Other super star clusters appear to have normal IMFs, so the dense cluster environment alone does not guarantee a flat IMF. Examples of normal IMFs are in NGC 1569−A (Ho & Filippenko 1996; Sternberg 1998), NGC 6946 (Larsen et al. 2001), and MGG−9 in M82 (McCrady et al. 2003).

Finding the light-to-mass ratio in a cluster is a difficult problem. The velocity dispersion in the cluster has to be measured along with the radius to get the mass, and the luminosity has to be measured. However, the velocity dispersion could vary with radius inside a young cluster, in which case the observed dispersion is a weighted integral over the position, and then the isothermal expression would not apply to the conversion between velocity dispersion and radius. The proper radius to use is uncertain because the cluster could be evaporating, out of equilibrium, non-isothermal, multi-component, non-isotropic, or non-centralized (Bastian & Goodwin 2006). Also, the core could be poorly resolved. The choice of aperture for the velocity dispersion measurement is difficult. Field-star corrections may be necessary for both the density profile and the velocity dispersion.

The IMFs of massive elliptical galaxies appear to be slightly flatter than in spiral galaxies (Pipino & Matteucci 2004; Nagashima et al. 2005b). Clusters of galaxies also suggest a history of top-heavy IMFs in the form of elliptical galaxy starbursts (Renzini et al. 1993; Loewenstein & Mushotsky 1996; Chiosi 2000; Moretti, Portinari, & Chiosi 2003; Tornatore et al. 2004; Romeo et al. 2005; Portinari et al. 2004; Nagashima et al. 2005a). Low surface brightness galaxies may have steeper-than-normal IMFs (Lee et al. 2004).

Taken together, these observations suggest a possible excess of high-mass stars in some starburst clusters or in early-phase starburst elliptical galaxies, and a possible deficit of high-mass stars in the most quiescent environments (low surface brightness galaxies). This trend is consistent with the existence of several routes to the formation of a high-mass star, with at least one of these routes more active in the type of environment that has a high star-formation rate. The exact physical processes that are involved with this "starburst route" are not observed yet, but enhanced accretion, protostellar coalescence, and high thermal temperatures would all work in this direction.

5. Mass segregation

A problem with IMF determinations for clusters is mass segregation, where the most massive stars are either born near the center or migrate toward the center after a random walk of scattering events. The IMF is often observed to be shallower in the central regions of clusters. The nearest large cluster, NGC 3603, has a relatively shallow IMF slope in the core and a relatively steep IMF slope near the edge (Sung & Bessell 2004) There is possibly a high-mass drop-off in this cluster too: the slope is -1.9 overall for $M > 40\ M_\odot$ (steeper than the Salpeter slope, which is -1.35). The Orion cluster has mass segregation too, prompting Hillenbrand & Hartmann (1998) to suggest it was there from birth, considering the young age of this cluster. At even younger age, the mm-wave continuum sources in Ophiuchus appear mass segregated (Elmegreen & Krakowski 2001).

A good example of a flat IMF that could either be top-heavy, as in some super star clusters, or mass segregated is in the Arches cluster. Yang et al. (2002) and Stolte et al. (2005) found an IMF slope of $\Gamma \sim -0.8$ there. The Arches cluster is in a region of intense tidal forcing (the Galactic center) and it seems plausible that the outer, low-density regions have been tidally stripped, leaving only the dense core. If the dense core was as mass segregated as other clusters which show the same flat slopes in their cores (e.g., de Grijs et al. 2002), then Arches would not have an unusual IMF.

The case of tidal stripping for the Arches cluster seems compelling after similarly flat IMFs have been observed in tidally stripped halo globular clusters. De Marchi, Pulone, & Paresce (2006) show a flat mass function in the galactic cluster NGC 6218. At four different radii, the mass function slopes in their figure are $+1.4$, $+1.3$, $+0.6$, and $+0.1$

(note the positive values, when the Salpeter slope on a comparable figure is −2.3). Flat mass functions are also seen in the globular clusters NGC 6712 (De Marchi et al. 1999) and Pal 5 (Koch et al. 2004). These latter two are expected to have undergone tidal stripping. Tidal stripping is suspected in NGC 6218 as well; the observed cluster mass is supposed to be only 20% of the original mass. Recent models by Baumgardt (2009) show how tidal stripping can leave a cluster with a flat IMF.

6. The low-mass IMF in clusters

The low-mass part of the IMF has been observed down to and beyond the brown-dwarf regime in nearby young clusters, where brown dwarfs are still bright on their pre-main sequence tracks. The count of low-mass stars is usually fairly high, making statistical fluctuations in the IMF much smaller than at higher masses. Generally the IMF turns over from its $\Gamma \sim 1.35$-type slope at intermediate to high mass and becomes somewhat flat with $\Gamma \sim 0$. The turnover occurs somewhere between 0.1 $M_\odot$ and 1 $M_\odot$ (Scalo 1986; Kroupa 2001; Chabrier 2003).

Several authors have noted variations in the relative abundance of low- and high-mass stars, or in the shape and extent of the flat low-mass part. For example, IC 348 (Preibisch, Stanke, & Zinnecker 2003; Muench et al. 2003; Luhman et al. 2003) and Taurus (Luhman 2000; Briceño et al. 2002) have brown-dwarf-to-star ratios that are ~2 times lower than the Orion Trapezium cluster (Hillenbrand & Carpenter 2000; Luhman et al. 2000; Muench et al. 2002), Pleiades (Bouvier et al. 1998; Luhman et al. 2000), M35 (Barrado y Navascués et al. 2001), and the Galactic field (Reid et al. 1999). There are many possible reasons for such fluctuations, but no observations yet favor one reason over another.

7. Theory

Stars form in dense molecular cores where self-gravity overcomes magnetic and pressure forces and where turbulent motions are too slow to disrupt the gas in a free-fall time. The origin of the cores is not fully understood. They are likely to have several formation mechanisms, including compression in shocks caused by turbulence and stellar outflows. Protostars form in the cores, but what happens after that is also unclear. If the protostars are rapidly converging in a cloud-wide collapse, then they can move together and interact strongly, creating tightly bound systems that disperse quickly (Bonnell et al. 2001). If the cores and protostars move slowly or have a low space density, then they will not interact. Strong interactions could make the stellar mass function different from the core mass function, while weak interactions might keep them about the same. This duality led to the suggestion above that there could be two high-mass IMFs appearing in different regions, depending on the degree of core and protostellar interactions.

Most simulations get the observed IMF, but then there are enough tunable parameters to assure this result if it is desired. Recent simulations have been probing the sensitivity of the results to the assumptions. Bate & Bonnell (2005) did two SPH simulations with no magnetic field, each having a different initial thermal Jeans mass. They found that the characteristic or turnover mass in the resulting IMF scaled directly with the input Jeans mass. Jappsen et al. (2005) considered variations in the equation of state. If the ratio of specific heats or adiabatic index γ is less than one at low density and greater than one at high density (meaning that the equation of state gets stiffer at high density), then the mean mass in the resulting IMF is comparable to the Jeans mass at this transition density (see also Larson 2005). If the transition density is higher, the Jeans mass becomes lower and there are more cores resulting. Martel et al. (2006) did SPH simulations with particle splitting, no magnetic fields, and an isothermal equation of state. They found that the characteristic mass in the resulting IMF depended on resolution: when the number of levels in the splitting hierarchy increased, and lower masses could be resolved, the mean core mass decreased in proportion.

Li et al. (2004) did magnetohydrodynamic (MHD) super-Alfvenic simulations on grids of various sizes and got a power-law IMF with a turnover at low mass, in reasonable agreement with observations. Tilley & Pudritz (2007) did MHD simulations in a 256^3 grid, testing the implications of different ratios of gravitational-to-magnetic energies. Their preferred model for the IMF had comparable thermal and magnetic pressures, and very large ratios of gravitational-to-magnetic energy density—on the order of 100. Padoan et al. (2005) did an adaptive mesh run without magnetic fields and found that brown dwarfs could form by turbulent fragmentation, even though their masses were much less than the initial Jeans mass. Nakamura & Li (2005) did a 2D MHD simulation and showed that magnetic diffusion is enhanced by turbulent compression; still, the slowness of magnetic diffusion lowers the efficiency of star formation compared to non-magnetic simulations.

An analytical model for the IMF in a turbulent medium was proposed by Padoan & Nordlund (2002). They suggested that a protostar mass is proportional to the density in a turbulence-compressed region times the cube of the compressed region thickness, L, and that the compression factor is proportional to the Alfven Mach number, as is typical for magnetic gas. This gives a mass $m \propto L^{4-\beta}$ where β is the slope of the energy power spectrum and is related to the size-linewidth power α by the equation $\beta = 2\alpha + 1$. Then they assumed that the size is distributed as $dN/dL \propto L^{-4}$, which comes from hierarchical fragmentation in a uniform medium. That is, $dN/dk \propto k^2$ for uniform partitions in the 3D phase space variable $k = 1/L$ (Di Fazio 1986). The result is a mass function $dN/dm \propto m^{-\Gamma-1}$ where $\Gamma = -3/(4-\beta) = -1.33$ for $\beta = 1.74$. The slope, Γ, compares well with the Salpeter slope, -1.35. In this model, higher Mach number regions give lower mass protostars.

Several important physical effects are missing from these models. Feedback that erodes disks and pre-collapse objects is usually not considered, although Li & Nakamura (2006) included protostellar winds as a source of turbulence in their 3D MHD simulations. Also missing are fully turbulent systems before star formation begins and anything like a turbulent environment that can affect the simulation boundary (i.e., variable total-grid mass, variable center of gravity, etc.). Heating and cooling are usually considered with only crude approximations, like polytrophic assumptions or constant temperatures. Generally only small numbers of stars form, so the modeled IMF is statistically inaccurate. In the case of SPH models, magnetic forces are always missing. For MHD on a single grid, there is a limited dynamic range for density. MHD simulations also do not treat the physics of detachment of the background magnetic field from stars. Nevertheless, it is only a matter of time before these limitations are overcome and simulations make the IMF in a realistic way.

The lack of magnetic fields in SPH simulations can severely affect the interpretation of the results. If the clump magnetic field is critical, or if the clump forms with a constant mass-to-flux ratio in a cloud where the average magnetic field is critical, then the field strength in the clump satisfies, $B_{\rm clump} \sim G^{1/2}\Sigma_{\rm clump}$ for clump mass column density $\Sigma_{\rm clump}$. The magnetic force per unit volume acting on the clump as it drags the field around is $\sim B^2_{\rm clump}/R_{\rm clump} \sim G\Sigma^2_{\rm clump}/R_{\rm clump}$. The gravitational force per unit volume acting on the clump by the rest of the cloud is $\sim G\Sigma_{\rm cloud} \times M_{\rm clump}/R^3_{\rm clump} =$

$G\Sigma_{\rm cloud}\Sigma_{\rm clump}/R_{\rm clump}$. Thus the ratio of the magnetic-to-gravitational force on a clump that drags the field around as the clump responds to the gravitational force of the surrounding cloud is

$$F_B/F_G \sim \Sigma_{\rm clump}/\Sigma_{\rm cloud} \gg 1 \quad . \tag{7.1}$$

Thus, clumps do not free fall in the cloud until either their magnetic-field lines are detached, or their fields diffuse out. Magnetic-field–free models that produce swarms of freely moving clumps and eventually protostars would seem to be unrealistic unless the whole clouds are collapsing too.

Similarly, we can calculate the likelihood of clump accretion from remote parts of the cloud. Magnetic fields should severely limit such accretion. The magnetic force per unit volume exerted on the ambient gas in a cloud is $\sim B^2_{\rm cloud}/R_{\rm cloud} \sim G\Sigma^2_{\rm cloud}/R_{\rm cloud}$. The gravitational force per unit volume on this ambient gas that is exerted by the clump is $\sim(GM_{\rm clump}/R^2_{\rm cloud}) \times M_{\rm cloud}/R^3_{\rm cloud}$. The magnetic to gravitational force ratio for accreted ambient cloud gas is

$$F_B/F_G \sim M_{\rm cloud}/M_{\rm clump} \gg 1 \quad . \tag{7.2}$$

Thus the ambient cloud gas cannot freely fall onto a single clump, or even onto a cloud core whose mass is significantly less than the mass of the whole cloud. These two effects from magnetic fields in clouds would seem to be a problem for magnetic-field–free simulations, where gravitational and turbulent motions turn into freely moving protostars which competitively accrete gas from the whole cloud. More likely, protostars accrete from their immediate neighborhoods or clumps, and they do not move rapidly in a cloud until their field lines are almost completely detached from the background.

There are also other magnetic effects that are not present in SPH simulations and which could be important for cloud structure and star formation. One large issue concerns dynamical communication between the cloud and the surrounding ISM. Magnetic fields connect the cloud, the cloud cores, and all of the pre-detached clumps to the external ISM. Magnetic stresses thereby transfer linear and angular momentum from inside the cloud to outside, and vice versa. These stresses are a source of damping for clump and cloud turbulent motions, and a possible source of energy into these motions from external turbulence (Elmegreen 1981). Internal feedback should be more influential on cloud structure when the cloud magnetic field is near the energy equilibrium value (the critical field). If the field is subcritical, it may have a more limited role (Padoan & Nordlund 1999).

8. Reflections

Computer simulations usually make IMFs like the observed IMF, but we should question whether they do this for the right reasons, because all of the different models have different assumptions and physical processes at work. The universality of the real IMF suggests an insensitivity to detailed processes. The IMF is the same inside and outside of star clusters, and it is about the same for starbursts and for slow star formation in galaxies. It is also nearly independent of metallicity, galaxy mass, and epoch in the Universe after some heavy elements form. With similar insensitivities, the simulations could also get the right result even if the assumed physics were oversimplified.

For example, hierarchical fragmentation alone gives a mass function $n(M)dM \sim M^{-2}dM$, which is very close to the Salpeter IMF ($M^{-2.35}dM$). What if the modeled IMF came mostly from fragmentation, regardless of the origin of this fragmentation? Such an IMF would be mostly the result of geometric effects. Then, if physical processes make star formation slightly more likely at intermediate mass (i.e., at about the Jeans mass M_J), the observed IMF could follow. That is, physical processes favoring M_J, in addition to fragmentation, could steepen the IMF from a function like M^{-2} to the observed function $M^{-2.4}$ for $M > M_J$. Similarly, a bias toward M_J would flatten the pure-fragmentation IMF from M^{-2} to $M^{-1.5}$ for $M < M_J$. Additional processes could act in dense clusters to make an excess of massive stars, or an excess of brown dwarfs. These additional processes might include the ablation of low-mass protostars, heightened accretion, coalescence, and multiple star interactions.

9. Conclusions

Observations suggest a more-or-less constant IMF in many diverse environments. There are hints of variations at the high- and low-mass ends of the IMF, suggesting perhaps a tri-modal IMF. Many things could cause these variations, such as protostellar coalescence, enhanced gas accretion, multiple-system star ejections, and so on, as discussed extensively in the literature. There are also possible false variations of the IMF from unknown star-formation histories, incorrect mass-to-light ratios, field-star contaminations, small number statistics, and so on.

The theory of gravo-turbulent fragmentation typically gets the observed IMF, but many uncertainties remain. Magnetic fields, feedback, boundary conditions and initial conditions are all concerns. Yet the diverse models usually get about the right IMF. If the simulations get the right IMF even under highly simplified conditions, it is fair to ponder what the simulations and reality have in common that always gives this IMF. Perhaps it is fragmentation alone, whether from gravity or turbulence, and independent of the proportion. Perhaps it is from accretion alone, as suggested by Bonnell et al. (2007). It could even be from the large number of independent parameters in both simulations and reality, which, through random variations, bring the system to a common mass function that is lognormal. More simulations are needed before these questions can be clarified.

REFERENCES

Alonso-Herrero, A., Engelbracht, C. W., Rieke, M. J., Rieke, G. H., & Quillen, A. C. 2001 *ApJ* **546**, 952.

Barrado y Navascués, D., Stauffer, J. R., Bouvier, J., & Martn, E. L. 2001 *ApJ* **546**, 1006.

Bastian, N. & Goodwin, S. P. 2006 *MNRAS* **369**, L9.

Bate, M. R. & Bonnell, I. A. 2005 *MNRAS* **356**, 1201.

Baumgardt, H. 2009. In *Globular Clusters: Guides to Galaxies* (eds. T. Richtler & Søren Larson). Springer; in press.

Belikov, A. N., Kharchenko, N. V., Piskunov, A. E., & Schilbach, E. 2000 *A&A* **358**, 886.

Bonnell, I. A. & Clarke, C. J. 1999 *MNRAS* **309**, 461.

Bonnell, I. A., Bate, M. R., Clarke, C. J., & Pringle, J. E. 2001 *MNRAS* **324**, 573.

Bonnell, I. A., Larson, R. B., & Zinnecker, H. 2007. In *Protostars and Planets V* (eds. B. Reipurth, D. Jewitt, & K. Keil). p. 149. University of Arizona Press.

Bouvier, J., Stauffer, J. R., Martin, E. L., Barrado y Navascués, D., Wallace, B., & Bejar, V. J. S. 1998 *A&A* **336**, 490.

Briceño, C., Luhman, K. L., Hartmann, L., Stauffer, J. R., & Kirkpatrick, J. D. 2002 *ApJ* **580**, 317.

Chabrier, G. 2003 *PASP* **115**, 763.

Chiosi, C. 2000, *A&A* **364**, 423.

De Grijs, R., Gilmore, G. F., Johnson, R. A., & Mackey, A. D. 2002 *MNRAS* **331**, 245.

De Marchi, G., Leibundgut, B., Paresce, F., & Pulone, L. 1999 *A&A* **343**, 9.

De Marchi, G., Pulone, L., & Paresce, F. 2006 *A&A* **449**, 161.

De Wit, W. J., Testi, L., Palla, F., & Zinnecker, H. 2005 The origin of massive O-type field stars: II. Field O stars as runaways. *A&A* **437**, 247.

Di Fazio, A. 1986 *A&A* **159**, 49.

Elmegreen, B. G. 1981 *ApJ* **243**, 512.

Elmegreen, B. G. 1983 *MNRAS* **203**, 1011.

Elmegreen, B. G. 1999 *ApJ* **515**, 323.

Elmegreen, B. G. 2004 *MNRAS* **354**, 367.

Elmegreen, B. G. 2005. In *Starbursts: From 30 Doradus to Lyman Break Galaxies* (eds. R. de Grijs & R. M. Gonzalez Delgado). Vol. 329, p. 57. Astrophysics & Space Science Library.

Elmegreen, B. G. & Krakowski, A. 2001 *ApJ* **562**, 433.

González Delgado, R. M. & Pérez, E. 2000 *MNRAS* **317**, 64.

Jappsen, A.-K., Klessen, R. S., Larson, R. B., Li, Y., & Mac Low, M.-M. 2005 *A&A* **435**, 611.

Hillenbrand, L. A. & Hartmann, L. 1998 *ApJ* **492**, 540.

Hillenbrand, L. A. & Carpenter, J. M. 2000 *ApJ* **540**, 236.

Ho, L. C. & Filippenko, A. V. 1996 *ApJ* **466**, L83.

Hunter, D. A., Baum, W. A., O'Neil, E. J. Jr., & Lynds, R. 1996 *ApJ* **456**, 174.

Khersonsky, V. K. 1997 *ApJ* **475**, 594.

Koch, A., Grebel, E. K., Odenkirchen, M., Martínez-Delgado, D., & Caldwell, J. A. R. 2004 *AJ* **128**, 2274.

Kroupa, P. 2001 *MNRAS* **322**, 231.

Krumholz, M. R., McKee, C. F., & Klein, R. I. 2005 *Nature* **438**, 332.

Larsen, S. S., Brodie, J. P., Elmegreen, B. G., Efremov, Y. N., Hodge, P. W., & Richtler, T. 2001 *ApJ* **556**, L801.

Larsen, S. S. & Richtler, T. 2000 *A&A* **354**, 836.

Larson, R. B. 1982 *MNRAS* **200**, 159.

Larson, R. B. 2005 *MNRAS* **359**, 211.

Lee, H.-C., Gibson, B. K., Flynn, C., Kawata, D., & Beasley, M. A. 2004 *MNRAS* **353**, 113.

Li, P. S., Norman, M. L., Mac Low, M.-M., & Heitsch, F. 2004 *ApJ* **605**, 800.

Li, Z.-Y. & Nakamura, F. 2006 *ApJ* **640**, L187.

Loewenstein, M. & Mushotsky, R. F. 1996 *ApJ* **466**, 695.

Luhman, K. L. 2000 *ApJ* **544**, 1044.

Luhman, K. L., Rieke, G. H., Young, E. T., Cotera, A. S., Chen, H., Rieke, M. J., Schneider, G., & Thompson, R. I. 2000 *ApJ* **540**, 1016.

Luhman, K. L., Stauffer, J. R., Muench, A. A., Rieke, G. H., Lada, E. A., Bouvier, J., & Lada, C. J. 2003 *ApJ* **593**, 1093.

Maíz-Apellániz, J. 2001 *ApJ* **563**, 151.

Martel, H., Evans, N. J. II, & Shapiro, P. R. 2006 *ApJS* **163**, 122.

Massey, P. & Hunter, D. A. 1998 *ApJ* **493**, 180.

McCrady, N., Gilbert, A., & Graham, J. R. 2003 *ApJ* **596**, 240.

Mengel, S., Lehnert, M. D., Thatte, N., & Genzel, R. 2002 *A&A* **383**, 137.

Moretti, A., Portinari, L., & Chiosi, C. 2003 *A&A* **408**, 431.

Muench, A. A., Lada, E. A., Lada, C. J., & Alves, J. 2002 *ApJ* **573**, 366.

Muench, A. A., Lada, E. A., Lada, C. J., Elston, R. J., Alves, J. F., Horrobin, M., Huard, T. H., Levine, J. L., Raines, S. N., & Román-Zíñiga, C. 2003 *AJ* **125**, 2029.

Nagashima, M., Lacey, C. G., Baugh, C. M., Frenk, C. S., & Cole, S. 2005a *MNRAS* **358**, 1247.

Nagashima, M., Lacey, C. G., Okamoto, T., Baugh, C. M., Frenk, C. S., & Cole, S. 2005b *MNRAS* **363**, L31.

Nakamura, F. & Li, Z.-Y. 2005 *ApJ* **631**, 411.

Oey, M. S., King, N. L., & Parker, J. Wm. 2004 *AJ* **127**, 1632.

Okumura, S., Mori, A., Nishihara, E., Watanabe, E., & Yamashita, T. 2000 *ApJ* **543**, 799.

Padoan, P., Kritsuk, A., Norman, M. L., & Nordlund, A. 2005 *Memorie della Societa Astronomica Italiana*, **76**, 187.

Padoan, P. & Nordlund, A. 1999 *ApJ* **526**, 279.

Parker, J. W., Hill, J. K., Cornett, R. H., Hollis, J., Zamkoff, E., Bohlin, R. C., O'Connell, R. W., Neff, S. G., Roberts, M. S., Smith, A. M., & Stecher, T. P. 1998 *AJ* **116**, 180.

Peretto, N., André, Ph., & Belloche, A. 2006 *A&A* **445**, 979.

Pipino, A. & Matteucci, F. 2004 *MNRAS* **347**, 968.

Portinari, L., Moretti, A., Chiosi, C., & Sommer-Larsen, J. 2004 *ApJ* **604**, 579.

Preibisch, T., Brown, A. G. A., Bridges, T., Guenther, E., & Zinnecker, H. 2002 *AJ* **124**, 404.

Preibisch, T., Stanke, T., & Zinnecker, H. 2003 *A&A* **409**, 147.

Reid, I. N., Kirkpatrick, J. D., Liebert, J., Burrows, A., Gizis, J. E., Burgasser, A., Dahn, C. C., Monet, D., Cutri, R., Beichman, C. A., & Skrutskie, M. L. 1999 *ApJ* **521**, 613.

Renzini, A., Ciotti, L., D'Ercole, A., & Pellegrini, S. 1993 *ApJ* **419**, 52.

Romeo, A. D., Sommer-Larsen, J., Portinari, L., & Antonuccio-Delogu, V. 2005 *MNRAS*, **371**, 548.

Sanner, J., Altmann, M., Brunzendorf, J., & Geffert, M. 2000 *A&A* **357**, 471.

Scalo, J. M. 1986 *Fund. Cos. Phys* **11**, 1.

Scalo, J. M. 1998. In *The Stellar Initial Mass Function* (eds. G. Gilmore, I. Parry, & S. Ryan). p. 201. Cambridge University Press.

Slesnick, C. L., Hillenbrand, L. A., & Massey, P. 2002 *ApJ* **576**, 880.

Smith, L. J. & Gallagher, J. S. 2001 *MNRAS* **326**, 1027.

Sternberg, A. 1998 *ApJ* **506**, 721.

Stolte, A., Brandner, W., Grebel, E. K., Lenzen, R., & Lagrange, A.-M. 2005 *ApJ* **628**, L113.

Sung, H. & Bessell, M. S. 2004 *AJ* **127**, 1014.

Tan, J. C. & McKee, C. F. 2004 *ApJ* **603**, 383.

Testi, L., Palla, F., & Natta, A. 1999 *A&A* **342**, 515.

Tilley, D. A. & Pudritz, R. E. 2007 *MNRAS* **382**, 73.

Tornatore, L., Borgani, S., Matteucci, F., Recchi, S., & Tozzi, P. 2004 *MNRAS* **349**, L19.

Walborn, N. R., Barbá, R. H., Brandner, W., Rubio, M., Grebel, E. K., & Probst, R. G. 1999 *AJ* **117**, 225.

Whitmore, B. C. 2003. In *A Decade of Hubble Space Telescope Science* (eds. M. Livio, K. Noll, & M. Stiavelli). p. 153. Cambridge University Press.

Yang, Y., Park, H. S., Lee, M. G., & Lee, S. G. 2002 *JKAS* **35**, 131.

Massive stars and star clusters in the Antennae galaxies

By BRADLEY C. WHITMORE

Space Telescope Science Institute, 3700 San Martin Drive, Baltimore, MD 21218, USA

Large numbers of young stars are formed in merging galaxies, such as the Antennae galaxies. Most of these stars are formed in compact star clusters (i.e., super star clusters), which have been the focus of a large number of studies. However, an increasing number of projects are beginning to focus on the individual stars as well. In this contribution, we examine a few results relevant to the triggering of star and star cluster formation; ask what fraction of stars form in the field rather than in clusters; and begin to explore the demographics of both the massive stars and star clusters in the Antennae.

1. Introduction

It is now well accepted that most star formation occurs in clustered environments, such as associations, groups and clusters (e.g., Lada & Lada 2003). In addition, it is clear that star formation is greatly enhanced in merging galaxies, making them an excellent place to study the formation of large numbers of young, massive stars—albeit with the disadvantage of having to work with stars at larger distances than the nearby groups and clusters in our own galaxy. In keeping with their galactic counterparts, most of the stars in merging galaxies also form in clusters, the brightest and most compact of which have been dubbed "super star clusters." Hence, understanding what triggers the formation of star clusters in mergers may be an important clue for understanding the formation of stars in general.

The excellent spatial resolution of the *Hubble Space Telescope* (*HST*) has rejuvenated the study of young star clusters in recent years (e.g., see reviews by Whitmore 2003; Larsen 2006). One of the most important results is that the brightest of the super star clusters have all the attributes expected of young globular clusters (e.g., Holtzman et al. 1992). An equally important result is that most of the groups and clusters do not appear to be bound, with roughly 90% being dispersed into the field each decade of log time (i.e., "infant mortality"; Whitmore 2003; Fall 2004; Fall, Chandar, & Whitmore 2005). Hence, understanding the destruction of clusters may be the key to understanding the demographics of both star clusters and field stars.

The Antennae galaxies (NGC 4038/39) are the nearest and youngest prototypical merger in the Toomre (1977) sequence. Hence, they may be our best chance for studying the formation of super star clusters and the massive stars within a major merger. While other galaxies will be briefly discussed at various parts of this review, the Antennae will be our centerpiece. Figure 1 shows an example of some of the super star clusters in the Antennae (two left panels). Knot S, shown in the upper left, will be the focus of several parts of this paper. The central cluster in Knot S contains at least $10^7 M_\odot$ alone, while the entire region contains well over $10^8\ M_\odot$. Note that Knot S consists of more than a single cluster. While it is difficult to distinguish individual supergiant stars in the outer region from faint clusters based on this image alone (this will be the main subject of Section 4), at least a dozen objects are clearly resolved, and hence are sizeable clusters in their own right. To provide some perspective, Figure 2 shows a superposition of what 30 Doradus ($M_V \approx -10$, $\approx 10^5\ M_\odot$) would look like at the distance of the Antennae.

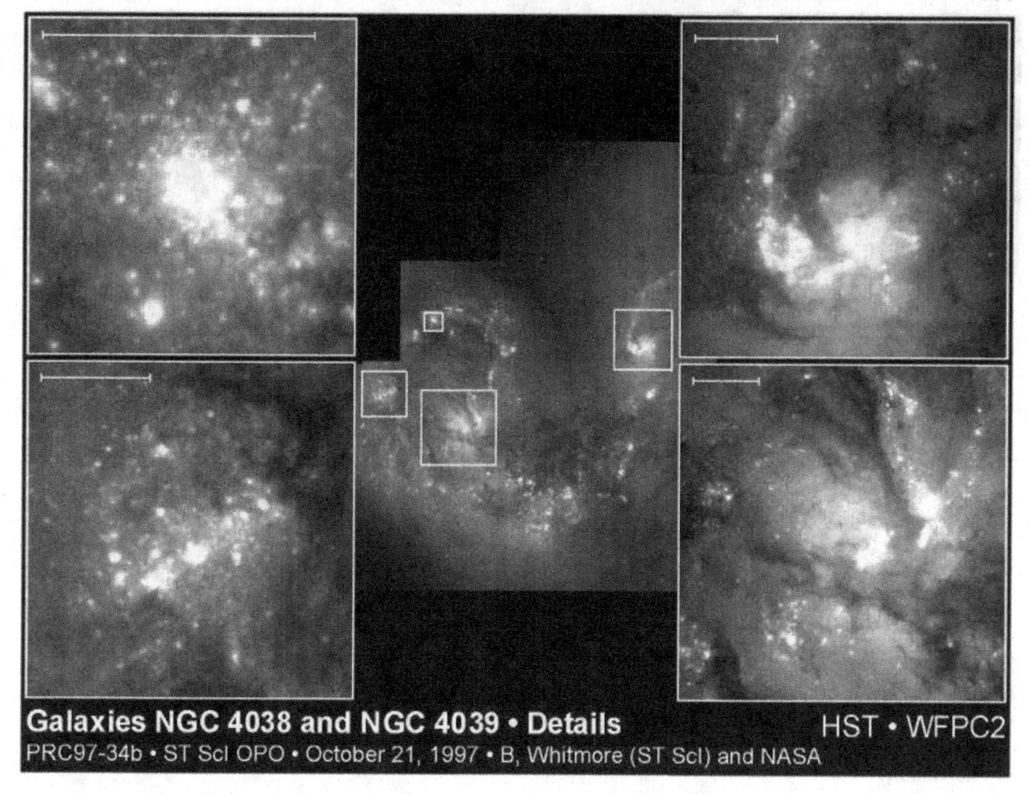

FIGURE 1. Blowup of two of the brightest knots of clusters in the Antennae galaxies (left) and the central regions of the two galaxies (right) from Whitmore et al. (1999).

While a great deal of attention has been paid to the study of super star clusters in external galaxies during the past decade, relatively little work has been done on the demographics of individual stars in these galaxies. Reasons include the larger distance, which makes it difficult to study anything but the brightest stars, and the high degree of crowding due to the large number of stars and the clustered nature of star formation. In general, it would seem that a detailed study of nearby star-formation regions, such as the Orion Nebula, would be more fruitful. However, there are two basic reasons why it is important to study individual stars in more distant galaxies as well. The first is the opportunity to study larger samples of stars (e.g., $\approx 10^7$ in Knot S) in a specific cluster. This would allow us to determine whether the most massive star in a cluster is determined by statistics or physics (see Weidner & Kroupa 2006; Elmegreen 2005; and Figer 2005 for discussions). Another motivating factor is to determine whether there are two modes of star formation (i.e., violent and quiescent; Gallagher 2004) which result in different stellar IMFs.

In this contribution we will first examine what has been learned about the triggering of star and star cluster formation in the Antennae. We will then address the question of whether essentially all stars form in clustered environments. We will also explore whether there is any evidence for an upper-mass cutoff for the stellar IMF in the Antennae. Finally, we will describe an effort to develop a general framework for understanding the demographics of both stars and star clusters.

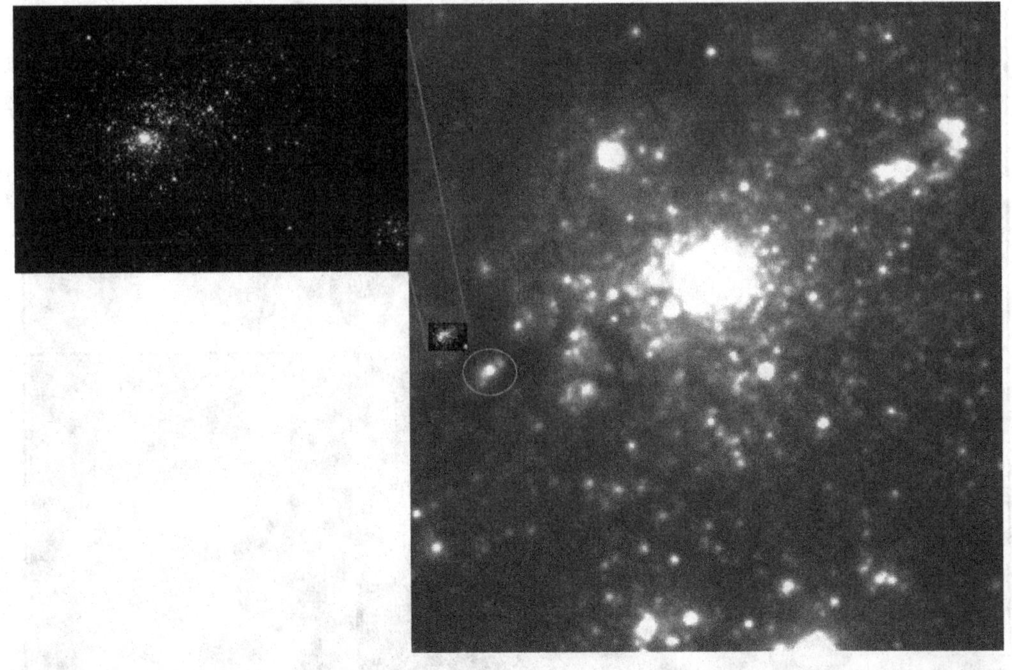

FIGURE 2. What 30 Doradus (upper left) would look like at the distance of the Antennae. While it would clearly be discernable as a cluster, it would be dwarfed by Knot S itself (the central object in the right-hand panel) and would also be superseded by about a dozen other clusters in the region.

2. What triggers the formation of star clusters (and hence stars) in the Antennae?

It is clear that shocks play an important role in triggering star formation. However, what is not clear is how they do this. One popular mechanism for triggering star formation in merging galaxies has been high-velocity cloud–cloud collisions (e.g., Kumai, Hashi, & Fujimoto 1993 suggest that collisions with relative velocities in the range 50–100 km s^{-1} are required).

Whitmore et al. 2005 obtained long-slit spectroscopy using STIS on *HST* to address this question. They found that the velocity fields are remarkably quiescent, with RMS dispersions less than about 10 km s^{-1}, essentially the same as in the disks of normal spiral galaxies (Figure 3). This does not support models that rely on high-velocity cloud-cloud collisions as the triggering mechanism, but is consistent with models where a high-pressure interstellar medium implodes the GMCs without greatly affecting their initial velocity distribution (e.g., Jog & Solomon 1992). This also supports earlier results (Zhang, Fall, & Whitmore 2001) that found essentially no correlation between star cluster formation and the velocity gradients and dispersions of Hα, H I, or CO. In retrospect, this is also evident from the existence of a large number of young clusters in the disk-like regions of NGC 4038, which still has a relatively quiescent, disk-like rotation curve (see Amram et al. 1992).

Another approach is to look for evidence of triggering by age-dating clusters and looking for a pattern of older star formation (the initial burst) surrounded by younger star formation (more recent bursts). Evidence for this has been seen around 30 Doradus (e.g., Walborn et al. 1999), with the youngest star formation at the tips of "pillars" pointing back towards the central object.

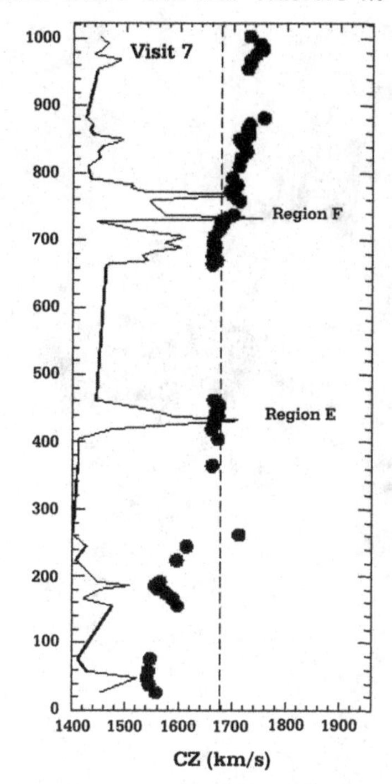

FIGURE 3. Hα velocities based on long-slit observations using the STIS detector on *HST*. Note how small the velocity dispersion is for a given subregion (e.g., ≈10 km s^{-1} for region F, once the large-scale gradient is removed). See Whitmore et al. (2005) for details.

Perhaps the central question here is not whether star formation can be triggered by previous star formation, which it clearly can, but how important this effect is (Whitmore 2003). Put another way, is local triggering more important, or global triggering? Few attempts have been made to try to quantify this. Figure 4 shows some evidence for sequential triggering around Knot S of the Antennae, with a clump of older clusters near the center (>10 Myr, circles), intermediate-age clusters further out (3–10 Myr, crosses), and a few very young clusters still further out (< 3 Myr, squares). We note that the clear clumping of the different ages shows that we are able to measure ages reasonably well, at least on a relative scale. If there was a very large amount of scatter in determining the ages we would find the different ages more randomly arranged in Figure 4.

In general, the fraction of luminosity in succeeding generations of star clusters appears to continuously decrease (i.e., each new generation does not produce a comparable generation, so that the process cannot continue in equilibrium). Hence, we conclude that triggered star formation is a significant, but not dominant, component of the overall star formation in the Antennae. More global processes, such as interactions and spiral arms, appear to be the primary drivers.

3. What fraction of stars are formed in clusters?

It is well recognized that for the Milky Way, most stars are formed in associations, groups and clusters (e.g., Lada & Lada 2003). De Wit et al. (2005) provide a recent demonstration of this. They use proper motions from *Hipparcos* to estimate that only 4±2% of the O and B stars in the Milky Way formed outside of groups or clusters (i.e.,

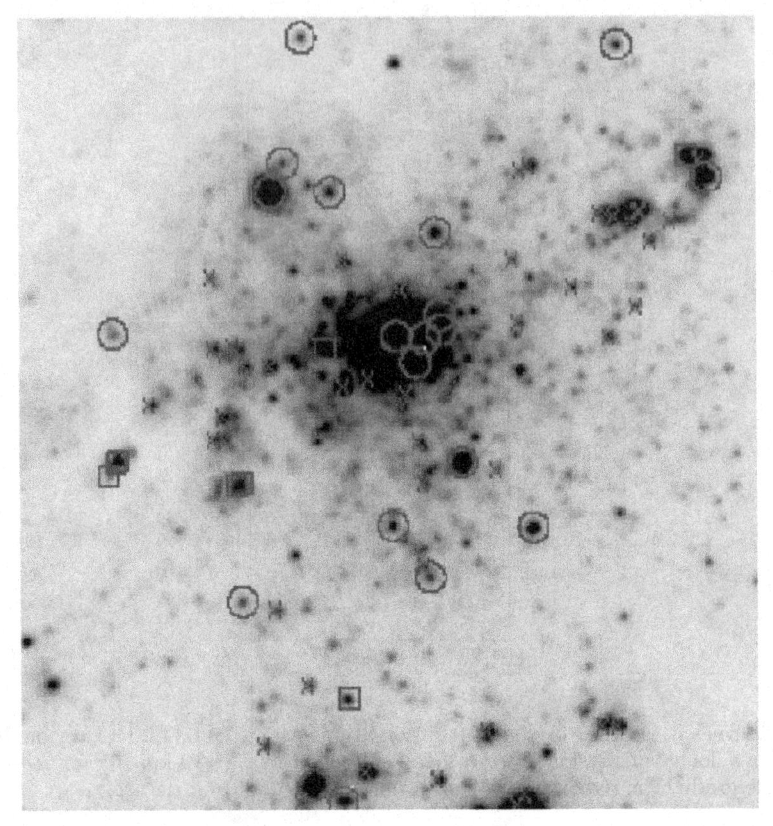

FIGURE 4. Age estimates of clusters in Knot S. Circles are for >10 Myr. Crosses are for 3–10 Myr. Squares are for <3 Myr.

most of the O and B stars in the field are consistent with being runaway stars from nearby groups.)

What have we learned on this subject from external galaxies, and in particular from the Antennae? Several early studies of star clusters in merger and starburst galaxies found that 10–50% of the UV light (i.e., young stars) are found in clusters (e.g., Meurer et al. 1995; Zepf et al. 1999; Whitmore & Zhang 2002). The initial fraction of stars in clusters is even higher than these estimates, since at least some clusters don't survive. In fact, as we shall see in Section 5 (also see Fall, Chandar & Whitmore 2005), we believe that roughly 80–90% of clusters disperse or are destroyed each decade of log time. Furthermore, our model, which incorporates this effect, predicts that if *all stars are formed in clusters*, the amount of UV light we should observe in clusters should be $\approx 8\%$ for the Antennae, in good agreement with observations ($\approx 9\%$; Whitmore & Zhang 2002). See Fall, Chandar, & Whitmore (2005) for a related calculation using total Hα flux, again concluding that the observations are consistent with the idea that essentially all stars are formed in groups or clusters.

A related question is: What are the relative fractions of stars formed in associations, open clusters, and super star clusters? This is a difficult question to answer for a variety of reasons. First, there is no clear dividing line between these types of groupings, (i.e., they probably represent a continuum rather than distinct modes). Second, the objects are barely resolved, and are often found in very crowded regions, making it difficult to reliably separate the objects into more than a single bin. In addition, it is not clear how diffuse an open cluster or association needs to be before it falls out of the sample because it cannot be detected. It is interesting to note, however, that we seem to be able to account for essentially all of the UV light in the Antennae by stars that originally formed in groups and clusters (either still existing within clusters we detect, or from stars where the cluster has already dispersed). This suggests that a large fraction of stars are not formed in very diffuse associations that would be too faint to be in our sample.

We should also keep in mind that even clusters that survive will lose a large fraction of their stars from their outer halos. For example, Whitmore et al. (1999) found that young clusters like Knot S have linear profiles, while older clusters have tidally truncated profiles, implying the removal of a large fraction of light from the outer regions (see Schweizer 2004 for a review on the sizes and radial profiles of clusters). In fact, we estimate that ≈50% of the light in Knot S falls beyond 50 pc from the center, a typical tidal radius for a globular cluster.

Bastian & Goodwin (2006) find similar profiles for the young clusters in M82, N1569, and N1705. They suggest that these profiles are compatible with N-body simulations of clusters with rapid removal of mass due to gas expulsion, hence supporting the basic interpretation that a large fraction of stars from clusters will eventually find themselves in the field. Fall, Chandar, & Whitmore (2005) make a similar argument to explain the high infant-mortality rate of clusters in the Antennae.

Comparisons between UV spectra from clusters, and from the diffuse field stars between clusters, provides another line of reasoning that supports this basic picture. For example, Chandar et al. (2005) find that the integrated spectrum of the field stars in several local starburst galaxies is consistent with formation of the stars within clusters which dissolve with typical timescales of 7–10 Myr.

4. What can we learn about the stellar content of the super star clusters in the Antennae?

In Whitmore et al. (1999), one of our primary difficulties was differentiating stars from clusters. This led us to conclude that the number of young star clusters in the Antennae was between 800 and 8000—a pretty big range! Our new ACS data, with its better spatial resolution, provides a better opportunity for making this determination and for studying the stars in their own right.

An important tool we are employing in this analysis is a maximum-likelihood SED-fitting software package named CHORIZOS, which is described in Maíz-Apellániz (2004). Ubeda, Maíz-Apellániz, & MacKenty (2007) employed CHORIZOS to analyze *HST* observations in six filter bands (F170, F336W, F555W, F814W, J, H) of NGC 4214, a nearby (3 Mpc) starburst dwarf galaxy. Their main conclusions are: 1) extinction is quite patchy, but relatively low around all but the youngest clusters, 2) 10 of the 12 clusters they studied have ages <10 Myr (note that this supports the infant-mortality discussion described in Section 1 and Section 5), 3) the blue-to-red supergiant ratios are consistent with theory, 4) the stellar IMF in the field is steeper than −2.8. This study is a good example of how researchers are starting to study both the stellar and cluster contents of external galaxies. In the current paper, we use CHORIZOS to estimate values of M_{bol} and T_e for candidate stars in the Antennae.

We first ask the question: How well can we distinguish clusters from stars in Knot S of the Antennae galaxies, based only on a concentration index (i.e., the luminosity of an object inside a 3 pixel radius compared to the luminosity inside a 1 pixel radius)? Figure 5 shows four luminosity ranges drawn from the sample of point-like objects around Knot S, starting with the brightest objects ($M_V < -10$; bottom left), and ranging to the fainter

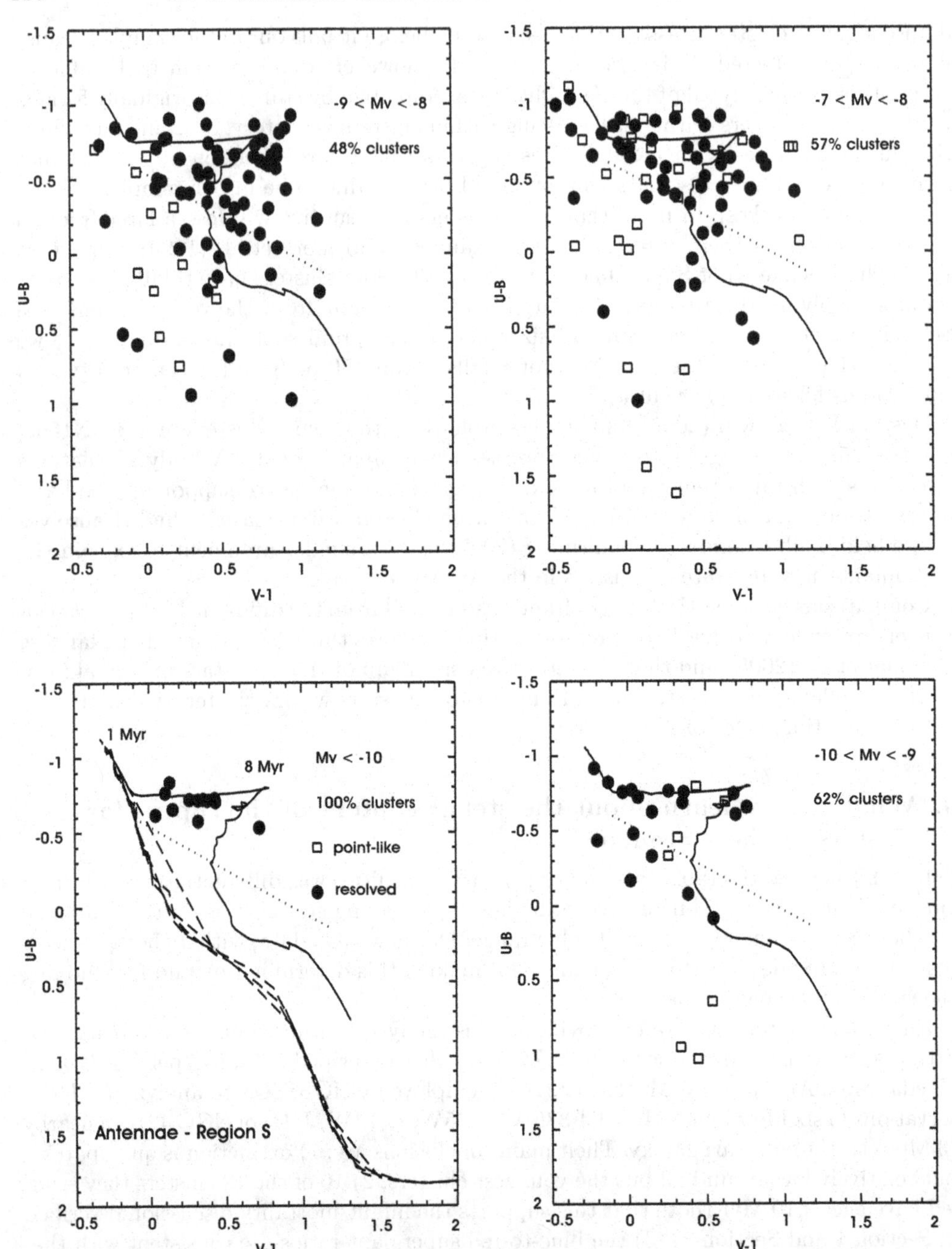

FIGURE 5. $U - B$ vs. $V - I$ color-color diagram for four luminosity ranges for a sample of objects around Knot S (from Whitmore, Chandar, & Fall 2007; see text for details).

objects ($-7 < M_V < -8$; in the upper right). The objects with profiles indistinguishable from stars are shown as open squares, while the resolved objects are solid circles. The data is plotted on a $U - B$ vs. $V - I$ color-color diagram with Bruzual & Charlot (2003) solar metallicity models superposed on all four panels using solid lines (young clusters are in the upper left and old clusters in the lower right; locations for 1- and 8-Myr clusters are shown on the bottom-left figure). Padova models of stars brighter than $M_V = -7$ are shown by the dashed lines in the bottom-left panel. The dotted line shows the reddening vector, and also acts as a rough dividing line between "cluster-space" (upper right) and "star-space" (lower left). This works because essentially all of the objects in this region are young, hence there are no clusters that populate the bottom part of the Bruzual & Charlot cluster models.

Several conclusions can be drawn from this figure. The first is that if we select only the brightest objects (i.e., $M_V < -10$), they are all consistent with being young clusters (<8 Myr) with relatively little extinction (i.e., they are very close to the Bruzual-Charlot models). This is reassuring, since the brightest stars might be expected to be fainter than $M_V \approx -9$ (i.e., the brightest stars in the Milky Way; Humphreys 1983). This is the value we—and several other researchers—have used to conservatively identify clusters in the past (e.g., Whitmore et al. 1999).

If we cut the sample at $M_V < -9$ (lower-right panel), things get a little more interesting. Near the bottom of the diagram we now have three point-like objects in Knot S in the part of the diagram appropriate for stars. These all happen to have values of $M_V \approx -9.1$, just slightly brighter than our boundary condition between stars and clusters. We also find three point-like objects in or near cluster-space. This is our second important result, that while the concentration index is useful for telling the difference between clusters and stars, it is only partially successful. It appears that some clusters (based on their position in color-color space) are so concentrated that they cannot be distinguished from stars based on their size alone. This is also apparent from the fainter bins (upper panels), where a majority of the point-like objects are found in star-space, but a fair fraction are also found in cluster-space.

Another interesting point is that while most of the resolved objects in the $-10 < M_V < -9$ diagram hug the Bruzual-Charlot models very nicely, about a half-dozen objects are just below the dotted line used to separate cluster- and star-space. We believe most of these are cases where there is a mixture of light from both a cluster and from one or two bright stars in the cluster (i.e., if you added the light from a cluster sitting on the Bruzual-Charlot cluster track and one of the three stars at the bottom of the diagram, which have roughly the same brightness, the result would be an object with an intermediate color). This does not happen for the brightest clusters (i.e., with $M_V < -10$) because these clusters have enough stars that one or two random bright stars cannot greatly affect the total color. Hence, a certain degree of "stochasticity" appears for young clusters with magnitudes around $M_V \approx -9$ (i.e., masses around $10^4\ M_\odot$). This effect has already been noted by other authors such as Cervino, Valls-Gabaud, & Mass-Hesse (2002). Ubeda, Maíz-Apellániz, & MacKenty (2007) also show a nice example of a cluster that appears to have both a blue and a red supergiant superposed.

Our fourth, and perhaps most important result, is that roughly 50% of the objects fainter than $M_V < -9$ are clusters, based on their position in the color-color diagram. This is actually a very conservative lower limit, since, as we just noted, some of the objects around the dividing line are likely to be clusters with one or two stars pulling them just below the dividing line. This has a number of important ramifications, the most important being that it shows that the number of faint clusters continues to rise in a power-law fashion. This provides a counter example to the claims that the initial cluster mass function in some galaxies may be a Gaussian (e.g., de Grijs, Parmentier, & Lamers 2005), since a Gaussian would require that essentially all the faint objects were individual stars. Another ramification is that the quantity of clusters in the Antennae numbers in the thousands, rather than the hundreds.

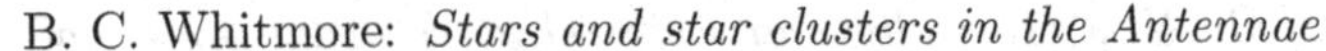

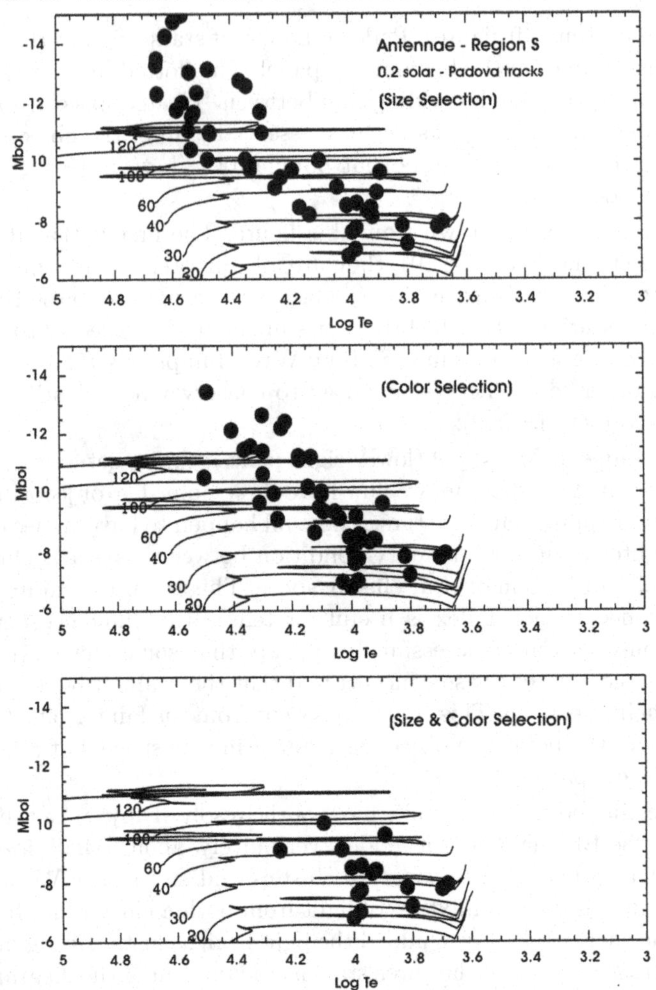

FIGURE 6. M_{bol} vs. Log T_e for candidate stars in Knot S, using only a size selection, only a color selection, or both a size and color selection. Note that using a combination of the size and color criteria does a good job of removing the clusters from the stars (i.e., the remaining objects are in the part of the diagram expected for stars). See text for details.

Hence, neither the concentration index (i.e., size), nor the position in the color-color diagram alone is completely successful in separating stars and clusters. What if we use a combination of both criteria? Figure 6 (M_{bol} vs. Log T_e diagrams for candidate massive stars around Knot S) shows that this appears to work fairly well. Using *either* the size or color criteria alone implies the existence of stars that are more massive than the stellar tracks (two upper panels in Figure 6). However, if we use *both* criteria simultaneously, all the remaining objects are consistent with being normal stars. We might note that this also suggests that there is an upper limit to the maximum mass of a star, since we would expect more massive stars in such a large sample of stars if the stellar IMF was a simple power law (see Weidner & Kroupa 2006; Elmegreen 2005; and Figer 2005 for detailed treatments of this issue). This result should be considered tentative, however, pending a more careful look at other knots in the Antennae galaxies, and the development of Monte Carlo simulations that will allow us to more quantitatively determine the statistical significance of the result.

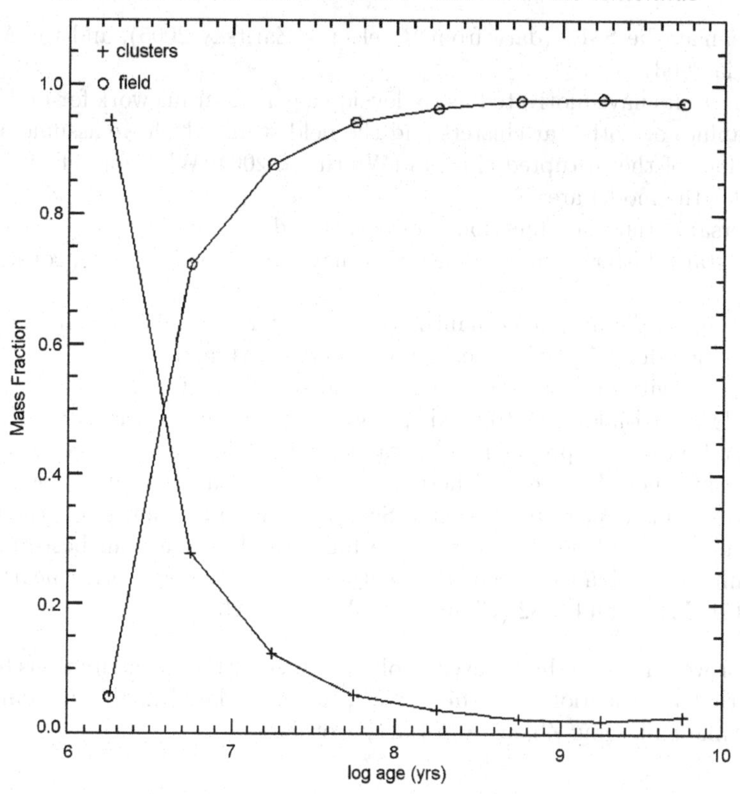

FIGURE 7. Fraction of mass in clusters and in field stars as a function of time for our Antennae model. See text for details.

5. The big picture—A general framework for understanding the demographics of stars and star clusters

The most extreme super star clusters, with magnitudes $M_V \approx -17$ and masses $\approx 10^8\, M_\odot$, are found in merging galaxies. One might therefore assume that there is something special about the physical environment in these galaxies that makes it possible to form such massive clusters there, but nowhere else. This suggests that there may be two modes of star cluster formation; one for relatively quiescent galaxies, such as normal spiral galaxies, and one for starbursting galaxies (e.g., Gallagher 2004). However, the discovery of super star clusters in spiral galaxies by Larsen & Richtler (1999)—and the subsequent demonstrations by Whitmore (2003; originally presented in 2000 as astro-ph/0012546), and Larsen (2002)—that there is a continuous correlation between the magnitude of the brightest cluster and the number of clusters (e.g., Figure 10 from Whitmore 2003), suggests that there may be a single universal mode of star cluster formation, with the correlation simply being due to statistics. Hence, mergers and starburst galaxies may have the brightest clusters only because they have the most clusters. Several recent papers (e.g., Hunter et al. 2003) have also realized that it is important to take this "size-of-sample" effect into consideration when interpreting results.

Similarly, there may be a universal power-law relationship for the disruption rate of clusters. For example, Fall et al. (2005) find that roughly 90% of the clusters in the Antennae are removed from the sample each decade of log time (i.e., a power law with index −1). Whitmore et al. (2007) show that this relationship appears to be the same for the Antennae, the SMC (data from Rafelski & Zaritsky 2005), and the Milky Way (Lada & Lada 2003).

These two results have motivated us to develop a general framework for understanding the demographics of both star clusters and the field stars, which we assume are formed as a by-product of the disrupted clusters (Whitmore 2004; Whitmore et al. 2007). The ingredients for the model are:

1. a universal initial mass function (power law, index −2);
2. various star (cluster) formation histories that can be coadded (e.g., constant, Gaussian, burst, ...);
3. various cluster disruption mechanisms (e.g., τ^{-1} for <100 Myr, i.e., infant mortality; constant mass loss for >100 Myr, i.e., 2-body relaxation); and
4. convolution with observational artifacts and selection effects.

This simple model allows us to predict a wide variety of properties for the clusters, field stars, and integrated properties of a galaxy. Of particular relevance for the present paper is the agreement between prediction and observations of what fraction of the UV light emitted by clusters (see discussion in Section 3). Figure 7 shows how the fraction of mass in clusters and in field stars varies as a function of time for our best-fitting model of the Antennae. We plan to extend this treatment to a number of other nearby galaxies including M51, M101, and M82 (Chandar & Whitmore 2006).

The author would like to thank several collaborators for their contributions to a variety of projects that are mentioned in this review, in particular, Rupali Chandar, Francois Schweizer, Mike Fall, Qing Zhang, and Barry Rothberg.

REFERENCES

Amram, P., Marcelin, M., Boulesteix, J., & Le Coarer 1992 *A&A* **266**, 106.
Bastian, N. & Goodwin, S. P. 2006 *MNRAS* **369**, 9.
Bruzual, A. G. & Charlot, S. 2003 *MNRAS* **344**, 1000.
Cervino, M., Valls-Gabaud, D., Luridiana, V., & Mass-Hesse, J. M. 2002 *A&A* **381**, 51.
Chandar, R., Leitherer, C., Tremonti, C. A., Calzetti, D., Aloisi, A., Meurer, G. R., & de Mello, D. 2005 *ApJ* **628**, 210.
Chandar, R. & Whitmore, B. C. 2009 (in preparation: Paper 2).
de Grijs, R., Parmentier, G., & Lamers, H. J. G. L. M. 2005 *MNRAS* **364**, 1054.
de Wit, W. J., Testi, L., Palla, F., & Zinnecker, H. 2005 *A&A* **437**, 247.
Elmegreen, B. G. 2005. In *Starbursts: from 30 Doradus to Lyman Break Galaxies* (eds. R. de Grijs & R. M. Gonzalez Delgado). ASSL Vol. 329, p. 57. Springer.
Fall, S. M. 2004. In *The Formation and Evolution of Massive Young Star Clusters* (eds. H. J. G. L. M. Lamers, L. J. Smith, & A. Nota). ASP Conf. Ser. 322, p. 399. ASP.
Fall, S. M., Chandar, R., & Whitmore, B. C. 2005 *ApJ* **631**, L133.
Figer, D. 2005 *Nature* **434**, 192.
Gallagher, J. 2004. In *The Formation and Evolution of Massive Young Star Clusters* (eds. H. J. G. L. M. Lamers, L. J. Smith, & A. Nota). ASP Conf. Ser. 322, p. 411. ASP.
Holtzman, J. A., et al. (the WFPC team) 1992 *AJ* **103**, 691.
Humphreys, R. M. 1983 *ApJ* **269**, 335.
Hunter, D. A., Elmegreen, B. G., Dupuy, T. T., & Mortonson, M. 2003 *AJ* **126**, 1836.
Jog, C. & Solomon, P. M. 1992 *ApJ* **387**, 152.
Kumai, Y., Hashi, Y., & Fujimoto M. 1993 *ApJ* **416**, 576.
Lada, C. J. & Lada, E. A. 2003 *ARA&A* **41**, 57.
Larsen, S. S. 2002 *AJ* **124**, 1393.
Larsen, S. S. 2006. In *Planets to Cosmology: Essential Science in* Hubble*'s Final Years* (ed. M. Livio). p. 35. Cambridge University Press.
Larsen, S. S. & Richtler, R. 1999 *A&A* **345**, 59.

Maíz-Apellániz, J. 2004 *PASP* **116**, 959.

Meurer, G. R., Heckman, T. M., Leitherer, C., Kinney, A., Robert, C., & Garnett, D. R. 1995 *AJ* **110**, 2665.

Rafelski, M. & Zaritsky, D. 2005 *AJ* **129**, 2701.

Schweizer, F. 2004. In *The Formation and Evolution of Massive Young Star Clusters* (eds. H. J. G. L. M. Lamers, L. J. Smith, & A. Nota. ASP Conf. Ser. 322, p. 411. ASP.

Toomre, A. 1977. In *The Evolution of Galaxies and Stellar Populations* (eds. B. M. Tinsley & R. B. Larson). p. 401. Yale.

Ubeda, L., Maíz-Apellániz, J., & MacKenty, J. 2007 *AJ* **3**, 932.

Walborn, N. R., Barba, R. H., Brandner, W., Rubio, M., Grebel, E., & Probst, R. 1999 *AJ* **117**, 225.

Weidner, C. & Kroupa, P. 2006 *MNRAS* **365**, 1333.

Whitmore, B. C. 2003. In *A Decade of* HST *Science* (eds. M. Livio, K. Noll, & M. Stiavelli). p. 153. Cambridge University Press.

Whitmore, B. C. 2004. In *The Formation and Evolution of Massive Young Star Clusters* (eds. H. J. G. L. M. Lamers, L. J. Smith, & A. Nota). ASP Conf. Ser. 322, p. 411. ASP.

Whitmore, B. C., Chandar, R., & Fall, S. M. 2007 *AJ* **133**, 1067.

Whitmore, B. C., Gilmore, D., Leitherer, C., Fall, S. M., Chandar, R., Blair, W. P., Schweizer, F., Zhang, Q., & Miller, B. W. 2005 *AJ* **130**, 2104.

Whitmore, B. C. & Zhang, Q. 2002 *AJ* **124**, 1418.

Whitmore, B. C., Zhang, Q., Leitherer, C., Fall, S. M., Schweizer, F., & Miller, B. W. 1999 *AJ* **118**, 1551.

Zepf, S. E., Ashman, K. M., English, J., Freeman, K. C., & Sharples, R. M. 1999 *AJ* **118**, 752.

Zhang, Q., Fall, M., & Whitmore, B. C. 2001 *ApJ* **561**, 727.

On the binarity of Eta Carinae

By THEODORE R. GULL

Exploration of the Universe Division, Code 667,
Goddard Space Flight Center, Greenbelt, MD 20771, USA

Multiple observations reinforce the binarity of Eta Carinae including the 5.54-year periodicity in x-rays, spectroscopic excitation of the Weigelt blobs and the behavior of the stellar line profiles. The *Hubble Space Telescope* (*HST*) STIS observations from 1998.0 to 2004.3 provide considerable new evidence of the binary system. We focus on the lines of He I, H I, Fe II and [N II] and provide initial visualizations of the binary system. Recent observations with VLTI/AMBER are consistent with a binary model.

1. Introduction

Eta Carinae (η Car) has intrigued astronomers for well over a century, beginning with its brightening to −1 magnitude in the late 1830s, rivaling Sirius as the brightest star in the sky for nearly two decades, then fading below naked-eye sensitivity. Observers in the Southern Hemisphere have pointed their telescopes in its direction since the 1820s; some navigational records exist even back to the late sixteenth century with visual magnitudes noted between 2nd and 4th magnitude (Frew 2004). Characterization of this peculiar star has been a challenge; D. Frew (private communication, 2003) noted that η Car was monitored by observers at Sydney Observatory in the nineteenth century in the suspicion that it was a binary system. Yet still today not all are convinced as direct evidence of the secondary star is not in hand.

η Car is of great interest, as at least one of the companions is at the end of its hydrogen-burning phase. Many observers noted the lack of oxygen in the nebular emission spectrum (see Davidson & Humphreys 1997 for a complete review as of that year). Davidson, Walborn & Gull (1982) found that while five ionic stages of nitrogen were present in ground-based and *International Ultraviolet Explorer* (*IUE*) spectra, no oxygen was detected. Davidson et al. (1986), using further *IUE* and ground-based observations, determined that helium was substantially overabundant. Verner, Bruhweiler & Gull (2005) modeled the *HST*/STIS spectra of the Weigelt blobs B and D (Weigelt & Ebersberger 1986), noted very weak lines of carbon and oxygen, and showed that carbon and oxygen had to be substantially depleted. The models of Maeder & Meynet (2003) for the evolution of massive stars with rotation indicate that mixing leads to nitrogen overabundance at the expense of carbon and oxygen. The abundances in the Weigelt blobs, the Strontium filament (Bautista et al. 2006), and the Homunculus (Gull et al. 2005) indicate that the ejecta from the 1840s event and a lesser event in the 1890s are deficient in carbon and oxygen and overabundant in nitrogen.

The amount of material thrown out by η Car in the Homunculus is estimated to be at least $10\,M_{\odot}$ (Smith et al. 2003). The Homunculus is a neutral bipolar nebulosity ejected during the 1840s event (Morse et al. 2001). Between the bipolar lobes is a skirt structure which includes the Weigelt blobs and the Strontium filament (Zethson et al. 2001; Hartman et al. 2004). The Strontium filament is a partially ionized gas cloud with many neutrals and singly ionized species including Fe I, Sr II, Ti II, Ni II, Sc II, and V II, but no H I, N I, N II, or O I. Bautista et al. (2006) modeled the unusually large [Ti II]/[Ni II] line ratios that led them to realize that Ti/Ni is nearly two decades larger than solar. Their conclusion was that the small amount of oxygen and carbon thrown out of η Car had to be chemically bound in molecules and dust grains that form at relatively high temperatures, leaving metals like titanium, strontium, scandium, and especially vanadium in gaseous form. Hence, the amount of material bound up in the dust grains around η Car is proportionately less, leading to a larger than normal gas-to-dust ratio. Likely the amount of ejecta in the Homunculus is significantly more than $10\,M_\odot$.

The argument for binarity was considerably strengthened by the discovery of Damineli (1996); Damineli et al. (1998) who noticed that the the He I λ10830 line and the H I Paschen lines changed with a 5.52-year periodicity, and then demonstrated that the nebular line emission of [Ar III], [Ne III] and many Fe II lines were modulated with the same period. Corcoran (2005), and references therein, monitored the x-ray flux with *RXTE*, finding a long-term buildup of the x-ray flux until just before a major drop, with recovery about 70 days later. He measured the period to be 2024 ± 2 days between the 1998.0 and 2003.5 minima. Pittard & Corcoran (2002) modeled the x-ray spectrum from *Chandra*, finding that the spectral properties were best explained by a wind-wind interaction between the visible star with $2 \times 10^{-4}\,M_\odot$/yr and 600 km s^{-1} wind and a companion star with $10^{-5}\,M_\odot$/yr with 3000 km s^{-1} wind. Their total wind for the system is substantially below the estimated $10^{-3}\,M_\odot$/yr by van Boekel et al. (2003), reinforcing the concern that mass-loss rates may be overestimated for massive stars.

More recently, Smith & Owocki (2006) have suggested that massive stars lose the bulk of their mass, not through the winds during the O-star phase, but in specific ejective events during the transition from O star to pre-supernova, e.g., during the various WR phases. It is likely at least one of the binary members of η Car has recently gone through one, and perhaps two or more, of these events. Given the recent connection between gamma-ray bursters, Population III, and other massive stars, the strong evidence of material thrown out in an eruptive pre-GRB event, we have many reasons to characterize the properties of this central source, η Car and its individual members.

2. The *HST*/STIS observations

Beginning in early 1998 soon after the 1998.0 x-ray drop, observations with *HST*/STIS were initiated. The high spatial resolution of *HST*, combined with moderate spectral resolving power of the first-order gratings, permitted full spectral coverage from 1600 to 10300 Å. Indeed, η Car is probably the object most frequently and completely observed with STIS during its operational lifetime. Multiple GI programs (K. Davidson, PI), STIS GTO programs (T. Gull, PI), and the Hubble η Car Treasury program (K. Davidson, PI) were committed to obtaining coverage with some overlap across one 5.54-period using the G230MB, G430M and G750M ($R = 8000$) gratings. Selective coverage extending from 1175 to 2380 Å ($R = 30,000$, E140M and E230M; Nielsen, Gull & Vieira 2005) and 2400 to 3160 Å ($R = 110,000$, E230H; Gull, Kober & Nielsen 2006) was accomplished during the broad spectroscopic maximum and the brief minimum. The echelle observations proved to be very informative about the structure of the Homunculus in line of sight as over 20 velocity components were identified, with two (-513 and -146 km s^{-1}) being well associated with the outer, neutral, or partially ionized Homunculus and the internal, ionized Little Homunculus (Gull et al. 2005). As seen in the Strontium filament, the Homunculus is a partially ionized structure with neutrals and ions not ionized by 8 eV photons; the Little Homunculus is ionized except during the spectroscopic minimum, when for a brief time even many lines of Ti II appear in strong absorption (Gull, Kober & Nielsen 2006).

The most important advantage of the *HST*/STIS observations has been separation of the central source from the nebulosity. The Weigelt blobs (Weigelt & Ebersberger 1986) are located with 0.1 to 0.3$''$ of η Car which prevents spectroscopic separation with ground-based instruments. *IUE* observations by Viotti et al. (1989), accomplished with a 3″ diameter or a 10″ × 18″ oval aperture, proved to be very difficult to analyze as the majority of UV flux is scattered by nebular structure close to the star and throughout the Homunculus. Stellar lines are contaminated by velocity-shifted scattered starlight and nebular emissions. The STIS longslit CCD observations were done with the 52″ × 0.1″ aperture. Extractions of the data were complicated by the small angle tilt of the spectral format with respect to the CCD rows and columns that lead to an artificially induced modulation. The STIS GTO data reduction routine was greatly modified with subpixel interpolation for extractions approaching the diffraction limit of *HST* angular resolution by Davidson, Ishibashi & Martin (2006). We used their data products to extract a spectrum with 0.15″ height centered on the central source.

3. Central source line profiles of interest

The spectroscopic dataset is large and packed with considerable information, especially in the ultraviolet accessible to STIS. Unfortunately that portion of the spectrum is dominated by nebular absorption lines from the Homunculus and Little Homunculus. Examples of the nebular information include the approximately 2000 nebular emission lines identified by Zethson (2001), the 2000 unique absorption lines in the 1300 to 2380 Å spectral region with multiple velocity components published by Nielsen, Gull & Vieira (2005) and the 1000+ absorption lines with over 30 velocity components presented by Gull, Kober & Nielsen (2006).

Our focus is on the broad lines originating from the central source, η Car. Most massive stars have strong, uncontaminated wind lines in the UV, but the foreground nebular lines and scattered starlight confuse the issue. Hillier et al. (2006) suggest that the companion stellar flux is dominated by the primary star until below 1400 Å. UV flux dropped by a factor of three in the STIS FUV spectra by 1300 Å from one week before the x-ray drop seen by *RXTE* to one week afterwards. *FUSE* observations, done up to a few days before the x-ray drop in 2003.5, showed that most of the flux below 1175 Å disappeared (Iping et al. 2005). Unfortunately, spacecraft safety issues prevented a *FUSE* observation during the 2003.5 minimum, but by early 2004 the FUV flux had returned. The *FUSE* spectra are heavily contaminated by nebular and interstellar absorptions, most noticeably originating from H_2 at very large column densities. To a lesser degree, H_2 plays a role in the STIS FUV spectra as Nielsen, Gull & Vieira (2005) identified ≈1000 absorption lines at -513 km s^{-1}, the velocity of the Homunculus in line of sight. These H_2 absorption lines, originating from very high rotational levels, are from vibrational levels populated by UV Lyman and Werner bands longward of 912 Å. They disappeared during the 2003.5 minimum, only to return in early 2004. Hence, we can state that the UV flux shortward of about 1600 Å is dramatically decreased within the Homunculus during the spectroscopic minimum. We cannot conclusively demonstrate that the FUV flux originates uniquely from the hot companion, and we have not identified unique line profiles in the FUV that would lead to stellar classification. The FUV radiation appears to be modified too much.

The nebular absorptions are from metals populated at a thermal temperature of 760 K for the -513 km s^{-1} Homunculus system and about 6400 K for the -146 km s^{-1} Little Homunculus system. Multiple intermediate-velocity systems are identifiable between -146 and -513 km s^{-1} that decrease in thermal properties with increasing velocity (distance). They are likely in the Homunculus wall and potentially the results of multiple spectroscopic minima since the 1840s event (Gull et al. 2005). Indeed a number of stellar wind lines in the mid-UV are so heavily contaminated with nebular structure that it is often difficult to separate the two contributions. As the line density decreases towards longer wavelengths to a minimum beginning around 4000 Å, the wind lines in the visible portion of the spectrum turn out to be more useful.

The stellar lines were examined individually across the entire dataset extending from 1998.0 to 2004.2, looking for changes in the profiles and velocities with time (in the following discussion, we rely heavily on Nielsen et al. 2007). Our primary search was for lines that showed 1) reproducibility at nearly the same phase in the 5.54-year period, and 2) systematic differences in features that could be attributed to the periodic variations. Davidson et al. (2005) focused on the changes in Balmer Hα and Hβ, and noted significant differences between spectra recorded in 1998 and in 2003 that were interpreted to be evidence for a long-term change in the physical state of η Car. Steiner & Damineli (2004) discovered that a line at λ4680 appeared in the late months of the broad maximum, and disappeared abruptly during the x-ray drop. They attributed this line to blue-shifted He II 4686 Å, possibly from the secondary star. Gull (2005) confirmed that the λ4680 line was present within the 0.1″-wide stellar spectrum, but not in the nebular radiation other than scattered starlight. Martin et al. (2006) analyzed the STIS spectra and found the emission to originate from the vicinity of the wind-wind interface, but determined that the shock structure did not provide sufficient energy for the He II 4686 Å line. Upper limits of other lines of He II were found to be consistent with predicted fluxes, especially that of He II 1640 Å, which is in the STIS FUV spectral region that is heavily modulated by nebular absorption lines.

Hillier et al. (2006) noted that the He I lines in the central source of η Car had peculiar velocity structures relative to the rest of the stellar wind lines. They appeared to be systematically blue-shifted throughout the orbit with the greatest blue shift being just before the spectroscopic minimum. Most characteristic He I wind lines proved to be contaminated by other wind lines, such as the He I λ5017 line which is heavily dominated by the Fe II λ5020 line (Nielsen et al. 2007). We isolated two singlet lines 6680 and 7283 Å and two triplet lines 7067 and 5877 Å that were relatively uncontaminated, at least on the blue side of the blue-shifted lines. The 5877 Å line is affected on the red side by IS Na I 5890, 96 Å absorptions. The narrow nebular emission-line contributions to each of these lines was minimized by the *HST*/STIS high spatial resolution.

We compared the He I line profile variations with other wind line profiles. Figure 1 provides a few examples of the many lines we examined. In addition to the He I λ7067 line, we selected H I Balmer δ λ4103 to follow the averaged changes throughout the wind structure, Fe II λ5170 for the changes in low ionization, outer structure and [N II] λ5756 for the outermost, low-density structure intermediate between the wind and the surrounding H II region. The He I λ7067 line (Figure 1, upper left) at phase 0.0, referenced to 1998.0 x-ray minimum, is noticeably shifted relative to the H I λ4103, the Fe II λ5170 and the [N II] λ5756 lines.

The [N II] line (Figure 1, lower right) shows little velocity shift throughout the orbit. Instead, the several narrow components increase or decrease with phase. All components drop during the spectroscopic minimum. Most noticeable is the increasing brightness of the most highly blue-shifted component until just before the spectroscopic minimum.

The H I line (Figure 1, upper right) shows modulation in emission with some variation in velocity. In addition to the major drop in the minimum ($\phi = 1.0$), there appears to be a shallow depression halfway through the period. More noticeable in this line is the strong absorption as the spectroscopic minimum begins, which then weakens to be nearly absent by halfway through the period, then abruptly reappears during the 2003.5 minimum.

The Fe II line (Figure 1, lower left) presents considerable changes in both the emission and absorption. Like the H I line, its strongest absorption is during the minimum, but

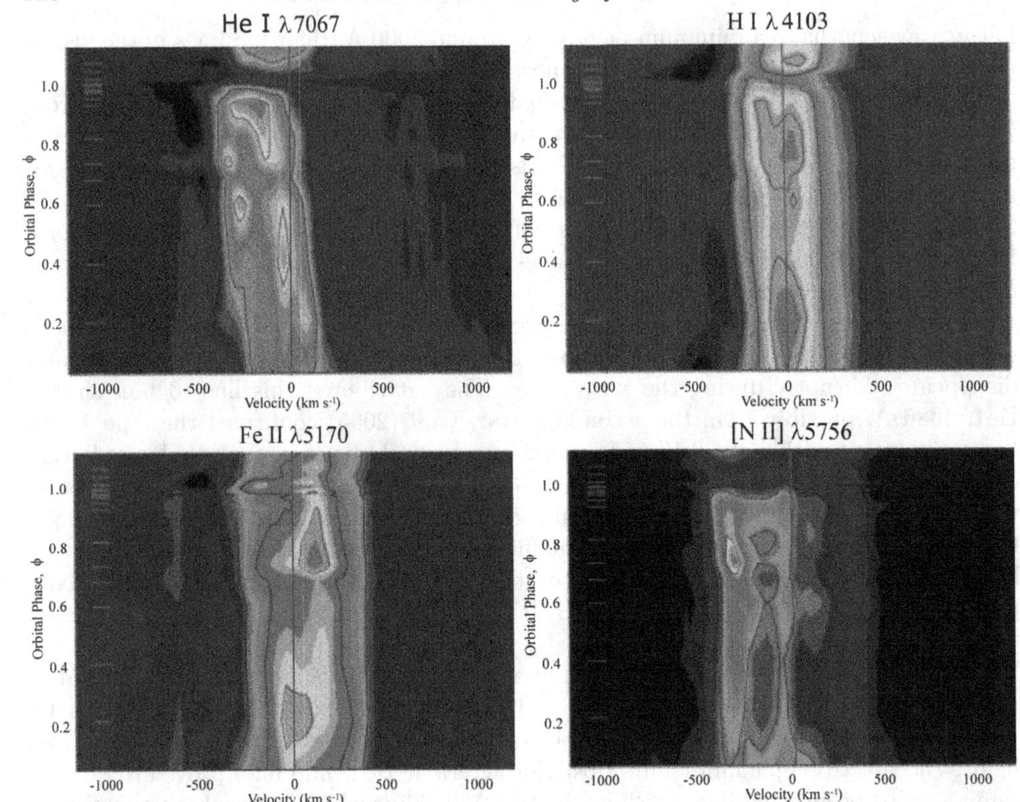

FIGURE 1. Comparison of selected line profiles across the STIS coverage of η Car from 1998.0 to 2004.3. Horizontal axes are velocity (km s^{-1}) and vertical axes are orbital phase referenced to 1998.0 with the x-ray derived 5.54-year period (Corcoran 2005). The He I λ7067 line (upper left) is systematically shifted to the blue with respect to all other lines. The H I Balmer γ λ4103 (upper right), the Fe II λ5170 and the [N II] λ5756 lines are representative of the stellar wind structure. (Contour plotting courtesy of Krister Nielsen.)

with generally quick recovery to a more shallow absorption throughout. More noticeable is the change in broad wind profile. While somewhat red-shifted early in the period, it drops in amplitude just past the halfway point in period, then strengthens again, but with the blue-shifted component depressed, coincident with an increase in [N II] emission on the blue-shifted side.

The He I line (Figure 1, upper left) behaves very differently from the three other examples. It starts out slightly blue-shifted, then increasingly blueshifts throughout the period. On these contour plots, the absorption becomes strong about $\phi = 0.6$, increasing continuously until just at the minimum. Like all three other lines, it too drops substantially during the minimum, *and* shifts to the red, almost overlapping that near $\phi = 0$.

4. A working model

The properties of these wind lines are very important clues to what is happening within the central source. Creating a systematically blue-shifted P-Cygni wind line in a symmetric wind is difficult at best to explain. Explaining predictable periodic behavior is still another problem. Explaining these properties with a typical binary system is also difficult. However, η Car has a very massive wind, so massive that much of the radiation from the primary star is trapped in the wind and reradiated. Hillier et al. (2001) note that up to 5 magnitudes of extinction appear to be right on the central source and is likely within the extended wind itself.

Curiously, the Weigelt blobs do not appear to be affected by this extinction. While STIS acquisition camera broad-band photometry showed a brightening by two or more (Martin & Koppelman 2004), the brightness and excitation of Weigelt B and D are virtually the same for spectra recorded under the same conditions 5.54 years apart. Moreover, the scattered starlight profile of H I Balmer α seen in the spectrum of the Weigelt blobs does not show the strong wind absorption seen in line of sight, nor in the starlight scattered off of the southeast lobe of the Homunculus (Gull et al. 2006b). A simple explanation is that the massive wind of the primary is not seen in the direction of the Weigelt blobs.

A second, important clue is provided by excitation models of the Weigelt blobs. Verner et al. (2002) using CLOUDY models could only account for the excitation of the Weigelt blobs during the spectroscopic minimum using the known spectral properties of η Car. A second model (Verner, Bruhweiler, & Gull 2005) showed that the flux of a massive star equivalent of an O or WR star must be added to account for the excitation leading to [Ar III], [Ne III], [Fe III], [Fe IV] and many Lyman-alpha-pumped lines of Fe II and Cr II. The UV properties of this companion star are quite consistent with the wind properties suggested by Pittard & Corcoran (2002). A concept of the primary extended wind enveloping part, but not all, of the orbit of the secondary begins to emerge. Basically for much of the orbital period, the secondary star, η Car B, with a less massive, but higher velocity wind, carves out a cavity in the dominant wind structure of η Car. The UV flux of the companion escapes in a preferred direction, ionizing the Weigelt blobs, which are only three noticeable clumps of an apparent necklace that surround η Car (Smith et al. 2004).

The Weigelt blobs appear to have originated from the 1890s event. Proper motion and velocity components suggest that they lie in the disk plane, located between the two lobes of the Homunculus (and the Little Homunculus). We use these clues to suggest that the orbital plane of the binary system lies in the disk plane with z-axis tilted along the bipolar axes of symmetry and clocked to about $-45°$ position angle, intermediate to the vectors defined by η Car to Weigelt D and η Car to Weigelt C.

The FUV radiation of this companion star also affects the very dominant primary wind of η Car. η Car itself ionizes the hydrogen in the primary wind, and a very small amount of helium—located in the inner regions close to η Car; about ten percent of a given He I line profile comes from a broad component approximately centered on the system velocity. The secondary star provides significant FUV flux that leads to a He II region, and some He III. The latter is mostly confined to the very hot, low-density secondary wind structure, except near periastron when the distance between the two stars closes to as little as 1.5 AU (Nielsen et al. 2007). The densities and FUV fluxes increase substantially, consistent with the He II λ4686 line appearing in the late stages of the period. The He II zone is a portion of the primary wind in the general direction of the secondary and moves around η Car following the motion of η Car B. We see positive velocities as the major axis is pointed at a relatively small angle from our direction, and the secondary spends the majority of its time close to apastron in an orbit determined by Pittard & Corcoran (2002) to be highly elliptical, $\epsilon = 0.9$. Other than just after periastron, the bulk of the He II zone has a positive velocity component relative to the observer.

We used the models of Smith, Norris & Crowther (2002) to estimate the amount of the primary wind in which helium would be singly ionized and find that it extends from the wind-wind interface, located about three-fourths the way from η Car to η Car B to about half way between the two stars and fills a volume roughly confined by two paraboloids.

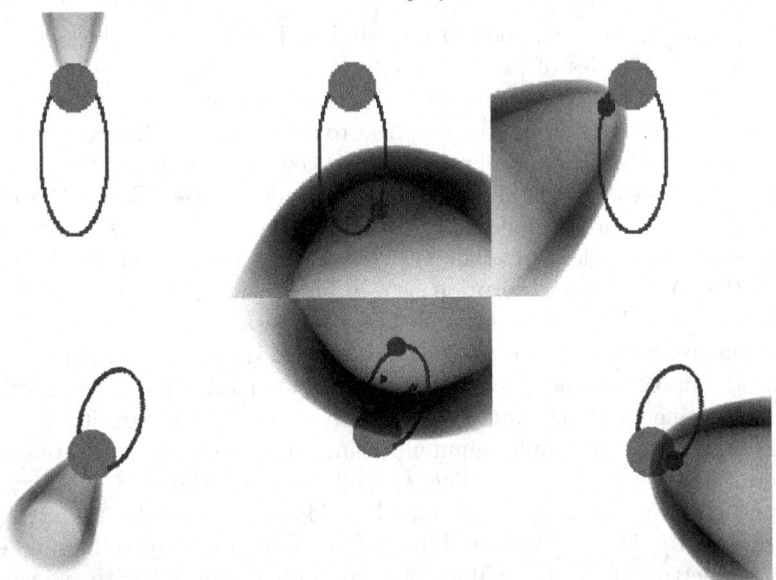

FIGURE 2. Three-dimensional visualization of the binary system with the He II zone. The wind and the UV radiation of the companion influence the primary wind. A dynamic cavity is carved out of the roughly spherically shaped primary wind by the less massive, but much faster secondary wind. The hot secondary wind is highly ionized, allowing the He II ionizing radiation to escape. A shell-like structure of He II results in the primary wind in the general direction of the secondary. Top row: looking down on the orbital plane (as in Figure 3). Bottom row: system seen from observer's viewpoint, assuming that the orbital plane lies in the skirt of the Homunculus and projects on the sky at a position angle of $-45°$. This visualization does not include distortion due to orbital motion. (Modeling accomplished with the help of Don Linder.)

We have constructed a datacube visualization of this structure in Figure 2. The top row is of a view looking down on the orbital plane and the lower row is from the observer's view assuming the orbital z-axis aligned to the axis of symmetry of the Homunculus and the major orbital axis pointing at an angle intermediate to the Weigelt blobs D and C. The larger star, depicted about the diameter measured by Weigelt et al. (2007) at a distance of 2300 pc, is η Car. The smaller star is η Car B in an orbit with $\epsilon = 0.9$ and semimajor axis of 20 AU. The gray-level structure is the approximate shape of an undistorted He II zone bounded by the wind-wind interface towards η Car B and by the limit of He II FUV photons towards η Car. For both rows, the left image is at $\phi = 0.0$, the middle is at $\phi = 0.25$, and the right image is at $\phi = 0.97$.

The He II zone shows the region of singly ionized helium, where recombination leads to He I emission and, because of the metastable levels, to absorption. The emission defines the He II zone locations and shifts with velocity along the He II zone, strongly oriented around the wind-wind interface. Indeed, the region is likely thinner as the wind-wind boundary is a very hot, dense gas, with density enhancement due to the shock, and if in pressure equilibrium, the gas behind cools, becoming even denser. We are modeling the wind-wind boundary, realizing that the gas follows the secondary star around the orbit, tracing a specific path. A possible mass ratio may be derivable from a consistent model. Indeed the absorption component seen in the He I lines is likely more confined than the He I emission as the absorption must be against background radiation. In this case, the background continuum source is η Car itself. For relatively optically thin lines, the absorption is thus tracing a portion of the He II zone that overlaps line of sight between η Car and the observer.

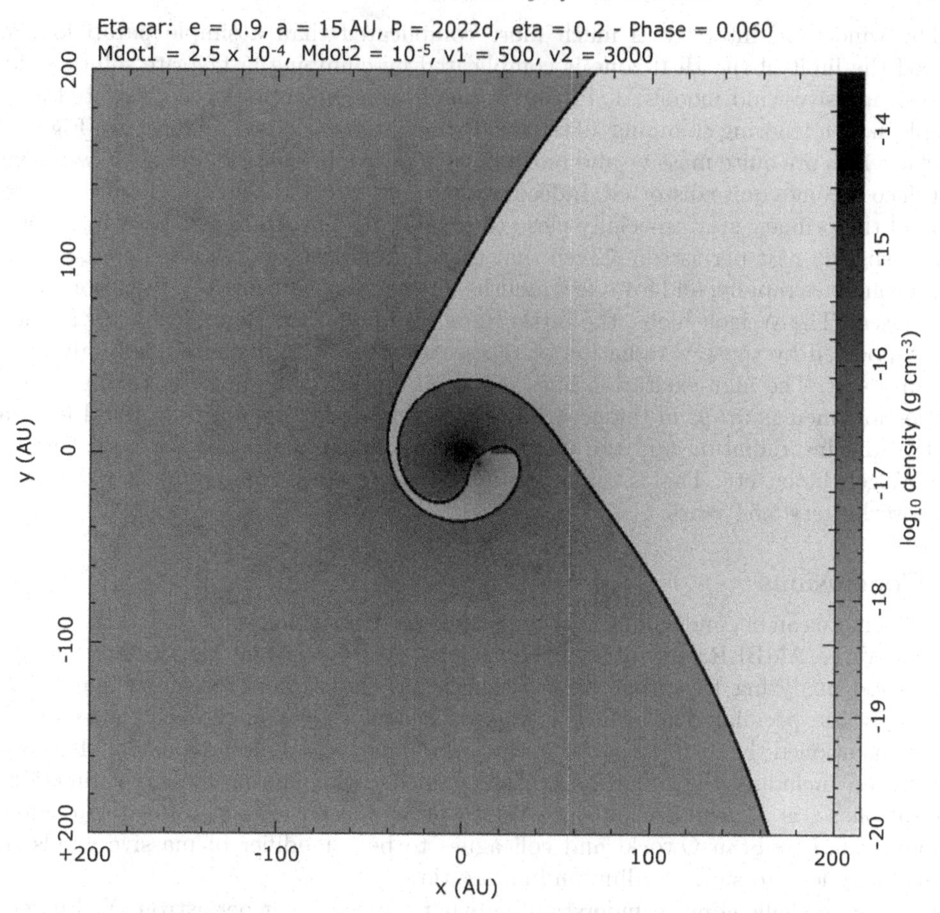

FIGURE 3. Two-dimensional toy model of the binary orbit. The vantage point is looking down on the orbital plane. The figure depicts the two stellar winds in their highly wrapped configuration just after periastron. Our best information suggests that the observer views the binary structure from the upper right, tilted about 45° out of the plane. (Graphics courtesy of Julian Pittard).

Recently Weigelt et al. (2007) observed the Br γ and He I 2.06-micron lines of η Car with the Very Large Telescope Interferometer and the AMBER spectrograph. As noted above, the continuum source measured about 4.6 milliarcseconds (m″). The H I Br γ line structure is about 5.5 times larger with the He I 2.06-micron line being intermediate at about 3.3 times larger. While weak Br γ absorption is present, the He I 2.06-micron line absorption is strong. The He I line profile is blue-shifted relative to the H I profile, quite consistent with that seen at visible wavelengths by *HST*/STIS. However, the He I absorption size appears to be comparable to the He I emission zone. This singlet line in absorption originates from a metastable level and has an f-value at least an order of magnitude greater than the transitions emitting the visible lines. The UV plane has only limited samples. Indeed, only an upper limit of the brightness of the secondary has been placed that is consistent with an O or WR star companion. More complete sampling of the UV plane is necessary to measure the two-dimensional geometry, and possibly three-dimensional geometry using velocity. One prediction by this model is that the absorption should lie on top of η Car, but the emission should be offset in the general direction towards η Car B.

The wind-wind interface is much more complicated than a simple paraboloid and indeed the limit of the He II zone is complicated by clumping, a favorite solution these days of massive-wind models. J. Pittard (private communication) has computed a very simple model, ignoring clumping, of the wind-wind interface in two-dimensions (Figure 3). As the winds are quite massive and move slowly relative to the secondary, the wind-wind interface becomes quite distorted. Indeed, shown in Figure 3, the wind structure can wrap around the primary star especially close to periastron. The example shown in Figure 3, is just slightly past periastron. Given that the wind structures are quite dense, gas piles up, recombines rapidly, and for a few months, little, if any, ionizing UV radiation escapes the system. The Weigelt blobs, the Little Homunculus and the Homunculus are suddenly not supported by the UV radiation of the secondary star, but maintained only by the primary star. The high-excitation lines relax, disappear, many ions recombine, and the Little Homunculus drops in temperature—but then the system starts to build up again as the ionizing radiation and the wind of the secondary plows out a new pathway to the Weigelt blobs, etc. This is the origin of the spectroscopic minimum that occurs like clockwork every 5.54 years.

5. Conclusions

Much more can be, and will be, written about the observations of *HST*/STIS, *Chandra*, *FUSE*, VLTI/AMBER, and other observatories of η Car. Many papers are yet to be completed analyzing the properties of the ejecta, the strange molecular species, and especially the peculiar dust. Much is yet to be done modeling the central source. We earlier mentioned the distortion of the wind-wind interface. Building a model of the wind structures, including the radiation transfer and deriving line profiles as a function of orbital phase, is a current challenge. Yet to be addressed is the problem of clumping demonstrated by Stan Owocki and colleagues to be a modifier of massive winds that indeed may lead to super-Eddington luminosities, etc.

Even more challenging is understanding what happens near periastron. Nielsen et al. (2007) note that the closest approach of the secondary is comparable to the terminal wind boundary of the primary. As periastron is approached, the primary wind to first order is being probed by the secondary wind. This probe may be actually a major distortion of the primary wind structure. Some have even suggested mass ejection occurs, others that we are simply seeing the wraparound of the massive winds for a brief time.

Perhaps this is what triggered the event of the 1890s: that a small change in forces affected a layer relatively deep within, or possibly that event was completely independent of the current orbit. Given the major mass loss of the 1840s, that event may have been what led to the large ellipticity of the current binary system. Calculations and models are needed to test this.

In January 2009, the next x-ray drop, spectroscopic minimum, periastron will occur. This provides an opportunity to test many things about η Car: What happens during periastron to the primary stellar wind? Is there mass ejection or simple wind wraparound? What is the effect on the ejecta? What can we learn, when and how frequently do we have to sample?

Many individuals and teams have contributed to the success of the STIS observations of η Car. Users of *HST* too often forget the instrument teams, the schedulers and the data handlers. Part of these observations were contributed from the STIS Guaranteed Observing Time. Others were from multiple Guest Observer Programs including the η Car Treasury Program. Beth Periello and Ray Lucas provided very enthusiastic support working directly with the author scheduling multiple visits in very narrow observing windows. Initial data reduction was accomplished through the STIS GTO software tools developed by Don Lindler. Considerable refinements addressing the artificial modulation caused by small angle tilts of the spectral format with respect to the CCD rows were done by K. Ishibashi, K. Davidson and J. Martin. The η Car Treasury Team, K. Davidson (PI), T. Gull, R. Humpheys, J. Morse, S. Johansson, J. Hillier, H. Hartman, A. Damineli, M. Bautista, N. Walborn, F. Hamann, K. Weis, O. Stahl and K. Ishibashi.

This discussion of η Car is the result of the efforts and inputs of many individuals: the *HST* η Car Treasury Team, the Goddard Eta Lunch Bunch, Gerd Weigelt, Stan Owocki, George Sonneborn, Rosina Iping, and many others.

Funding and observations were from the Hubble Space Telescope Project at Goddard Space Flight Center and the Space Telescope Science Institute.

REFERENCES

Bautista, M., et al. 2006 *MNRAS* **370**, 1991.
Corcoran, M. F. 2005 *AJ* **129**, 2018.
Damineli, A. 1996 *ApJ* **460**, L49.
Damineli, A., et al. 1998 *A&AS* **133**, 299.
Davidson, K., et al. 1986 *ApJ* **305**, 867.
Davidson, K. & Humphreys, R. 1997 *ARA&A* **35**, 1.
Davidson, K., et al. 2005 *AJ* **129**, 900.
Davidson, K., Ishibashi, K., & Martin, J. C. 2006 http://etacar.umn.edu/.
Davidson, K., Walborn, N. R., & Gull, T. R. 1982 *ApJ* **254**, L47.
Frew, D J. 2004 *JAD* **10**, 6.
Gull, T. R. 2005. In *The Fate of the Most Massive Stars* (eds. R. Humphreys and K. Stanek). ASP Conference Series, Vol. 332, p. 281. ASP.
Gull, T. R., et al. 2005 *ApJ* **620**, 442.
Gull, T. R., Kober, G. Vieira, & Nielsen, K. E. 2006a *ApJS* **163**, 173.
Gull, T. R., et al. 2006b *BAAS* **207**, 114.07.
Hartman, H., Gull, T., Johansson, S., Smith, N., & the HST Eta Carinae Treasury Project Team 2004 *A&A* **419**, 215.
Hillier, J. D., et al. 2001 *ApJ* **553**, 837.
Hillier, J. D., et al. 2006 *ApJ* **642**, 1098.
Iping, R. C., et al. 2005 *ApJ* **633**, L37.
Maeder, A. & Meynet, g. 2003 *A&A* **411**, 543.
Martin, J. C. & Koppelman, M. D. 2004 *AJ* **127**, 2352.
Martin, J. C., et al. 2006 *ApJ* **640**, 474.
Morse, J. A., et al. 2001 *ApJ* **548**, L207.
Nielsen, K. E., Gull, T. R., & Vieira, G. 2005 *ApJS* **157**, 138.
Nielsen, K. E., et al. 2007 *ApJ* **660**, 669.
Pittard, J. M. & Corcoran, M. F. 2002 *A&A* **383**, 636.
Smith, L. J., Norris, R. P. F., & Crowther, P. A. 2002 *MNRAS* **337**, 1309.
Smith, N., et al. 2003 *AJ* **125**, 1458.
Smith, N., et al. 2004 *ApJ* **605**, 405.
Smith, N. & Owocki, S. 2006 *ApJ* **645**, L45.
Steiner, J. E. & Damineli, A. 2004 *ApJ* **612**, L133.
van Boekel, R., et al. 2003 *A&A* **410**, L37.
Verner, E., et al. 2002 **ApJ** *581*, 1154.
Verner, E., Bruhweiler, F., & Gull, T. 2005 *ApJ* **624**, 973.
Viotti, R., et al. 1989 *ApJS* **71**, 983.
Weigelt, G. & Ebersberger, J. 1986 *A&A* **163**, L5.
Weigelt, G., et al. 2007 *A&A* **464**, 87.
Zethson, T. 2001 *Ph.D. Thesis*, Lund University.
Zethson, T., et al. 2001 *AJ* **122**, 322.

Parameters and winds of hot massive stars

By ROLF P. KUDRITZKI AND MIGUEL A. URBANEJA

Institute for Astronomy, University of Hawaii, 2680 Woodlawn Drive, Honolulu, HI 96822, USA

In recent years a new generation of model atmosphere codes, which include the effects of metal line blanketing of millions of spectral lines in NLTE, has been used to re-determine the properties of massive stars through quantitative spectral analysis methods applied to optical, IR and UV spectra. This has resulted in a significant change of the effective temperature scale of early-type stars and a revision of mass-loss rates. Observed mass-loss rates and effective temperatures depend strongly on metallicity, both in agreement with theoretical predictions. The new model atmospheres, in conjunction with the new generation of 10-m-class telescopes equipped with efficient multi-object spectrographs, have made it possible to study blue supergiants in galaxies far beyond the Local Group in spectroscopic detail to determine accurate chemical composition, extinction and distances. A new distance determination method, the flux-weighted gravity–luminosity relationship, is discussed as a very promising complement to existing stellar distance indicators.

Observationally, there are still fundamental uncertainties in the determination of stellar mass-loss rates, which are caused by evidence that the winds are inhomogeneous and clumped. This may lead to major revisions of the observed rates of mass loss.

1. Introduction

Hot massive stars are cosmic engines of fundamental importance, not only in the local, but also in the early universe. A first generation of very massive stars has very likely influenced the formation and evolution of the first building blocks of galaxies. The spectral appearance of Lyman break galaxies and Lyα-emitters at high redshift is dominated by an intrinsic population of hot massive stars. Gamma-ray bursters are very likely the result of terminal collapses of very massive stars and may allow the tracing of the star-formation history of the universe to extreme redshifts.

It is obvious that the understanding of important processes of star and galaxy formation in the early universe is intimately linked to our understanding of the physics of massive stars. The observational constraints of the latter are provided by quantitative spectroscopic diagnostics of the population of hot massive stars in the local universe. It is the goal of this contribution to provide an overview of the dramatic progress which has been made in this field over the last five years.

There are two factors which have contributed to this progress—new observational facilities such as the optical/infrared telescopes of the 10-m class on the ground, and observatories in space allowing for spectroscopy in the UV (*HST*, *FUSE*, *GALEX*), IR (*ISO*, *Spitzer*) and at x-ray wavelengths (*XMM*, *Chandra*), and the enormous advancement of model atmosphere and radiative-transfer techniques. As for the latter, it is important to realize that modeling the atmospheres of hot stars is a tremendous challenge. Their physics are complex and very different from standard stellar atmosphere models. They are dominated by a radiation field with energy densities larger than, or of the same order as, the energy density of the atmospheric matter. This has two important consequences. First, severe departures from Local Thermodynamic Equilibrium of the level populations in the entire atmosphere are induced, because radiative transitions between ionic energy levels become much more important than inelastic collisions. Second, supersonic hydrodynamic outflow of atmospheric matter is initiated by line absorption of photons transferring outwardly directed momentum to the atmospheric plasma. This latter effect

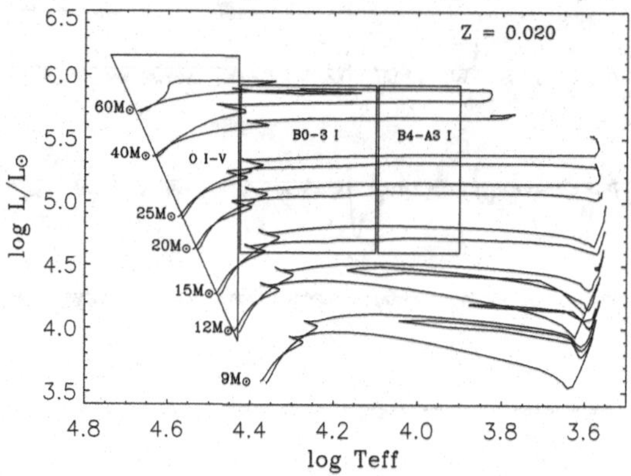

FIGURE 1. Evolutionary tracks of massive stars in the HRD with (thick lines) and without (thin lines) stellar rotation (Meynet & Maeder 2000), and the domains of spectral types discussed in this review.

is responsible for the existence of the strong stellar winds observed and requires the use of non-local thermodynamic equilibrium (NLTE) model atmospheres, which include the hydrodynamic effects of stellar winds (see Kudritzki 1998 for a detailed description of the physics of hot-star atmospheres).

The winds of hot massive stars are fundamentally important. Their energy and momentum input into the ISM is significant, creating circumstellar shells, stellar wind bubbles, and initiating further star formation. They affect stellar evolution by modifying evolutionary timescales, chemical profiles, surface abundances, and stellar luminosities. They also have substantial effects on the structure of stellar atmospheres. They dominate density stratification and radiative transfer through their transonic velocity fields, and they significantly modify the amount of emergent ionizing radiation (Gabler et al. 1989, 1991, 1992; Najarro et al. 1996).

While the basic concepts for the hydrodynamic atmospheres of hot stars and the spectroscopic diagnostics of their parameters and stellar winds have been developed in the 1980s and 1990s (see reviews by Kudritzki & Hummer 1990; Kudritzki 1998; Kudritzki & Puls 2000), the development of the most recent generation of model atmospheres, which for the first time accounts self-consistently for the effects of NLTE metal line blanketing (see Section 2), has led to a dramatic change in the diagnostic results. A new effective temperature scale has been obtained for the O-star spectral types, which apparently are significantly cooler than originally assumed, and thus, on average, have a lower luminosity and mass, and also provide less ionizing photons than previously thought. We will discuss these effects in detail in Sections 2–4. Note that we will focus our review on "normal" hot massive stars in a mass range between 15 and 100 $M_\odot$ in well-established evolutionary stages, such as dwarfs, giants, and supergiants (Figure 1). We will not discuss objects with extreme winds, such as Wolf-Rayet stars, luminous blue variables in outbursts, etc. For those, we refer the reader to the contributions in this volume by Paul Crowther and Nathan Smith.

The new model atmospheres, in conjunction with the new generation of 10-m-class telescopes equipped with efficient multi-object spectrographs, allow the study, in great spectroscopic detail, of blue supergiants in galaxies far beyond the Local Group to determine accurate chemical composition, extinction and distances. We will report recent results in this new field of "extragalactic stellar astronomy" in Section 5. In Section 6, we will discuss a new distance-determination method based on stellar photospheric spectroscopy, the flux-weighted gravity–luminosity relationship.

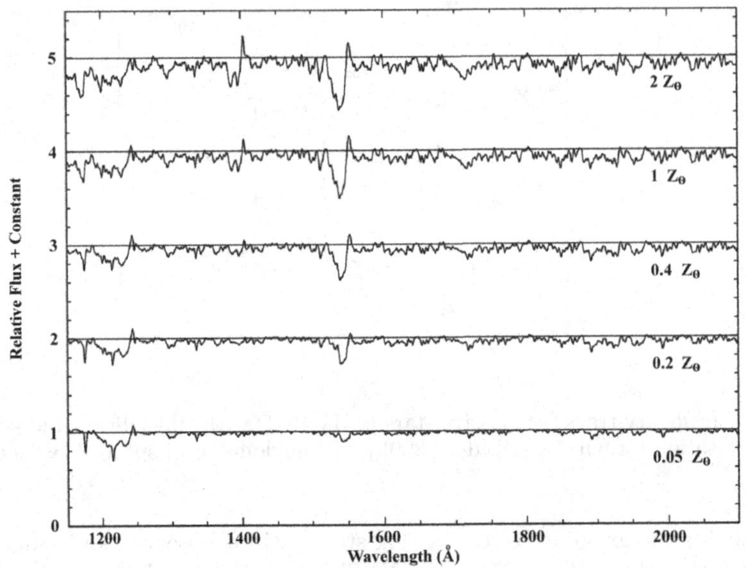

FIGURE 2. Model UV spectra of an integrated population of massive stars in a starburst galaxy displayed as a function of metallicity. The model atmosphere code used was WMBasic (Pauldrach, Lennon, & Hoffmann 2001) and the population synthesis was done with Starburst99 (Rix et al. 2004).

The diagnostics of stellar winds, in particular their mass-loss rates, are also affected by the proper accounting for line-blanketing effects. While the general scaling relations of stellar wind parameters—with stellar luminosity, mass, radius, and metallicity as described in Kudritzki & Puls (2000)—remain qualitatively unchanged, important quantitative changes have been found. There are still puzzling uncertainties with regard to the observed rates of mass loss related to the inhomogeneous structure of stellar wind outflows. We will present and discuss the most recent results in Section 7.

As indicated above, the new generation of model atmosphere codes has important applications for the interpretation of spectra of star-bursting galaxies in the early universe. Figure 2 is a nice example. This work by Rix et al. (2004) has been used to constrain star-formation rates and metallicities in high-redshift Lyman-break galaxies. Another example is the work by Barton et al. (2004), which uses stellar-atmosphere–model predictions by Bromm, Kudritzki, & Loeb (2001) and Schaerer (2003) to explore the possible detection of the first generation of very massive stars at redshifts around ten with present-day 8-m-class telescopes and with future (diffraction-limited) optical/IR telescopes of 30 m aperture, such as the Giant Segmented Mirror Telescope (GSMT; see also report by the GSMT Science Working Group, Kudritzki et al. 2003a). As it turns out, not only would the GSMT be able to detect such objects, it would also be able to constrain the initial mass function (IMF) of massive stars in the early universe from the relative comparison of L_α and He II 1640 recombination lines caused by the population of the first stars.

2. The effects of metal line blanketing

The inclusion of the opacity of millions of spectral lines in NLTE has two major effects. First, it changes the spectral energy distribution in the UV because of strong metal line absorption in the outer atmosphere ("line blanketing"). An example is given in Figure 3. While the UV flux is reduced, the physical requirement of radiative equilibrium and

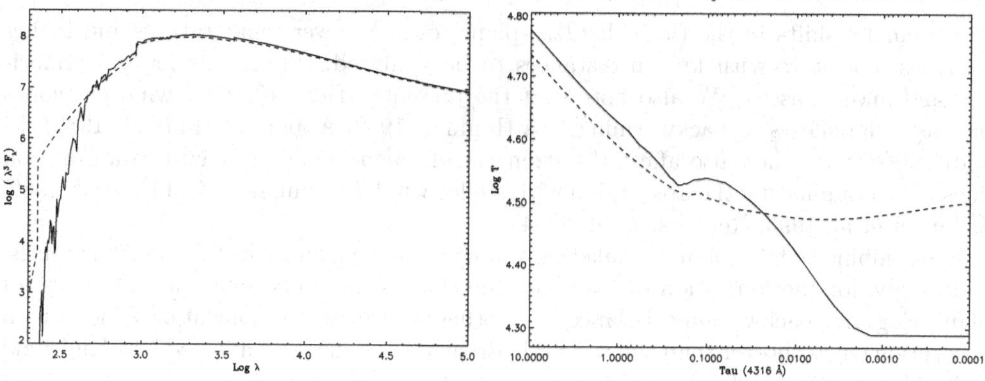

FIGURE 3. Effects of metal line blanketing. Left: Emergent flux of a NLTE hydrodynamic O-star model of $T_{\rm eff} = 40,000$ K with (solid) and without (dashed) metal line blanketing. Right: Local kinetic temperature as a function of monochromatic optical depth at blue wavelength for the same two models. The photospheric backwarming (as discussed in the text) as the result of blanketing is clearly seen. The NLTE code FASTWIND (Puls et al. 2005) was used for the calculations.

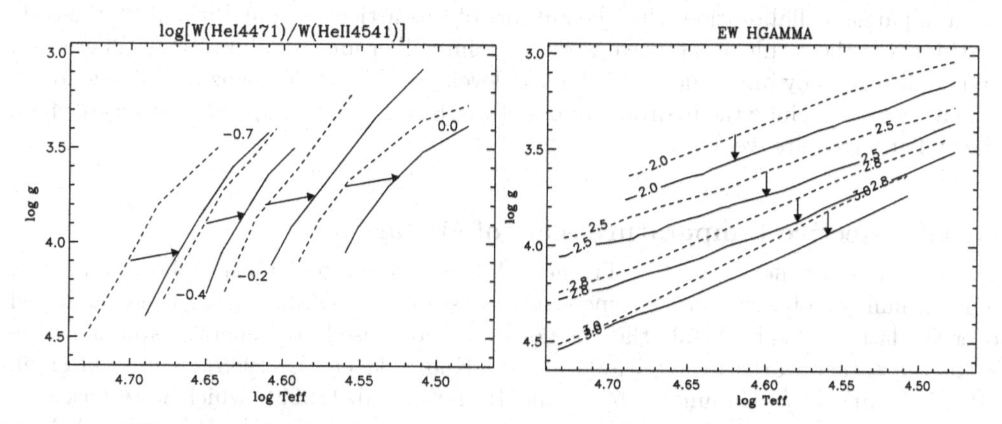

FIGURE 4. Effects of metal line blanketing on the stellar diagnostics. The isocontours of log $[W_\lambda(\text{He I } 4471)/W_\lambda(\text{He II } 4542)]$ (left) and $W_\lambda(H_\gamma)$ (right) in the ($\log g$, $\log T_{\rm eff}$)-plane are plotted. Dashed isocontours are unblanketed models, solid are blanketed. Vectors indicate the shifts caused by the effects of NLTE metal line blanketing. As for previous figures, the calculations were done with the NLTE code FASTWIND (Puls et al. 2005).

conservation of the total flux leads to an increase of the emergent flux at optical and IR wavelengths relative to the unblanketed case. This increase comes from the fact that a significant fraction of the photons absorbed in the outer layers are emitted back to the inner photosphere providing additional energy input and, thus, heating the deeper photospheres. This second "backwarming" effect increases the local kinetic temperature (see Figure 3), which then leads to stronger photospheric continuum emission and modifies diagnostically important ionization equilibria such as He I/II, which are used for the determination of $T_{\rm eff}$. Figure 4 demonstrates how the He I/II ionization equilibrium is shifted towards lower $T_{\rm eff}$ because of the backwarming effect. At the same time, the pressure-broadened wings of the Balmer lines (the standard diagnostic for $\log g$) become weaker, because the millions of metal lines increase the radiative acceleration $g_{\rm rad}$ and decrease the effective gravity $g_{\rm eff} = g - g_{\rm rad}$. As a result, higher gravities are needed to fit the Balmer lines (see Figure 4), in addition to the lower temperatures obtained from the helium ionization equilibrium. In summary, the use of blanketed models leads to systematic shifts in the ($\log g, \log T_{\rm eff}$)-plane towards lower temperatures and higher gravities. For stars with known distances to be analyzed, this means lower luminosities and lower masses. We also note that the presence of dense stellar wind envelopes increases the effects of backwarming (see Hummer 1982; Abbott & Hummer 1985). In addition, stellar winds also affect the strengths of diagnostically crucial He I absorption lines by contaminating the absorption with stellar wind line emission (Gabler et al. 1989; Sellmaier et al. 1993; Repolust et al. 2004).

The combined effects of line blanketing and backwarming also affect the ionizing fluxes. Amazingly, for the ionization of hydrogen the changes are very small, as the effects of blanketing and backwarming balance each other. However, the ionization of ions with absorption edges shorter than the one for hydrogen is significantly affected (see Kudritzki 2002; Martins et al. 2005a).

Over the last few years very powerful and user-friendly software packages have been developed to calculate NLTE model atmospheres with metal line blanketing, which are now available to the astronomical community and have been intensively applied for spectral diagnostics of hot massive stars. At Munich University Observatory the codes WMBasic (Pauldrach, Lennon, & Hoffmann 2001) and FASTWIND (Puls et al. 2005) were developed in partial collaboration with the authors of this article, and at Pittsburgh the code CMFGEN is the result of intensive work by John Hillier and collaborators (Hillier et al. 2003). Ivan Hubeny and Thierry Lanz have developed TLUSTY (Lanz & Hubeny 2003), which does not include the hydrodynamics of winds, but can be applied in all cases where the stellar winds are weak.

3. The effective temperature scale of O stars

With the new generation of NLTE metal line-blanketed model atmospheres available, quite a number of very detailed spectroscopic studies of O stars have been published over the last five years. While the diagnostic methods used are generally still the same as in the earlier works by Kudritzki (1980), Kudritzki et al. (1983), Kudritzki et al. (1989), Kudritzki & Hummer (1990), and Herrero et al. (1992), which were based on first-generation hydrostatic NLTE model atmospheres with only hydrogen and helium opacity, the results coming out of the new work are substantially different because of the tremendous improvements of the model atmospheres. Since the effects of metal line blanketing depend on the stellar metallicity, we will discuss O stars in the solar neighborhood of the Milky Way and in the Magellanic Clouds separately.

3.1. *O stars in the Milky Way*

Milky Way O stars have been studied by Pauldrach, Lennon, & Hoffmann (2001), Herrero et al. (2002), Martins et al. (2002), Bianchi & Garcia (2002), Garcia & Bianchi (2004), Repolust et al. (2004), Markova et al. (2004), Martins et al. (2005a), Mokiem et al. (2005), and Bouret et al. (2005). The results of all these studies with regard to the effective temperature scale is quite dramatic. An example is given by Figure 5, which displays the shifts in the ($\log g$, $\log T_{\rm eff}$)-plane of individual O stars, when the classical analysis of hydrogen and helium lines based on unblanketed model atmospheres is replaced by blanketed models. As to be expected from the previous section, the use of blanketed models leads to cooler effective temperatures, which simply is caused by the fact that blanketed models have higher intrinsic local photospheric temperatures, and thus models with lower $T_{\rm eff}$ are required to fit the observed helium ionization equilibrium. In addition, blanketed models have a lower local gas pressure and, thus, a higher gravity is generally needed to fit the pressure-broadened Balmer lines.

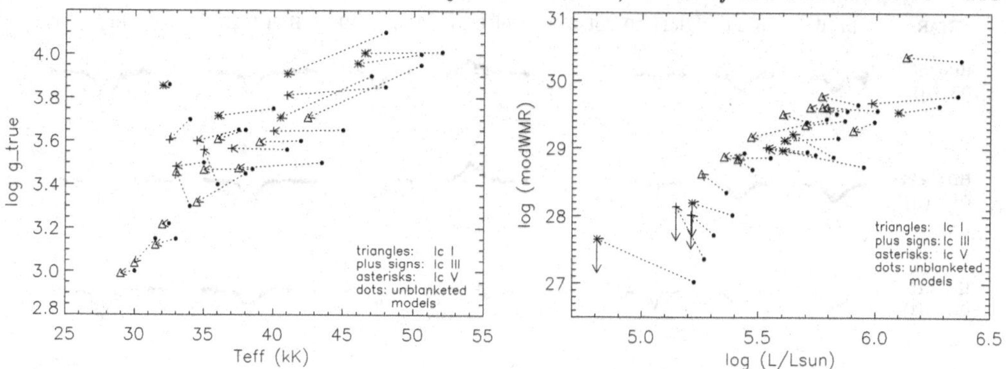

FIGURE 5. Effects of metal line blanketing on the stellar diagnostics. Left: Shifts of the location of individual O stars in the ($\log g$, $\log T_{\rm eff}$)-plane caused by the use of metal line-blanketed models in the analysis of the hydrogen and helium optical line spectrum. Right: Shifts of modified stellar wind momenta (see Section 7) as a function of luminosity (Repolust et al. 2004).

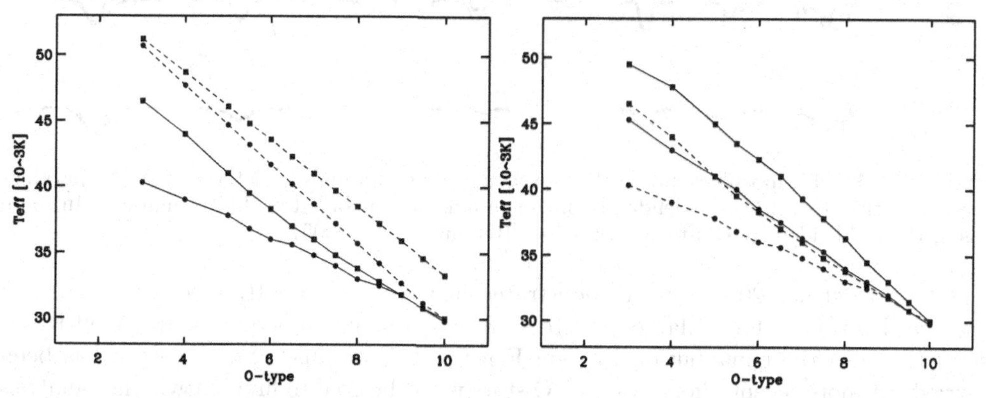

FIGURE 6. Left: Effective temperature as a function of spectral type for Milky Way O stars. Supergiants are plotted as circles, and giants and dwarfs as squares. The new scales are represented by the solid curves, the old scales by the dashed curves. For discussion see text. Right: The new, effective temperature scale of O stars in the SMC (solid) relative to the Milky Way (dashed; Massey et al. 2005).

While systematic effects towards somewhat lower effective temperatures and higher gravities were always expected in previous work based on unblanketed models, the large shifts as displayed in Figure 5 come as a surprise. We note that the effects are strongest for very hot objects and low-gravity supergiants, which have very strong and dense winds additionally affecting the helium ionization equilibrium by the mechanisms discussed above. In some cases, we encounter effective temperature changes of the order of 10 to 15%. Based on the work by Repolust et al. (2004), Massey et al. (2005) have introduced a new, effective temperature scale for Milky Way O stars, which is displayed in Figure 6 and compared to the old scale by Vacca et al. (1996). We realize that, particularly for supergiants, the differences between the two scales are dramatic. A very similar new, effective temperature scale has also been independently introduced by Martins et al. (2005a), based on a comprehensive study of a large sample of O stars.

3.2. *Quantitative IR spectroscopy of O stars*

Massive stars are frequently found in star-forming regions heavily obscured by interstellar dust. In such a situation, IR spectroscopy is the only way to obtain information. In the Galactic center, Najarro et al. (1994), Najarro et al. (1997), Figer et al. (1998),

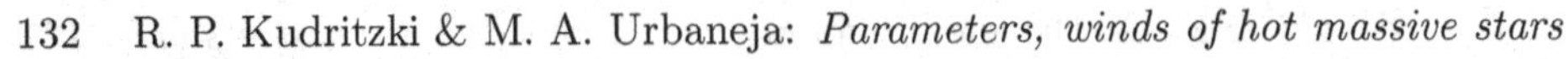

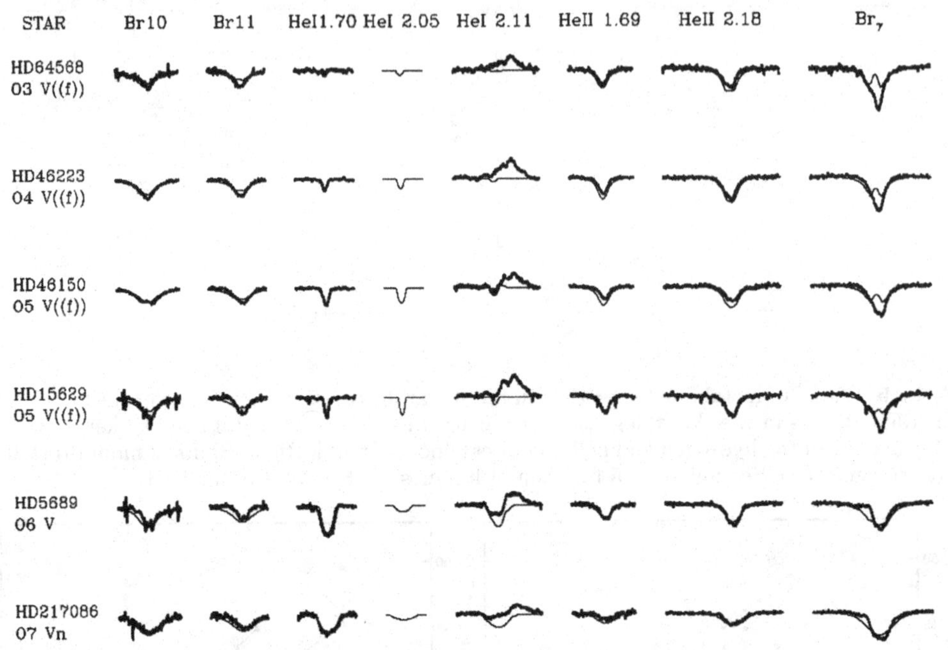

FIGURE 7. Model atmosphere fits of IR hydrogen and helium lines to determine stellar parameters. Note that He I 2.11μ is blended by nitrogen lines not included in the calculations. In some cases, Br_γ is blended by H II-region emission (Repolust et al. 2005).

and Najarro et al. (2004) have demonstrated how quantitative IR spectroscopy can be used to determine the stellar parameters of very extreme supergiants and Wolf-Rayet stars (see also the contribution by Don Figer in this volume). Similar work has been carried out more recently for "normal" O stars with the goal to find out whether analysis of the hydrogen and helium lines in the IR yields accurate information about stellar parameters consistent with the ones obtained from the analysis of the optical spectrum. Hanson et al. (2005) and Repolust et al. (2005) have used Subaru H- and K-band spectra of high S/N to study a large sample of O stars with very encouraging results showing that IR spectroscopy leads to effective temperatures, gravities, and mass-loss rates practically identical with those determined from optical spectra (Figure 7 gives an example). In parallel, Lenorzer et al. (2004) have carried out a systematic model-atmosphere study of the IR spectral diagnostics for normal O stars, focusing on both stellar lines as well as H II-region emission lines. It is obvious from this new work that for future investigations of massive stars in dense and highly obscured star-forming regions, quantitative IR spectroscopy has enormous potential.

3.3. *Effects of metallicity: The effective temperature scale of O stars in the Magellanic Clouds*

From the discussion in Section 2, it is clear that metallicity must have an influence on the strengths of blanketing effects, and thus, the effective temperature scale of O stars. The Magellanic Clouds are the ideal laboratory to investigate metallicity dependence, because of their lower metallicity relative to the Milky Way. Massey et al. (2004) and Massey et al. (2005) and, independently, Mokiem et al. (2004) and Mokiem et al. (2006), have carried out a systematic and comprehensive spectroscopic study of O stars in the clouds. Figures 6 and 8 describe the work done by Massey and collaborators. From a detailed fit of the optical hydrogen and helium lines and *HST* and *FUSE* UV spectra,

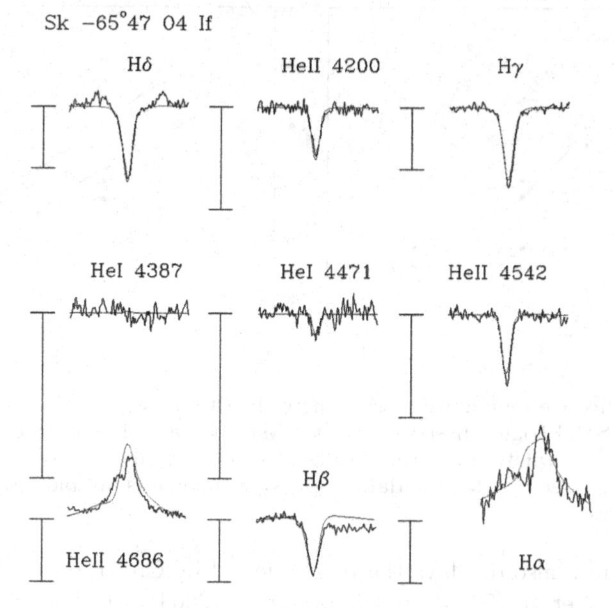

FIGURE 8. Model atmosphere fits of optical hydrogen and helium lines to determine stellar parameters of an O star in the LMC (Massey et al. 2004).

they were able to determine stellar parameters and wind properties. With regard to the effective temperature scale, the result is striking, and in agreement with expectations based on model atmosphere theory. For a given spectral type defined by the relative strengths of He I and He II lines, O stars at lower metallicity are significantly hotter than their galactic counterparts—a result which is also clearly confirmed by the work of Mokiem and collaborators. For the determination of IMFs in other galaxies based on spectral classification, or for the investigation of the stellar content of galaxies or H II regions based on nebular emission-line analysis, this is an important effect to take into account. The metallicity dependence of the effective temperature scale might also be important for the study of integrated spectra of starburst galaxies.

3.4. *Systematic effects depending on the analysis method*

In the work discussed so far, effective temperatures and gravities have been determined from the fit of the He I/II ionization equilibrium of optical or IR helium lines and the Balmer lines. Since O stars are UV bright, and hundreds of photospheric lines can be identified in high-resolution UV spectra (*IUE*, *HST*, *ORFEUS*, *FUSE*), it is tempting to use this spectral range to obtain independent information about the stellar parameters. As it turns out, it is difficult to constrain gravities solely through the UV. However, there are ionization equilibria such as Fe IV/V/VI or C III/IV, which can be used for a determination of temperatures. For Milky Way O stars, such work has been carried out by Pauldrach, Lennon, & Hoffmann (2001), Bianchi & Garcia (2002), Garcia & Bianchi (2004), and Bouret et al. (2005). As shown in Figure 9, significantly lower $T_{\rm eff}$ is obtained in some cases from the UV lines are compared to optical/IR He I/II lines. In particular, the results by Bianchi & Garcia (2002) and Garcia & Bianchi (2004) have resulted in very low effective temperatures.

A similar result has been obtained by Heap, Lanz, & Hubeny (2006) for Small Magellanic Cloud (SMC) O stars, mostly of luminosity class V. Figure 9 shows their results, compared to the effective temperature scale obtained by Massey et al. (2004) and Massey

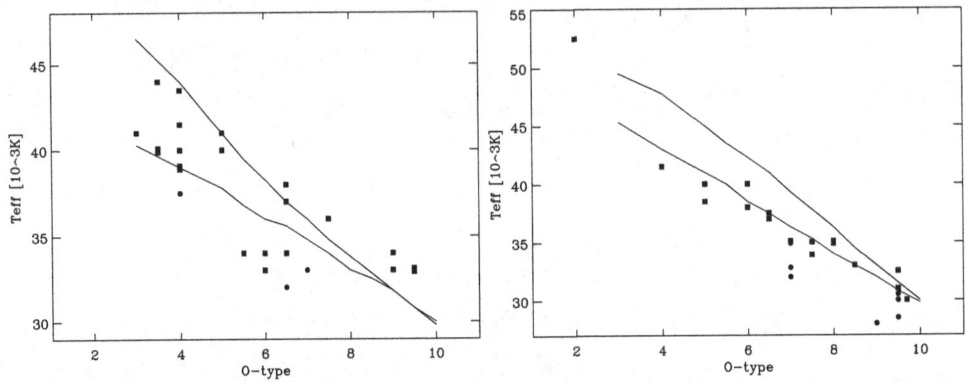

FIGURE 9. O star effective temperature scales according to Massey et al. (2005) for the Milky Way (left) and the SMC (right) based on the analysis of the He I/II ionization equilibrium at optical wavelengths. Overplotted are the results of spectral analyses using UV metal lines as temperature indicator (see text for the data sources). Supergiants are plotted as circles, giants and dwarfs as squares.

et al. (2005). Figure 9 also displays the results found by Crowther et al. (2002), Hillier et al. (2003), Bouret et al. (2003), and Evans et al. (2004), which show a less extreme, but similar trend.

This indication of a systematic effect depending on the analysis method used deserves a very careful and systematic future investigation. What is needed is a comprehensive simultaneous UV, optical, and IR analysis of all the O stars studied so far. The observational material seems to be available, or can easily be obtained. Of course, this will be a time-consuming effort, requiring a lot of detailed spectroscopic and model atmosphere work. However, it is extremely important that the reasons for these obvious discrepancies are well understood.

4. The effective temperature scale of B and A supergiants

As shown in Figure 1, massive O stars evolve into B, and later A supergiants. Based on unblanketed, hydrostatic NLTE model atmospheres, an effective temperature scale was introduced for galactic B supergiants by McErlean et al. (1998, 1999). More recently, Trundle et al. (2004), Trundle & Lennon (2005), and Crowther, Lennon, & Walborn (2006) have analyzed a large sample of Milky Way and SMC B supergiants with the improved models and have revised the effective temperature scale for the earliest spectral types of luminosity class Ia. The results are shown in Figure 10. Note that if the spectral classification scheme introduced by Lennon (1997) is used, the effective temperature scale should not depend on metallicity. This is indeed the case, as shown by a comparison of the SMC sample (Trundle & Lennon 2005) with the MW sample (Crowther, Lennon, & Walborn 2006).

Kudritzki, Bresolin, & Przybilla (2003) have recently introduced an effective temperature scale for late B and early A supergiants of luminosity class Ia and of solar metallicity, which is also displayed in Figure 10. A discussion of the metallicity dependence can be found in Evans & Howarth (2003).

5. Extragalactic stellar astronomy with B and A supergiants

As shown in Figure 1, O stars with masses between 15 to 40 $M_\odot$ evolve into spectral B and A supergiants. Although not as massive and luminous as the most massive O stars,

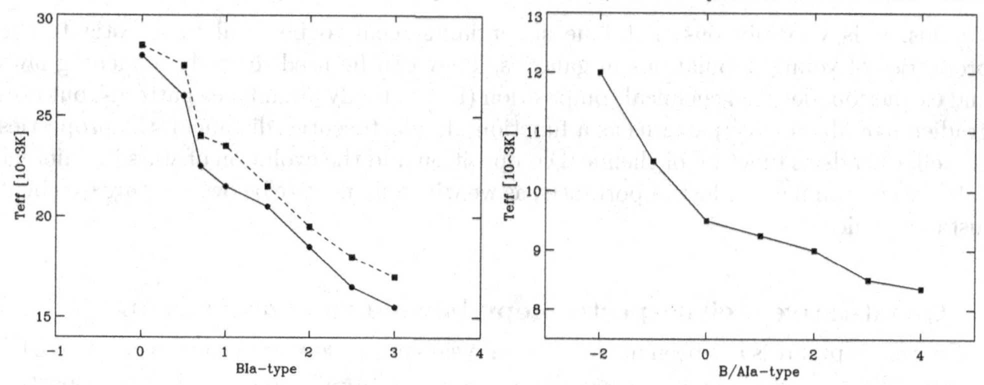

FIGURE 10. The effective temperature scales of early B supergiants (left, the dashed curve shows the old scale based on unblanketed NLTE models) and late B and early A supergiants (right) of luminosity class Ia. Note that in the right plot on the abscissa, the value of −2 corresponds to spectral type B8.

these stars are the brightest "normal" stars at visual light, with absolute magnitudes $-7.0 \geqslant M_V \geqslant -9.5$ (see Bresolin 2003; by "normal" we mean stars evolving peacefully without showing signs of eruptions or explosions, which are difficult to handle theoretically and observationally). While O stars emit most of their radiation in the extreme and far UV because of their high atmospheric temperature, late B and early A supergiants are cooler, and their bolometric corrections are much smaller because of Wien's law, so that their brightness at visual light reaches a maximum value during stellar evolution. It is this enormous intrinsic brightness at visual light which makes them extremely interesting for extragalactic studies far beyond the Local Group.

During their smooth evolution from the left to the right in the Hertzsprung-Russell Diagram (HRD), massive stars are crossing the temperature range of late B and early A supergiants in a timescale on the order of several 10^3 years (Meynet & Maeder 2000). During this short evolutionary phase, stellar winds with mass-loss rates of the order $10^{-6}\,M_\odot\,\mathrm{yr}^{-1}$ or less do not have enough time to significantly reduce the mass of the star, so the mass remains constant (Kudritzki & Puls 2000). In addition, as Figure 1 shows, the luminosity stays constant as well. The fact that the evolution of these objects can very simply be described by constant mass, luminosity, and a straightforward mass-luminosity relationship makes them a very attractive stellar distance indicator, as we will explain later in this review.

As evolved objects, the blue supergiants are older than their O-star progenitors, with ages between 0.5 to 1.3×10^7 years (Meynet & Maeder 2000). All galaxies with ongoing star formation or bursts of this age will show such a population. Because of their age, they are spatially less concentrated around their place of birth than O stars, and can frequently be found as isolated field stars. Together with their intrinsic brightness, this makes them less vulnerable as distance indicators against the effects of crowding, even at the larger distances where less luminous objects, such as Cepheids and RR Lyrae, start to have problems.

With regard to the crowding problem, we also note that the short evolutionary time of 10^3 years generally makes it very unlikely that an unresolved blend of two supergiants with very similar spectral types would be observed. On the other hand, since we are dealing with spectroscopic distance indicators, any contribution of unresolved additional objects of different spectral type would be detected immediately, as soon as it significantly affects the total magnitude.

Thus, it is very obvious that blue supergiants seem to be ideal to investigate the properties of young populations in galaxies. They can be used to study reddening laws and extinction, detailed chemical composition (i.e., not only abundance patterns, but also gradients of abundance patterns as a function of galactocentric distance), the properties of stellar winds as function of chemical composition and the evolution of stars in different galactic environments. Most importantly, as we will demonstrate below, they are excellent distance indicators.

6. Quantitative stellar spectroscopy beyond the Local Group

Enormous progress has been made in recent years in the development of accurate NLTE spectral diagnostics of B and A supergiants. Using high-resolution and high-S/N spectra, Urbaneja (2004), Urbaneja et al. (2005b), Trundle et al. (2004), Trundle & Lennon (2005), and Crowther, Lennon, & Walborn (2006) have studied early B supergiants in Local Group galaxies and the Milky Way to determine stellar properties, chemical composition and abundance gradients. For late B and early A supergiants, Przybilla & Butler (2001) and Przybilla et al. (2001a,b, 2006) have developed very detailed model atoms for the NLTE radiative transfer diagnostics, which allow for extremely accurate determination of effective temperatures (1%), gravities (0.05 dex), and chemical abundances (0.1 dex). Detailed abundance studies of A supergiants in many Local Group galaxies using these NLTE methods were carried out by Venn (1999) and Venn et al. (2000, 2001, 2003) and Kaufer et al. (2004). The analysis technique is similar for O stars, except that different ionization equilibria are used for the determination of effective temperatures (Si II/III/IV for early B supergiants and O I/II, N I/II, Mg I/II, S II/III for late B and early A supergiants; see Kudritzki 2003 for a more detailed description).

6.1. *Chemical composition*

For extragalactic applications beyond the Local Group, spectral resolution becomes an issue. The important points are the following: Unlike the case of late-type stars, crowding and blending of lines are not severe problems for hot massive stars—as long as we restrict our investigation to the visual part of the spectrum. In addition, it is important to realize that massive stars have angular momentum, which usually leads to high rotational velocities. Even for A supergiants, which have already expanded their radius considerably during their evolution and, thus, have slowed down their rotation, the observed projected rotational velocities are still on the order of 30 km s^{-1} or higher. This means that the intrinsic full half-widths of metal lines are on the order of 1 Å. In consequence, for the detailed studies of supergiants in the Local Group, a resolution of 25,000—sampling a line with five data points—is ideal. This is indeed the resolution which has been applied in most of the work referred to above.

However, as we have found out empirically (Przybilla 2002), degrading the resolution to 5,000 (FWHM = 1 Å) has only a small effect on the accuracy of the diagnostics—as long as the S/N remains high (i.e., 50 or better). Even for a resolution of 2,500 (FWHM = 2 Å), it is still possible to determine $T_{\rm eff}$ to an accuracy of 2%, $\log g$ to 0.05 dex and individual element abundances to 0.1.

Bresolin (2003), Bresolin et al. (2001, 2002a, 2002b, 2004, 2006), Urbaneja et al. (2003, 2005a, 2008), Kudritzki, Bresolin, & Przybilla (2003) and Kudritzki et al. (2008) have used the Focal Reducer and low-dispersion Spectrograph (FORS) of the Very Large Telescope (VLT) with a resolution of 1,000 (FWHM = 5 Å) to study blue supergiants far beyond the Local Group. The accuracy in the determination of stellar properties at this rather low resolution is still remarkable. The effective temperature (for late B and

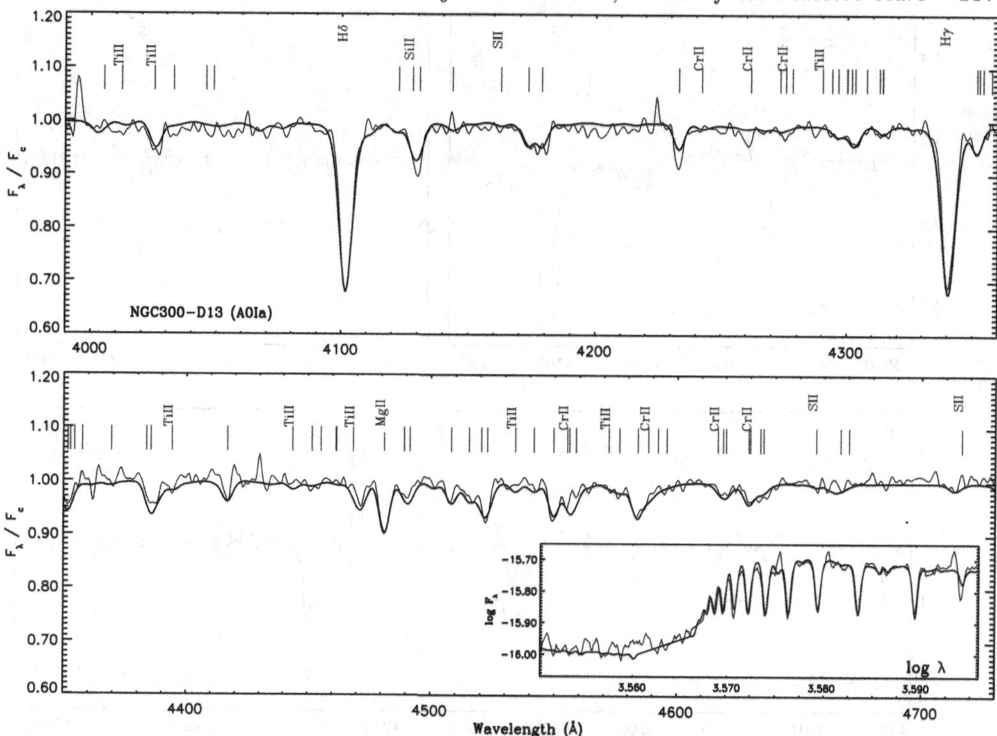

FIGURE 11. Quantitative spectral analysis of an A supergiant in the Sculptor spiral galaxy NGC 300 at a distance of 2 Mpc. The inset shows the fit of the Balmer jump, which is used to determine $T_{\rm eff}$ (Kudritzki et al. 2008; see also Bresolin et al. 2002a).

A supergiants are now determined from the Balmer jump, rather than from ionization equilibria) is accurate to roughly 4%, and the determination of gravity based on fitting the broad Balmer lines remains unaffected by the lower resolution and is still good to 0.05 dex. Abundances can be determined with an accuracy of 0.2 dex. In Figure 11, 12, and 13, we show examples of the detailed spectral fits that can be accomplished. Figure 13 demonstrates how important information about the metallicity gradients of the young stellar population in spiral galaxies can be obtained directly from the spectral analysis of blue supergiants.

6.2. *Extragalactic distance determinations with the Flux-Weighted Gravity–Luminosity Relationship*

The best-established stellar distance indicators, Cepheids and RR Lyrae, suffer from two major problems—extinction and metallicity dependence—both of which are difficult to determine with sufficient precision for these objects. Thus, in order to improve distance determinations in the local universe, and to assess the influence of systematic errors, there is definitely a need for alternative distance indicators, which are at least as accurate, but are not affected by uncertainties arising from extinction or metallicity. Blue supergiants are ideal objects for this purpose because of their enormous intrinsic brightness, which makes them available for accurate quantitative spectroscopic studies even far beyond the Local Group, using the new generation of 8-m-class telescopes and the extremely efficient multi-object spectrographs attached to them (see previous subsection). Quantitative spectroscopy allows us to determine the stellar parameters, and thus the intrinsic energy distribution, which can then be used to measure reddening and the extinction

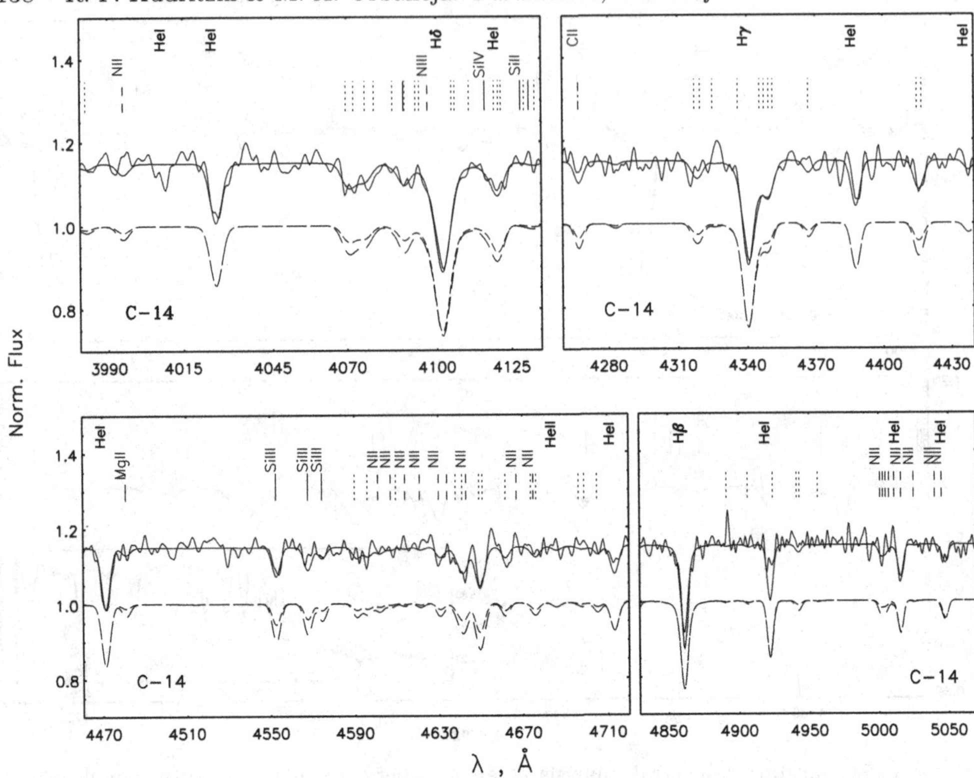

FIGURE 12. Quantitative spectral analysis of an early B supergiant in the Sculptor spiral galaxy NGC 300 (Urbaneja et al. 2005a).

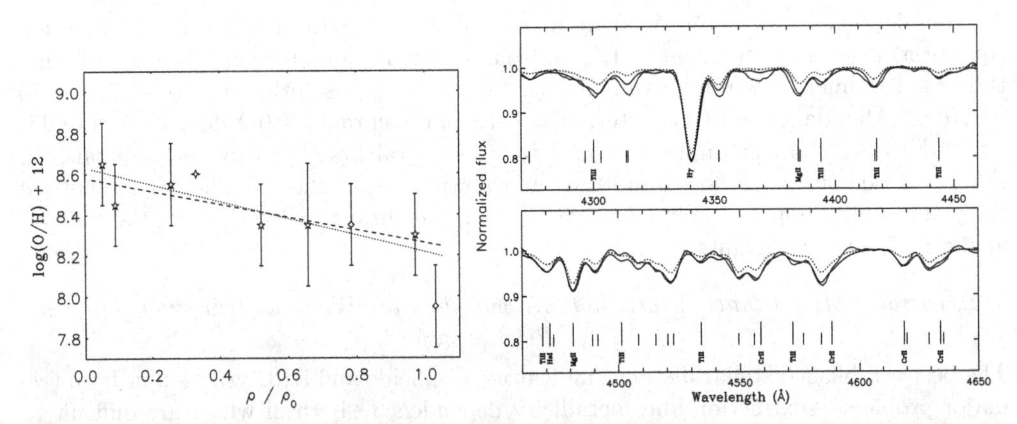

FIGURE 13. Left: the metallicity gradient of the young stellar population in the Sculptor spiral galaxy NGC 300, as determined for quantitative stellar spectroscopy (Urbaneja et al. 2005a; Bresolin et al. 2002a,b). Right: Quantitative spectral analysis of an A supergiant in the spiral galaxy NGC 3621 at a distance of 7 Mpc (Bresolin et al. 2001).

law. In addition, metallicity can be derived from the spectra. We emphasize that a reliable *spectroscopic* distance indicator will always be superior, since an enormous amount of additional information comes for free, as soon as one is able to obtain a reasonable spectrum.

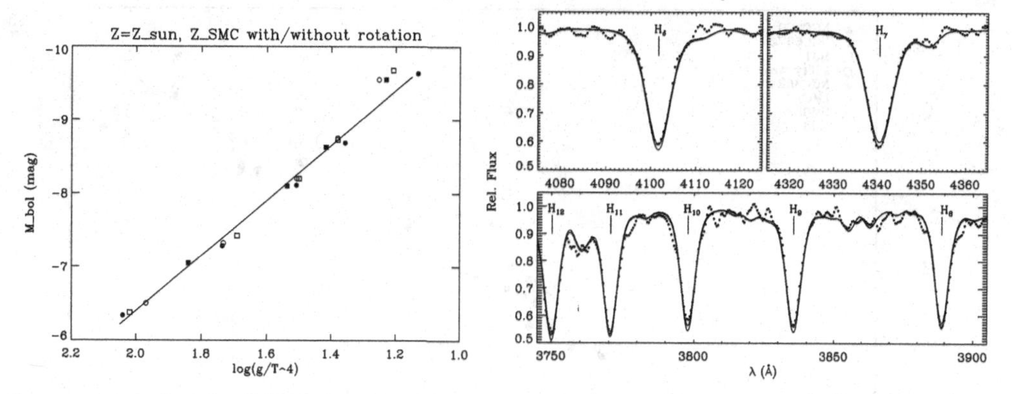

FIGURE 14. Left: The FGLR of stellar evolution models from Meynet et al. (1994) and Meynet & Maeder (2000). Circles correspond to models with rotation, squares represent models without the effects of rotation. Solid symbols refer to galactic metallicity and open symbols represent SMC metallicity. The solid curve corresponds to $a = -3.75$ in the FGLR. Note that gravities are in cgs units and the temperatures are in units of 10^4 K. Right: Fit of the higher Balmer lines of an A supergiant in the Sculptor galaxy NGC 300, using two atmospheric models with $T_{\rm eff} = 9500$ K and $\log g = 1.60$ (thick line) and 1.65 (thin line), respectively. The data were taken with FORS at the VLT. For further discussion see Kudritzki, Bresolin, & Przybilla (2003).

A very promising spectroscopic distance determination method based on simple stellar physics is the "flux-weighted gravity–luminosity relationship" (FGLR), which was introduced by Kudritzki, Bresolin, & Przybilla (2003). When discussing Figure 1 in Section 1, we noted that massive stars evolve through the domain of blue supergiants with constant luminosity and constant mass. This has a very simple, but very important consequence for the relationship of gravity and effective temperature along each evolutionary track. From

$$L \propto R^2 T_{\rm eff}^4 = \text{const.}\ ;\ M = \text{const.}\ , \tag{6.1}$$

follows immediately that

$$M \propto g\, R^2 \propto L\, (g/T_{\rm eff}^4) = \text{const.} \tag{6.2}$$

This means that each object of a certain initial mass on the Zero-Age Main Sequence (ZAMS) has its specific value of the "flux-weighted gravity" $g/T_{\rm eff}^4$ during the blue supergiant stage. This value is determined by the relationship between stellar mass and luminosity, which to a good approximation is a power law:

$$L \propto M^x\ . \tag{6.3}$$

Inspection of evolutionary calculations with mass loss, cf. Meynet et al. (1994) and Meynet & Maeder (2000), shows that $x = 3$ is a good value in the range of luminosities considered, although x changes towards higher masses. With the mass–luminosity power law, we then obtain:

$$L^{1-x} \propto \left(g/T_{\rm eff}^4\right)^x\ , \tag{6.4}$$

or with the definition of bolometric magnitude $M_{\rm bol} \propto -2.5 \log L$

$$-M_{\rm bol} = a \log\left(g/T_{\rm eff}^4\right) + b\ . \tag{6.5}$$

This is the FGLR of blue supergiants. Note that the proportionality constant a is given by the exponent of the mass–luminosity power law through

$$a = 2.5x/(1-x)\ . \tag{6.6}$$

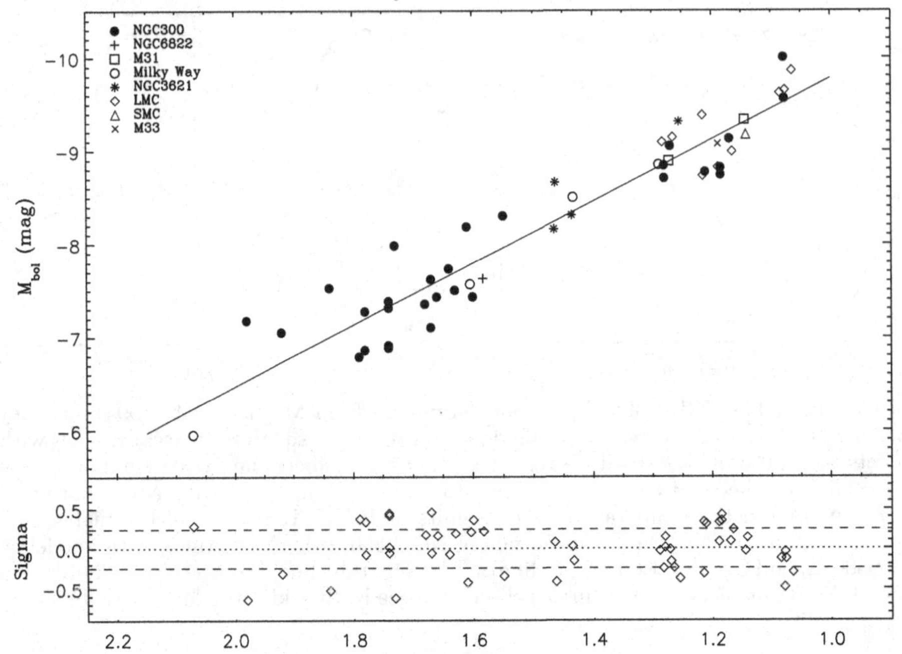

FIGURE 15. The FGLR of B and A supergiants in Local Group galaxies and in the spiral galaxies NGC 300 and NGC 3621 at a distance of 2 and 7 Mpc, respectively. The abscissa is the same as in Figure 14 (left). For a discussion, see text (Kudritzki, Urbaneja, Bresolin et al. 2008; see also Kudritzki, Bresolin, & Przybilla 2003).

and $a = -3.75$ for $x = 3$. Mass loss will depend on metallicity and therefore affect the mass–luminosity relation. In addition, stellar rotation through enhanced turbulent mixing might be important for this relation. In order to investigate these effects, we have used the models of Meynet et al. (1994) and Meynet & Maeder (2000) to construct the stellar evolution FGLR, which is displayed in Figure 14. The result is very encouraging. All different models with or without rotation, and with significantly different metallicity form a well defined, very narrow FGLR.

In order to verify the existence of the theoretically predicted FGLR, Kudritzki, Bresolin, & Przybilla (2003) re-analyzed a large sample of late B and early A supergiants in several Local Group galaxies (high-resolution spectra) and in the spiral galaxies NGC 300 (2 Mpc) and NGC 3621 (7 Mpc) using VLT FORS low-resolution spectra as described above. Figure 14 demonstrates how precisely the gravities can be determined, even at low resolution. The effective temperatures were obtained from the spectral types using the relation displayed in Figure 10. Taking into account the valid criticism by Evans & Howarth (2003) about the metallicity dependence of the temperature vs. spectral-type relationship, Kudritzki et al. (2008) used the information about the Balmer jump in the VLT/FORS spectrophotometric data (see Figure 11) to independently determine effective temperature and metallicity. This new procedure led to the FGLR shown in Figure 15. A least-square fit yields $a = -3.31$ and $b = 13.09$ with a one-σ scatter of 0.24 mag. Fixing the slope to the theoretical value $a = -3.75$, we obtain as a zero point $b = -13.73$ with $\sigma = 0.25$.

As displayed in Figure 15, the FGLR is an extremely tight relationship with a scatter comparable to the observed period–luminosity relationships of Cepheids. We conclude that blue supergiants provide a great potential as excellent extragalactic distance indicators. The quantitative analysis of their spectra—even at only moderate resolution—allows the determination of stellar parameters, stellar wind properties, and chemical composition with remarkable precision. In addition, since the spectral analysis yields intrinsic energy distributions over the whole spectrum from the UV to the IR, multi-color photometry can be used to determine reddening, extinction laws, and extinction. This is a great advantage over classical distance indicators—for which only limited photometric information is available—when observed outside the Local Group. Spectroscopy also allows dealing with the effects of crowding and multiplicity, as blue supergiants, due to their enormous brightness, are less affected by such problems than, for instance, Cepheids, which are fainter.

Applying the FGLR method on objects brighter than $M_V = -8$ mag and using multi-object spectrographs at 8- to 10-m-class telescopes, which allow for quantitative spectroscopy down to $m_V = 22$ mag, we estimate that with 20 objects per galaxy, we will be able to determine distances out to distance moduli of $m - M \sim 30$ mag with an accuracy of 0.1 mag. We emphasize that these distances will not be affected by uncertainties in extinction and metallicity, because we will be able to derive the corresponding quantities from the spectrum.

7. Winds of hot massive stars

All hot massive stars have winds which are driven by radiation. As emphasized in the introduction, these winds are fundamentally important for the spectral diagnostics of massive stars, for their evolution, and for the galactic environment. Over the last decades, very detailed and refined methods have been developed for the diagnostic of stellar winds and for modeling their hydrodynamics. A comprehensive review describing these methods and summarizing the basic properties of stellar winds was published a few years ago by Kudritzki & Puls (2000; see also Kudritzki 2000, which was published in this Symposium Series). The mass-loss rates and terminal velocities of these winds are related to the physical parameters of massive hot stars through simple relationships. The stellar-wind momentum $\dot{M}v_\infty$ is related to radius R, luminosity L and metallicity Z through:

$$\dot{M}v_\infty R^{1/2} \propto L^{1.8}(Z/Z_\odot)^{-0.8} \quad . \tag{7.1}$$

This the wind momentum–luminosity relationship (WLR), which was first introduced by Kudritzki, Lennon, & Puls (1995). (Note that the left-hand side is called the "modified stellar-wind momentum.") The physics leading to this relationship has been described in detail in Kudritzki (1998) and Kudritzki (2000). Proportionality constants for the WLR have been determined by Puls et al. (1996) for O stars and by Kudritzki et al. (1999) for B and A supergiants (see also Kudritzki & Puls 2000).

The terminal velocities v_∞ of winds of hot massive stars are related to the photospheric escape velocities $v_{\rm esc}^{\rm phot}$ and metallicity through

$$v_\infty \propto v_{\rm esc}^{\rm phot}\,(Z/Z_\odot)^{-0.15} \quad , \tag{7.2}$$

$$v_{\rm esc}^{\rm phot} = \left(2GM\,(1-\Gamma)\,R^{-1}\right)^{1/2} \quad . \tag{7.3}$$

The proportionality constant in Eq. (7.2) is about 2.6 (± 0.5) for O stars and early B supergiants, but becomes smaller for lower temperatures and is about 1.0 for A supergiants (see Kudritzki & Puls 2000, but see also the new results discussed below). Γ is the usual distance to the Eddington limit.

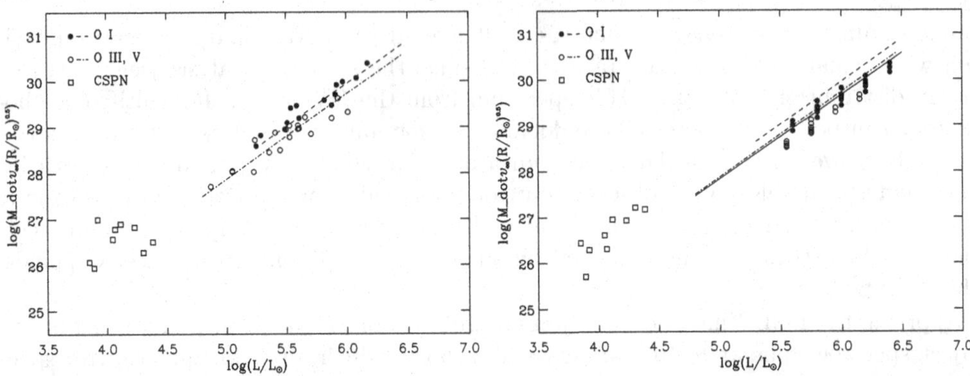

FIGURE 16. Left: Observed wind momenta of O stars and CSPN as a function of luminosity. Right: Calculated stellar-wind momenta for O stars and CSPN using the theory of line-driven winds as developed by Kudritzki (2002). Symbols refer to model calculations. The dashed lines are the regression curves obtained from the observations displayed in the left diagram. The solid line represents the theoretical approach by Vink et al. (2000).

7.1. *O-star wind momenta based on new diagnostics with metal line-blanketed model atmospheres*

The proportionality constants for the WLR were based on spectral diagnostics of H_α, which is very sensitive to the strengths of stellar winds and, therefore, regarded as a very good indicator of mass-loss rates (see Kudritzki & Puls 2000). However, the original work was based on the use of NLTE model atmospheres, which neglected metal line blanketing. It was, therefore, crucially important to repeat the H_α studies of stellar winds with the improved model atmospheres described in the sections before.

Repolust et al. (2004) and Markova et al. (2004) carried out comprehensive studies of O stars in the Milky Way. Figure 5 indicates the general trend with regard to the diagnostics of wind momenta. Since the analysis with metal line-blanketed NLTE atmospheres yields lower temperatures and, therefore, lower luminosities, the WLR is simply shifted towards lower luminosities. Figure 16 shows the observed WLR for O stars. As in Kudritzki & Puls (2000), we include the results of similar diagnostics of Central Stars of Planetary Nebulae (CSPN; for details see Kudritzki, Urbaneja, & Puls 2006) for a demonstration that the WLR very obviously extends to these low-mass hot stars in a post-AGB evolutionary stage as well. Repolust et al. (2004) and Markova et al. (2004) discuss the difference between the WLR of dwarfs and supergiants, and argue that the theory cannot reproduce this difference. They conclude that the H_α mass-loss rates of supergiants are affected by inhomogeneous stellar-wind clumping and in reality are probably close to the ones of dwarfs. To check this hypothesis, we compare with the theoretical wind momenta for O stars and CSPN calculated by Kudritzki (2002), who used a new theoretical approach for the theory of line-driven winds. The result is also shown in Figure 16. While we find a clear offset between the WLRs of dwarfs and supergiants, the theoretical wind momenta are a little too small in both cases, which would mean that clumping might be important for both dwarfs and supergiants. We will discuss the effects of clumping at the end of this review.

With the new stellar parameters derived with metal line-blanketed models, there is also a change in the values of photospheric escape velocities. After a first check of all the new results, the factor relating terminal velocity with escape velocity appears to be 3.1 rather than 2.6.

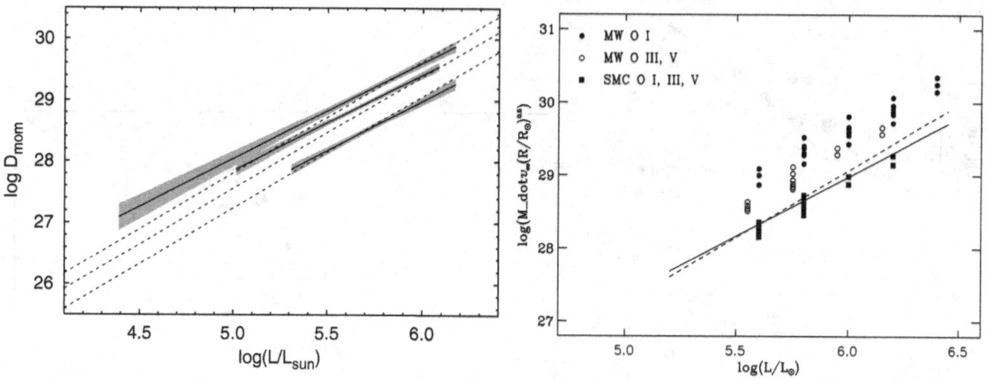

FIGURE 17. Left: Observed wind momenta of O stars and CSPN as a function of luminosity. The shaded area represent observational results in the MW, LMC, and SMC and the solid lines are the observed WLR regression curves. The dashed curves represent theoretical predictions by Vink et al. (2001) for MW, LMC and SMC metallicities, respectively (Mokiem et al. 2007). Right: Calculated stellar wind momenta for O stars and CSPN using the theory of line-driven winds as developed by Kudritzki (2002). Symbols refer to model calculations. The solid curve is the observed regression for the SMC from the left diagram. The dashed line represents the theoretical approach for the SMC by Vink et al. (2001).

7.2. *Metallicity dependence*

Since the mechanism of driving stellar winds is the absorption of photospheric photon momentum through many thousands of spectral lines, it is clear that the strengths of the winds is expected to depend on metallicity. First predictions of the metallicity dependence of both mass-loss rates and terminal velocities were made by Abbott (1982) and Kudritzki et al. (1987), and later confirmed and extended by Leitherer et al. (1992). Improved calculations were carried out more recently by Vink et al. (2001) and Kudritzki (2002). Observationally, Puls et al. (1996) analyzed O stars in the Clouds relative to the Milky Way and found a first observational indication for the power-law dependence of wind momenta on metallicity. Moreover, Kudritzki & Puls (2000) showed that O stars in the metal-poor SMC have lower terminal velocities than their MW counterparts. More recently, Massey et al. (2004), Massey et al. (2005), Evans et al. (2004), Hillier et al. (2003), and Crowther et al. (2002) have clearly confirmed these results. Figure 17 shows the most recent results obtained by Mokiem et al. (2007), which indicate excellent agreement with the predictions of the theory.

7.3. *B supergiants*

Because of changes in ionization leading to different sets of spectral lines absorbing photospheric photon momentum and driving the stellar winds, the WLR is expected to be spectral-type dependent. Indeed Kudritzki et al. (1999), in their study of winds of B and A supergiants, found a strong variation of the WLR with spectral type. While early-B supergiants seemed to have wind momenta only somewhat weaker than their O-stars' counterparts, mid-B supergiants showed much weaker winds, whereas A supergiants had stellar-wind momenta comparable to the early-B spectral types, but a steeper slope of the WLR.

With the new line-blanketed models available, Trundle et al. (2004), Trundle & Lennon (2005), Urbaneja (2004), Evans et al. (2004) and Crowther, Lennon, & Walborn (2006) have re-investigated the wind properties of B supergiants. In general, the results by Kudritzki et al. (1999) are confirmed. There is still, on average, a difference in the strengths of wind momenta between early and mid-B spectral types. However, while mass-loss rates

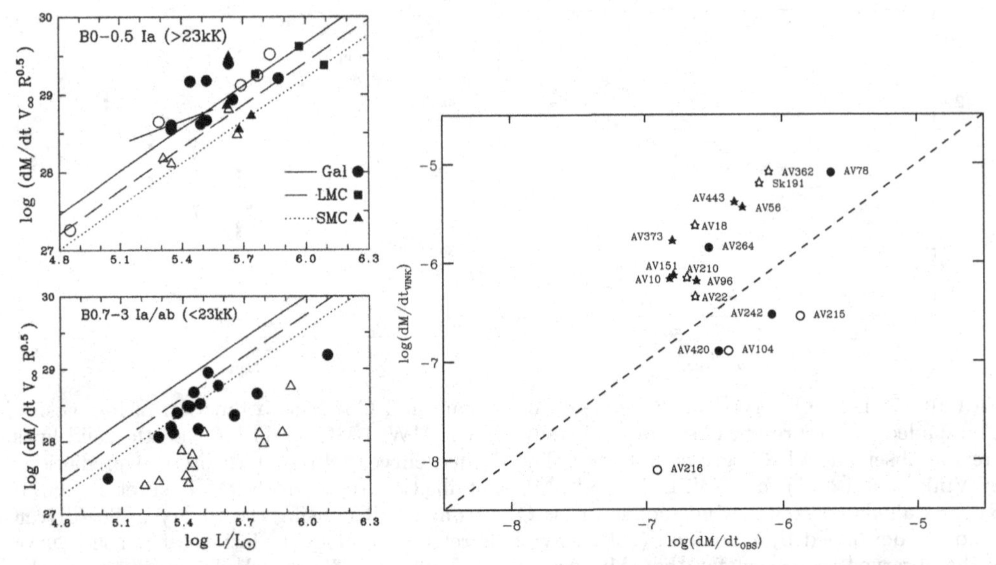

FIGURE 18. Left: Observed wind momenta of BIa supergiants as a function of luminosity for objects hotter than 23 kK (B0–B0.5; upper panel) and cooler than 23 kK (B0.7–3; lower panel) and for objects in the Galaxy (circles), LMC (squares), and SMC (triangles). Overplotted are wind momenta predicted by the computations by Vink et al. (2000, 2001) for the metallicities of the three galaxies (Crowther, Lennon, & Walborn 2006). Right: Mass-loss rates as predicted by Vink et al. (2001) for SMC B-supergiants compared to observations. Circles represent early-type supergiants, stars correspond to mid types.

for early spectral types are very similar to those obtained with the unblanketed models by Kudritzki et al. (1999), on average the new values for mid-B supergiants are about a factor of three higher—but even with this increase they remain significantly lower than those for the early-B types (see Figure 18). There is a clear metallicity dependence of wind momenta.

As shown in Figure 18, the theory of line-driven winds fails to reproduce these observations. Vink et al. (2000, 2001) predict a strong increase of mass-loss rates below 24 kK due to changes of ionization, which is definitely not observed, for the early-type supergiants show mass-loss rates much stronger than predicted by the theory. While this can be explained by stellar-wind clumping, we believe that in general, more theoretical work for B-supergiant winds is needed.

The ratio between terminal and escape velocities changes as a function of effective temperature showing a similar trend as found by Prinja & Massa (1998), but quantitatively different. For temperatures above 24 kK the ratio is 3.4, between 20 kK and 24 kK it is 2.5, and below 20 kK 1.9 is found, though in all temperature ranges there is a large scatter around these average values.

7.4. *Winds at very low metallicity*

As indicated in the introduction and discussed in several talks throughout this symposium, there is growing evidence that the evolution of galaxies in the early universe is heavily influenced by the formation of the first generations of very massive stars. Thus, it is important to understand the nature of radiation-driven winds at very low metallicity. As has been shown by Kudritzki (2002), radiative-line forces show a different dependence on optical depth and electron density at very low metallicity, which requires significant modifications of the theoretical description. With those implemented, it can be shown that the power-law dependence on the metallicity of mass-loss rates, wind momenta, and velocities breaks down and a much stronger dependence on metallicity is found. For details we refer the reader to the paper by Kudritzki (2002), which also provides ionizing fluxes and predicted UV spectra as a function of metallicity.

7.5. *The effects of rotation and instabilities*

There are many mechanisms which might lead to an enhancement of mass loss during stellar evolution. Centrifugal forces provided by stellar rotation are a classical mechanism already taken into account by Pauldrach, Puls, & Kudritzki (1986) and Friend & Abbott (1986). Kudritzki & Puls (2000), Petrenz & Puls (2000), and Owocki (2005) give overviews about the possible effects. In particular, at low metallicities, when massive O stars are hotter and have much smaller radii, and the effects of radiation-driven winds become smaller, stellar rotation might become a crucial mechanism not only for stellar mass loss, but also for stellar evolution (see Marigo et al. 2003; Maeder et al. 2005; Meynet et al. 2005; Hirschi et al. 2005; Ciappini et al. 2006; Meynet et al. 2006).

Continuum-driven winds and their instabilities as very likely encountered in LBV outbursts are another important physical mechanism, at least for objects very close to the Eddington limit (see Shaviv 2001; Shaviv 2005; Owocki et al. 2004; Owocki 2005; Smith & Owocki 2006; and Smith, these proceedings).

Stellar pulsations have also been frequently discussed as enhancing stellar mass loss, but as shown by Baraffe et al. (2001), they become less important at low metallicities.

7.6. *The problem of the "weak-wind stars"*

Bouret et al. (2003) and Martins et al. (2004) conducted UV and optical studies of O stars in the SMC with luminosities $L \leqslant 10^{5.5}\,L_{\odot}$ which indicated much smaller stellar-wind momenta than expected from the theory of radiation-driven winds and from simple extrapolation of the WLR from higher luminosities. More recently, Martins et al. (2005b) have investigated similar "weak-wind stars" in the Milky Way, and found that O dwarfs with $L \leqslant 10^{5.2}\,L_{\odot}$ indeed seem to have much weaker winds than predicted by the theory, with a discrepancy of the order of a factor of hundred. It is always simple to speculate about overly strong stellar winds. One can invent additional wind-driving mechanisms, or blame the neglect of clumping in the diagnostics. However, it is very difficult to explain winds that are too weak, because the radiative force at a given luminosity is always there, and cannot simply be switched off. Thus, it is natural to be suspicious about the accuracy of the spectroscopic diagnostics of these objects.

The determination of mass-loss rates comes mostly from the analysis of UV metal lines, since H_{α} provides only the upper limits in most of the cases. While the diagnostics of those lines have been done most carefully, with the state-of-the-art model atmospheres described before, the mass-loss rates determined depend crucially on ionization calculations which could severely be affected by the soft x-ray emission of shocks in the stellar-wind flow. The authors are aware of this problem and have included effects of shock emission, but it is open at this point whether this treatment of shocks is sufficient. Other possible options—less likely, from our point of view—to explain the weak-wind stars are discussed by Martins et al. (2005b) and Mokiem et al. (2007).

7.7. *Stellar wind clumping*

While H_{α} is, in principle, a perfect tool to measure mass-loss rates (see Kudritzki & Puls 2000 for discussion and references), the results might be affected by stellar-wind clumping. It has long been known that line-driven winds are intrinsically unstable (Owocki et al. 1988, 2004). This might lead to inhomogeneous clumped winds such as described by

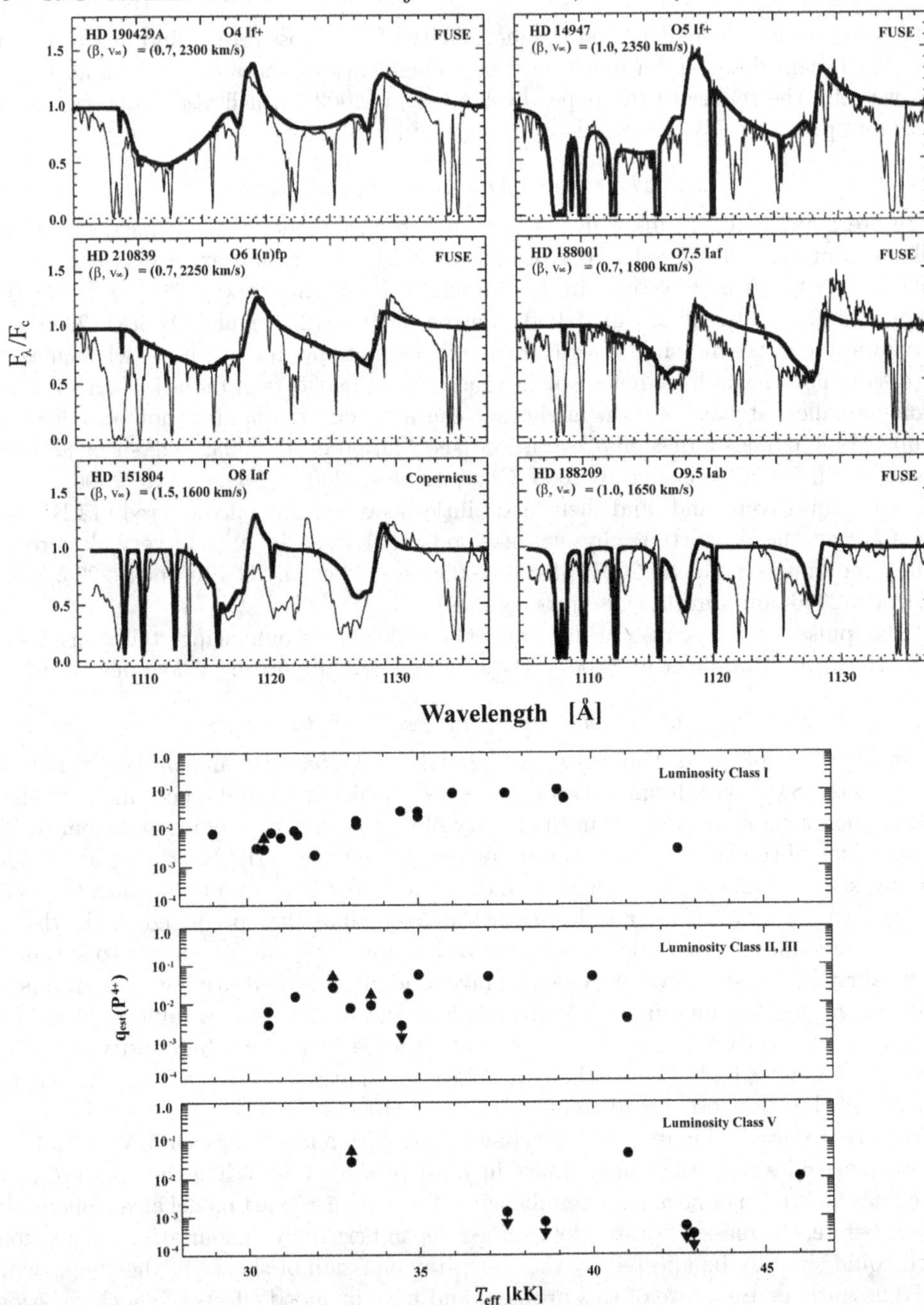

FIGURE 19. Upper panels: Radiative transfer fits of the far UV PV lines of MW O stars. Lower panels: Ionization fraction of PV as determined from fits of H_α and PV as function of effective temperature (Fullerton et al. 2006).

Owocki & Runacres (2002), with regions of enhanced density ρ_{cl} and regions where the density is much lower. In a very simple description, introducing clumping factors f_{cl} similar to PN diagnostics, the relationship between the average density of the stellar wind flow ρ_{av} and the density in the clumps is then given by $\rho_{cl} = \rho_{av} f_{cl}$. The same relationship holds for the occupation numbers n_i of ions.

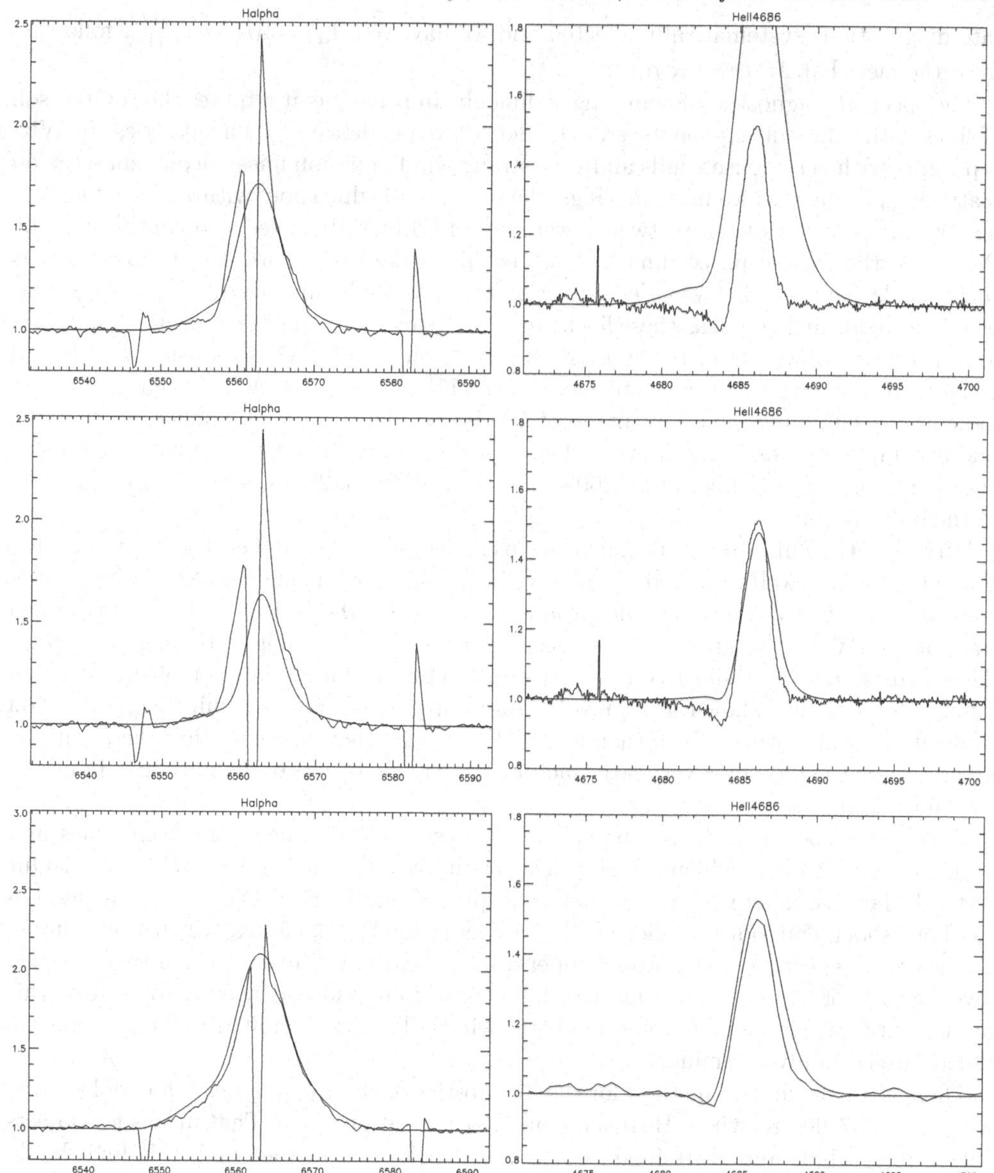

FIGURE 20. Diagnostic of clumping factors in the winds of CSPN. The first two rows show model fits of the H_α (left) and He II 4686 (right) profile of IC 418 with $f_{cl} = 1$ (top row) and $f_{cl} = 50$ indicating enormous clumping. The bottom row shows the same for He2−108 and $f_{cl} = 1$, demonstrating that for this object the wind is very likely homogeneous. Note that the H_α and N II nebular emission lines have been (imperfectly) subtracted (Kudritzki, Urbaneja, & Puls 2006).

Line opacities κ depend on density through $\kappa \propto n_i \propto \rho^x$ and for very small, optically thin clumps the average optical line depth in the wind is given by $\tau_{av} \propto n_i^{av} \propto n_i^{cl} f^{-1} \propto \rho_{av}^x f^{x-1}$. For a dominating ionization stage we have $x = 1$, and the clumping along the line of sight cancels and does not affect the diagnostics. However, bound hydrogen is a minor ionization stage in hot stars depending on recombination from ionized hydrogen with $n_i(H) \propto n_E n_P \propto \rho^2$. Thus, if f_{cl} is significantly larger than one, the H_α mass-loss rate diagnostic is systematically affected and we have $\dot{M}(H_\alpha) = \dot{M}(true) f_{\rm cl}^{1/2}$, following from the fact that $\dot{M}(true) \propto \rho_{\rm av}$.

The spectral diagnostics of clumping is difficult. In principle, it requires the comparison of lines with different exponents x in the density dependence of their opacities. In WR-type stars with very dense winds and very strong wind-emission lines, incoherent electron scattering produces wide emission wings, the strength of which goes with $x \sim 1$. Clumping factors on the order of ten to twenty were found (Hillier 1991; see also contribution by Paul Crowther, these proceedings). This technique does not work for O-type stars, as their winds have much lower density. Also the UV P-Cygni lines of dominating ions usually provide little help, as these lines are mostly saturated and the ionization equilibria are uncertain. However, in the most recent work on massive O stars using *FUSE* and Copernicus spectra, the PV resonance line at 1118 Å and 1128 Å has been used as an indicator of clumping. The advantage of PV is the low cosmic abundance, so that the line is completely unsaturated even when in a dominating ionization stage. Substantial clumping was found (Hillier et al. 2003; Bouret et al. 2003, 2005), with clumping factors of the order of ten.

Very recently, Fullerton et al. (2006) have carried out a comprehensive study of Milky Way O stars with well-observed *FUSE* PV-line profiles producing detailed PV radiative-transfer line fits to determine the product $q(P^{4+})\dot{M}$, where $q(P^{4+})$ is the ionization fraction of PV in the ground state (see Figure 19). Comparing with mass-loss rates derived from H_α (or radio free-free emission), they produced the plot of $q(P^{4+})$ as a function of $T_{\rm eff}$, also shown in Figure 19. Assuming that PV is a dominating ionization stage at all temperatures (and, therefore, $q(P^{4+}) = 1$), they conclude that the H_α mass-loss rates are too high by a very large factor (up to many orders of magnitudes), implying enormous filling factors.

This very important work is based on the best available diagnostic techniques and poses a very serious problem. The crucial assumption is, of course, $q(P^{4+}) = 1$. Our test calculations, done with the model atmosphere code FASTWIND show that models (without shock emission) predict $q(P^{4+}) = 1$ only for $T_{\rm eff} \leqslant 35$ kK, which would imply a much lower effect than at hotter temperatures. Clearly, for future work a very detailed investigation of the PV ionization (including shock emission) is needed to address this fundamental problem of clumping in O-star winds. The results found by Fullerton et al. (2006) are certainly alarming.

There is an alternative method for the diagnostics of clumping, at least for cool O stars with $T_{\rm eff} \leqslant 37,000$ K, where He II is a dominant ionization stage. That means for objects with strong winds and He II 4686 in emission and formed in the wind, this line should have a density dependence close to $x = 1$. Its relative strength to H_α should allow us to constrain $f_{\rm cl}$.

Kudritzki, Urbaneja, & Puls (2006) have applied this technique to study clumping in the winds of CSPN with very interesting results (see Figure 20), yielding clumping factors varying in a range from 50 to 1.

REFERENCES

ABBOT, D. C. 1982 *ApJ* **259**, 282.
ABBOT, D. C. & HUMMER, D. G. 1985 *ApJ* **294**, 286.
BARAFFE, I., HEGER, A., & WOOSLEY, S. E. 2001 *ApJ* **550**, 890.
BARTON, E., DAVE, R., SMITH, J., ET AL. 2004 *ApJ* **604**, L1.
BIANCHI, L. & GARCIA, M. 2002 *ApJ* **581**, 610.
BOURET, J. C., LANZ, T., & HILLIER, J. D. 2005 *A&A* **438**, 301.

Bouret, J. C., Lanz, T., Hillier, J. D., et al. 2003 *ApJ* **595**, 1182.

Bresolin, F. 2003 *Lect. Notes Phys.* **635**, 149.

Bresolin, F., Gieren, W., Kudritzki, R. P., Pietrzynski, G., & Przybilla, N. 2002a *ApJ* **567**, 227.

Bresolin, F., Kudritzki, R. P., Mendez, R. H., & Przybilla, N. 2001 *ApJ* **548**, L149.

Bresolin, F., Kudritzki, R. P., Najarro, F., Gieren, W., & Pietrzynski, G. 2002b *ApJ* **577**, L107.

Bresolin, F., Pietrzynski, G., Gieren, W., Kudritzki, R. P., Przybilla, N., & Fouque, P. 2004 *ApJ* **600**, 182.

Bresolin, F., Pietrzynski, G., Urbaneja, M. A., Gieren, W., Kudritzki, R. P., & Venn, K. A. 2006 *ApJ* **600**, 182.

Bromm, V., Kudritzki, R. P., & Loeb, A. 2001 *ApJ* **552**, 464.

Ciappini, C., Hirschi, R., Meynet, G., et al. 2006 *A&A* **449**, L27.

Crowther, P. A., Hillier, D. J., Evans, C. J., Fullerton, A. W., de Marco, O., & Willis, A. J. 2002 *ApJ* **579**, 774.

Crowther, P. A., Lennon, D. J., & Walborn, N. R. 2006 *A&A* **446**, 279.

Evans, C. J., Crowther, P. A., Fullerton, A. W., & Hillier, D. J. 2004 *ApJ* **610**, 1021.

Evans, C.J. & Howarth, I. 2003 *MNRAS* **345**, 1223.

Figer, D. F., Najarro, F., Morris, M., et al. 1998 *ApJ* **506**, 384.

Friend, D. & Abbott, D. C. 1986 *ApJ* **202**, 153.

Fullerton, A., Massa, D., & Prinja, R. 2006 *ApJ* **637**, 1025.

Gabler, R., Gabler, A., Kudritzki, R. P., & Mendez, R. H. 1992 *A&A* **265**, 656.

Gabler, R., Gabler, A., Kudritzki, R. P., Puls, J., & Pauldrach, A. W. A. 1989 *A&A* **226**, 162.

Gabler, R., Kudritzki, R. P., & Mendez, R. H. 1991 *A&A* **245**, 587.

Garcia, M. & Bianchi, L. 2004 *ApJ* **606**, 497.

Hanson, M. M., Kudritzki, R. P., Kenworthy, M. A., Puls, J., & Tokunaga, A. T. 2005 *ApJS* **161**, 154.

Heap, S. R., Lanz, T., & Hubeny, I. 2006 *ApJ* **638**, 409.

Herrero, A., Kudritzki, R. P., Vilchez, J. M., et al. 1992 *A&A* **261**, 209.

Herrero, A., Puls, J., & Najarro, F. 2002 *A&A* **396**, 949.

Hillier, J. D. 1991 *A&A* **247**, 455.

Hillier, J. D., Lanz, T., Heap, S. R., et al. 2003 *ApJ* **588**, 1039.

Hirschi, R., Meynet, G., & Maeder, A. 2005 *A&A* **443**, 581.

Hummer, D. G. 1982 *ApJ* **257**, 724.

Kaufer, A., Venn, K. A., Tolstoy, E., Pinte, C., & Kudritzki, R. P. 2004 *AJ* **127**, 2723.

Kudritzki, R. P. 1980 *A&A* **85**, 174.

Kudritzki, R. P. 1998. In *Stellar Astrophysics for the Local Group* (eds. A. Aparicio, A. Herrero & F. Sanchez). p. 149. Cambridge University Press.

Kudritzki, R. P. 2000. In *Unsolved Problems in Stellar Evolution* (ed. M. Livio). p. 202. Cambridge University Press.

Kudritzki, R. P. 2002 *ApJ* **577**, 389.

Kudritzki, R. P. 2003 *Lect. Notes Phys.* **635**, 123.

Kudritzki, R. P., Bresolin, F., & Przybilla, N. 2003 *ApJ* **582**, L83.

Kudritzki, R. P., Cabanne, K. P., Husfeld, D., et al. 1989 *A&A* **226**, 235.

Kudritzki, R. P. and the GSMT Science Working Group 2003 *GSMT Science Working Group Report*, http://www.aura-nio.noao.edu/.

Kudritzki, R. P. & Hummer, D. G. 1990 *ARA&A* **28**, 303.

Kudritzki, R .P., Lennon, D. J., & Puls, J. 1995. In *Science with the VLT* (eds. J. R. Walsh & J. Danziger). p. 246. Springer-Verlag.

Kudritzki, R. P., Pauldrach, A. W. A., & Puls, J. 1987 *A&A* **173**, 293.

Kudritzki, R. P. & Puls, J. 2000 *ARA&A* **38**, 613.

Kudritzki, R. P., Puls, J., Lennon, D. J., et al. 1999 *A&A* **350**, 970.

Kudritzki, R. P., Simon, K. P., & Hamann, W. R. 1983 *A&A* **118**, 245.

Kudritzki, R. P., Urbaneja, M. A., Bresolin, F., et al. 2008 *ApJ* **681**, 118.

Kudritzki, R. P., Urbaneja, M. A., & Puls, J. 2006. In *Planetary Nebulae in our Galaxy and Beyond* (eds. M. J. Barlow & R. H. Mendez). IAU Symp. Ser. 234, p. 119. Cambridge University Press.

Lanz, T. & Hubeny, I. 2003 *ApJ* **465**, 359,

Leitherer, C., Robert., C., & Drissen, L. 1992 *ApJ* **401**, 596.

Lennon, D. J. 1997 *A&A* **317**, 871.

Lenorzer, A., Mokiem, M. R., de Koter, A., & Puls, J. 2004 *A&A* **422**, 275.

Maeder, A., Meynet, G., & Hirschi, R. 2005. In *The Fate of the Most Massive Stars* (eds. R. Humphreys & K. Stanek). ASP Conf. Series 332, p. 3. ASP.

Marigo, P., Chiosi, C., & Kudritzki, R. P. 2003 *A&A* **399**, 617.

Markova, N., Puls, J., Repolust, T., & Markov, H. 2004 *A&A* **413**, 693.

Martins, F., Schaerer, D., & Hillier, D. J. 2002 *A&A* **382**, 999.

Martins, F., Schaerer, D., & Hillier, D. J. 2005a *A&A* **436**, 1049.

Martins, F., Schaerer, D., Hillier, D. J., et al. 2005b *A&A* **441**, 735.

Martins, F., Schaerer, D., Hillier, D. J., & Heydari-Malayeri, M. 2004 *A&A* **420**, 1087.

Massey, P., Bresolin, F., Kudritzki, R. P., Puls, J. & Pauldrach, A. W. A. 2004 *ApJ* **608**, 1001.

Massey, P., Puls, J., Pauldrach, A. W. A., Bresolin, F., Kudritzki, R. P., & Simon, T. 2005 *ApJ* **627**, 477.

McErlean, N. D., Lennon, D. J., & Dufton, P. L. 1998 *A&A* **329**, 613.

McErlean, N. D., Lennon, D. J., & Dufton, P. L. 1999 *A&A* **349**, 553.

Meynet, G., Ekström, S., & Maeder, A. 2006 *A&A* **447**, 623.

Meynet, G. & Maeder, A. 2000 *A&A* **361**, 101.

Meynet, G., Maeder, A., & Ekström, S. 2005. In *The Fate of the Most Massive Stars* (eds. R. Humphreys & K. Stanek). ASP Conf. Series 332, p. 228. ASP.

Meynet, G., Maeder, A., Schaller, G., Schaerer, D., & Charbonnel, C. 1994 *A&AS* **103**, 97.

Mokiem, M. R., de Koter, A., Evans, C., et al. 2006 *A&A* **456**, 1131.

Mokiem, M. R., de Koter, A., Evans, C., et al. 2007 *A&A* **465**, 1003.

Mokiem, M. R., de Koter, A., Puls, J., et al. 2005 *A&A* **441**, 711.

Mokiem, M. R., Martin-Hernandez, N. L., Lenorzer, A., de Koter, A., Tielens, A. G. G. M. 2004 *A&A* **419**, 319.

Najarro, F., Figer, D. F., Hillier, D. J., & Kudritzki, R. P. 2004 *ApJ* **611**, L105.

Najarro, F., Hillier, D. J., Kudritzki, R. P., et al. 1994 *A&A* **285**, 573.

Najarro, F., Krabbe, A., Genzel, R., Lutz, D., Kudritzki, R. P., & Hiller, D. J. 1997 *A&A* **325**, 700.

Najarro, F., Kudritzki, R. P., Cassinelli, J. P., Stahl, O., & Hillier, D. J. 1996 *A&A* **306**, 892.

Owocki, S. 2005. In *The Fate of the Most Massive Stars* (eds. R. Humphreys & K. Stanek). ASP Conf. Series 332, p. 169. ASP.

Owocki, S. P., Castor, J. I., & Rybicki, G. B. 1988 *ApJ* **335**, 914.

Owocki, S. P., Gayley, K., & Shaviv, N. 2004 *ApJ* **616**, 520.

Owocki, S. P. & Runacres, M. C. 2002 *A&A* **381**, 1015.

Pauldrach, A. W. A., Lennon, M., Hoffmann, T. 2001 *A&A* **375**, 161.

Pauldrach, A. W. A., Puls, J., & Kudritzki, R. P. 1986 *A&A* **164**, 86.

Petrenz, P. & Puls, J. 2000 *A&A* **358**, 956.

Prinja, R. K. & Massa, D. 1998. In *Boulder-Munich II: Properties of Hot, Luminous Stars* (ed. I. D. Howarth). ASP Conf. Series 131, p. 218. ASP.

Przybilla, N. 2002 *Ph.D. Thesis*, Ludwig-Maximilians-Universtität München.

Przybilla, N. & Butler, K. 2001 *A&A* **379**, 995.

Przybilla, N., Butler, K., Becker, S. R., & Kudritzki, R. P. 2001a *A&A* **369**, 1009.

Przybilla, N., Butler, K., Becker, S. R., & Kudritzki, R. P. 2006 *A&A* **445**, 1099.

Przybilla, N., Butler, K., & Kudritzki, R. P. 2001b *A&A* **379**, 936.

Puls, J., Kudritzki, R. P., Herrero, A., et al. 1996 *A&A* **305**, 171.

Puls, J., Urbaneja, M. A., Venero, R., et al. 2005 *A&A* **435**, 669.

Repolust, T., Puls, J., Hanson, M. M., Kudritzki, R. P., & Mokiem, M. R. 2005 *A&A* **440**, 261.

Repolust, T., Puls, J., & Herrero, A. 2004 *A&A* **415**, 349.

Rix, S., Pettini, M., Leitherer, C., Bresolin, F., Kudritzki, R. P., & Steidel, C. 2004 *ApJ* **615**, 98.

Schaerer, D. 2003 *A&A* **397**, 527.

Sellmaier, F., Puls, J., Kudritzki, R. P., Gabler, A., & Gabler, R. 1993 *A&A* **273**, 533.

Shaviv, N. 2001 *ApJ* **594**, 1093.

Shaviv, N. 2005. In *The Fate of the Most Massive Stars* (eds. R. Humphreys & K. Stanek). ASP Conf. Series 332, p. 180. ASP.

Smith, N. & Owocki, S. 2006 *ApJ*, **645**, L45.

Trundle, C. & Lennon, D. J. 2005 *A&A* **434**, 677.

Trundle, C., Lennon, D. J., Puls, J., & Dufton, P. L. 2004 *A&A* **417**, 217.

Urbaneja, M. A. 2004 *Ph.D. Thesis*, University of La Laguna, Instituto Astrofisica de Canarias, Spain.

Urbaneja, M. A., Herrero, A., Bresolin, F., Kudritzki, R. P., Gieren, W., & Puls, J. 2003 *ApJ* **584**, L73.

Urbaneja, M. A., Herrero, A., Bresolin, F., Kudritzki, R. P., Gieren, W., Puls, J., Przybilla, N., Najarro, F., & Pietrzynski, G. 2005a *ApJ* **622**, 862.

Urbaneja, M. A., Herrero, A., Kudritzki, R. P., Najarro, F., Smartt, S. J., Puls, J., Lennon, D. J., & Corral, L. J. 2005b *ApJ* **635**, 311.

Urbaneja, M. A., Kudritzki, R. P., Bresolin, F., et al. 2008 *ApJ* **684**, 118.

Vacca, W. D., Garmany, C. D., & Shull, J. M. 1996 *ApJ* **460**, 914.

Venn, K. A. 1999 *ApJ* **518**, 405.

Venn, K. A., Lennon, D., Kaufer, A., McCarthy, J. K., Przybilla, N., Kudritzki, R. P., & Lemke, M. 2001 *ApJ* **547**, 765.

Venn, K. A., McCarthy, J. K., Lennon, D., Przybilla, N., Kudritzki, R. P., & Lemke, M. 2000 *ApJ* **541**, 610.

Venn, K. A., Tolstoy, E., Kaufer, A., Skillman, E. D., Clarkson, S. M., Smartt, S. J., Lennon, D. J., & Kudritzki, R. P. 2003 *ApJ* **547**, 765.

Vink, J. S., de Koter, A., & Lamers, H. J. G. L. M. 2000 *A&A* **362**, 295.

Vink, J. S., de Koter, A., & Lamers, H. J. G. L. M. 2001 *A&A* **369**, 574.

Unraveling the Galaxy to find the first stars

By JASON TUMLINSON

Yale Center for Astronomy and Astrophysics, Department of Physics, P.O. Box 208121, New Haven, CT, USA; jason.tumlinson@yale.edu

I review our knowledge of metal-free stars in the early universe—based on highly detailed new data on the most metal-poor stars in the Galaxy, and interpreted with new models of nucleosynthesis and chemical evolution. Present data supports the theoretical prediction that metal-free gas did not form stars of $M \sim 1\,M_{\odot}$, but indicates a characteristic mass $M \sim 10\,M_{\odot}$, lower than the $M \sim 100\,M_{\odot}$ suggested by *ab initio* simulations. This field is expected to grow dramatically in the next decade with future instruments and large surveys of metal-poor Galactic stars.

---

1. Introduction and motivations

Despite theoretical progress and rapid advances in the discovery of high-redshift galaxies, many questions about the first stars remain open: What was their IMF? How long did the epoch of metal-free stars last? Did they contribute much to reionization? Do their remnants include compact objects, and if so, where are they? The highly interdisciplinary "first stars" field is just beginning to confront observations in the form of reionization tests with the Gunn-Peterson effect and the *Wilkinson Microwave Anisotropy Probe* (*WMAP*), and chemical abundances from extremely metal-poor Galactic stars. These indicators have yielded some indirect constraints on the nature of the first stars. I believe that neither high-z galaxies nor the "second stars" will tell the whole story on their own. Present and future observational advances yield an abundance of data—much of it still unexploited or poorly understood—about the earliest stages of star formation in the Galaxy. Thus, the most promising theoretical approach is to connect high-z signatures of the first stars *in situ* and the Galactic "fossil record" preserved in the second stellar generation, an approach that has been called "near-field cosmology" (Freeman & Bland-Hawthorn 2002).

This review of nucleosynthesis by the first, metal-free stars will summarize recent progress in uncovering the nature of stars of this type in the universe by using their surviving chemical signatures. In the last decade, the field of the "first stars" has become a recognizably distinct sub-field of astrophysics, with its own dedicated investigators and meetings, and I am pleased to be able to review and represent it at this meeting and in the proceedings. This area is distinguished by the strong and unusual links it draws between the theory of the high-redshift universe and observational Galactic astronomy. I hope the logic and power of this strong link will emerge clearly throughout this review, and will grow stronger as our understanding of the first stars improves. A general review of the field is provided by Bromm & Larson (2004).

From the point of view of this meeting on massive stars, this topic is ironically positioned, because the empirical information that we have on the first massive stars comes solely from studying the atmospheres of low-mass stars in the Galactic halo. This indirect method presents its own problems in terms of selection biases, systematic errors, and observational details, which I will leave aside here. The interested reader is referred to the excellent review by Beers & Christlieb (2005), who describe the observational aspects in detail. For our purposes here, it is sufficient to deal only with the rich and numerous abundance data on its own terms.

We study nucleosynthesis by metal-free stars for reasons that go far beyond our innate desire to understand the properties of the first massive stars. These four reasons seem to be the most important to me:

- *To understand stellar evolution at low and zero metallicity.* Metal-free stars are intrinsically interesting to fans of stars as a natural limiting case of the physical processes that drive stellar evolution. Metal-free stars burn hotter and live shorter lives than more metal-enriched stars, and as a consequence they may rotate faster, lose more or less mass, differ in their post-main-sequence stages, and explode differently. These differences may provide leverage for tests of the physical models for these various phenomena. It is reasonable to hope that theoretical studies of metal-free stars, which differ in these features to varying degrees, can even help illuminate mysterious features of stellar evolution at higher metallicity.
- *To constrain the IMF in primordial gas.* This is one of the jackpot questions of first-stars research: What was the mass function of stars formed from metal-free gas? A lot of sophisticated theoretical work and numerical simulations have gone into developing physical models of early star-forming "minihaloes" with $M = 10^6$–$10^7\,M_\odot$ (Abel et al. 2002; Bromm et al. 2001; Yoshida et al. 2003). These models generally agree in their prediction that the restricted cooling properties of primordial gas imply a $\sim 100\,M_\odot$ mass scale for protostellar cores, and correspondingly larger final stellar masses. The hope is that we can exploit the variation of nucleosynthetic yields with initial stellar mass to extract information about the mass function from abundance data. Initial steps in this direction are promising, and the growth of abundance datasets and refinements of theoretical yields suggest that further progress is forthcoming.
- *To reconstruct the early mass assembly of galaxies.* A population of stars may carry in their abundances and kinematics a memory of the environment in which they formed. In the standard ΛCDM theory of hierarchical galaxy formation, the Galaxy is constructed from the accretion of hundreds of subunits over a wide range of redshift, most of which are completely disrupted and some of which are still identifiable as "star streams" or satellites in the Galactic halo. The hope of large-area surveys such as RAVE and SEGUE is to identify this early accretion history of the Galaxy by studying these chemical and kinematic signatures of CDM substructure. The optimistic view is that we can develop these techniques and the surveys sufficiently to unravel the Galactic mass assembly history from a time before the earliest epochs at high redshift that we can directly probe.
- *To understand the origin of important and rare elements.* Identifying the physical processes and formation sites of the chemical elements is a headline goal of astrophysics, and for many people the main reason to study metal-poor stellar abundances. At the lowest metallicities, we see the periodic table in its least mixed form, where the contributions of individual stars and enrichment events are most easily isolated. We have associated the various element groups loosely with their progenitor stars—the light elements, the α region, the Fe peak, and the neutron-capture elements, in rough order of decreasing confidence. The last group, the heavy nuclei formed in the rapid and slow neutron-capture processes, are particularly interesting because the group contains such critical elements as barium, and trivial elements like gold, but they have not yet been isolated to either a specific physical mechanism or production site. By studying these abundances in relatively undiluted form, we can hope to explain the origin of all elements on the periodic table.

2. Definitions

This review will sharpen and maintain a distinction in the literature between "nucleosynthesis" and "chemical evolution." These terms are often used interchangeably. They should not be, as they denote distinct physical processes that are, nevertheless, closely related. Nucleosynthesis refers to the creation of the elements by nuclear reactions in stars. At the points where nuclear reactions cease, so does the formal process of nucleosynthesis. Nucleosynthesis occurs in hot, dense astrophysical environments such as supernova explosions and AGB star atmospheres, and generally occurs over the short characteristic timescales of the relevant nuclear reactions.

Chemical evolution is the set of astrophysical processes that distribute these elements throughout space, and as such is a catchall term for processes generally known by other names. These include supernova blast waves, Galactic superwinds, interstellar mixing, dust formation and destruction, and molecular cloud formation. These processes occur on a wide range of densities and timescales, so modeling them is physically complex and mathematically complicated.

Of course, the details of stellar evolution—rotation, mass loss, post-MS evolution, supernovae, etc., and the physical processes that drive chemical evolution—may depend sensitively on chemical abundances, and nucleosynthesis by a single star may depend on its initial chemical abundances, so these two basic concepts are closely coupled. This is sometimes frustrating, because it means that we cannot disentangle their effects. At the same time, this coupling means that we can use information about one to understand the other.

Naturally, we believe that as stars increase in metallicity, they represent the aggregation of many prior generations. This means that abundances in metal-poor stars are closer to the raw products of nucleosynthesis. Two interesting and ironic implications are that we may never know the detailed nucleosynthetic yields of a single massive star with half-solar metallicity, and that the stars about which we can obtain the most detailed information are the first, metal-free stars in the universe. It is their products that we see in least diluted form.

3. Mysteries in the data

Whatever the metal-poor stellar abundances are saying about the first few generations of chemical evolution, they are now saying a lot of it. Observational surveys to discover metal-poor stars in the Galaxy and follow-up observations to obtain high-quality abundances have been underway for more than three decades, and in recent years have obtained data at a much faster rate than understanding. As of this writing, theory is still lagging behind.

Rather than review all aspects of the abundance data, this contribution will focus on three surprising aspects of the data that hint at the nature of the first stars. These are (1) the trends and small scatter in Fe-peak abundances at [Fe/H] -4 to -2, (2) the high incidence and large scatter in the r-process abundances over the same range of metallicity, and (3) the surprising and unusual abundance patterns seen in the only two stars known with [Fe/H] < -5. These are described here, in turn with references to more complete analyses and interpretation.

Trends in mainstream EMP stars. Even at [Fe/H] ~ -3, abundances in the mainstream (i.e., unremarkable for no CNO or s-process enhancement, binarity, etc.) depart from their values at higher metallicity (McWilliam et al. 1995; Cayrel et al. 2004). The changes are most evident in the Fe-peak elements Cr, Mn (just below Fe), and Co and Zn (just above

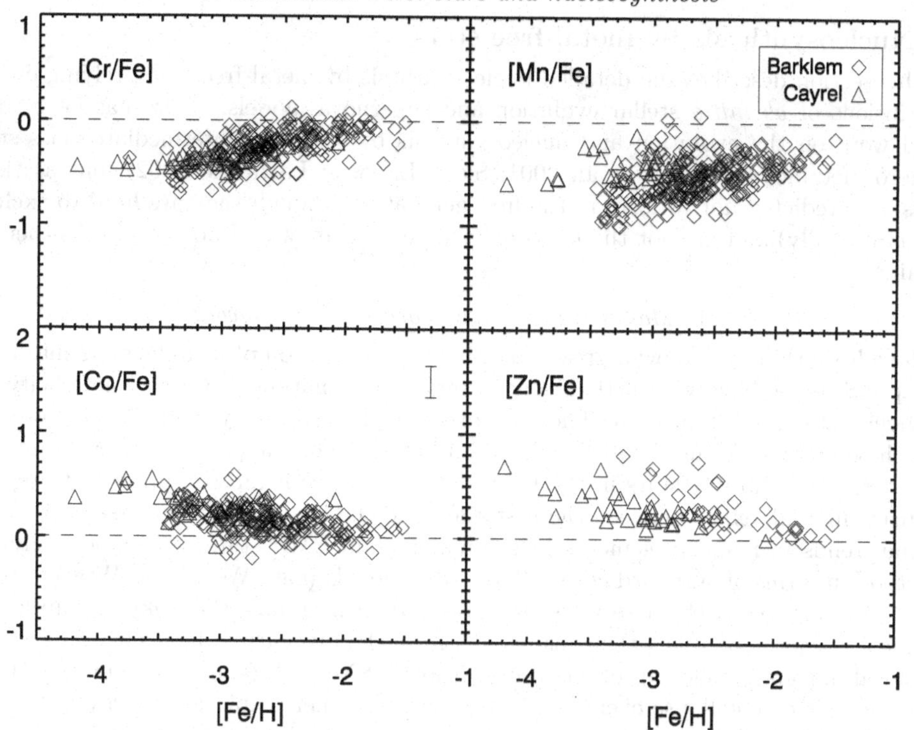

FIGURE 1. Trends in Fe-peak abundance data from Cayrel et al. (2004) and HERES (Barklem et al. 2005).

Fe). What is the supernova progenitor and what feature unique to zero or low metallicity affects the change?

Trends in the neutron-capture elements. We believe that below [Fe/H] ~ -2.5, all neutron-capture abundances are provided purely by the r-process (except perhaps a contribution to Sr-Y-Zr from a weak s-process associated with massive stars—see Travaglio et al. 2004). The r-process elements are thought to be produced in the neutrino-driven winds associated with core-collapse events (Woosley et al. 1994), but the physical site and/or mass range of the process is still unknown. Enormous scatter in the data, increasing in magnitude as [Fe/H] declines, strongly suggests an almost complete decoupling between the r-process and iron (Barklem et al. 2005). This trend, which persists to the lowest metallicities we can probe, is very helpful for constraining the first stars (see Section 6).

"Hyper-metal-poor" (HMP) stars. The two most Fe-poor objects in the universe were discovered by the Hamburg/ESO survey and have now been exhaustively studied at high resolution with 8-m telescopes. HE 0107–5240 has [Fe/H] $= -5.3$ (Christlieb et al. 2002) and HE 1327–2326 has [Fe/H] $= -5.6$ (Frebel et al. 2005; Aoki et al. 2006). Even though these stars have 1/300,000 the solar iron abundance, and we call them hyper-metal-poor stars in the nomenclature of Beers & Christlieb (2005), these objects are not at all metal poor! In fact, they both show extreme enhancements of the light elements C, N, O, with [C, N, O/Fe] = 2–4. These extraordinary objects not only show extremely non-solar relative abundance patterns, they also are very much unlike the "mainstream" EMP stars (see Figure 12 of Aoki et al. 2006). If these stars carry primordial abundances, and therefore preserve nucleosynthesis by the first generation, the first stars were unusual indeed.

4. Nucleosynthesis by metal-free stars

This section describes the detailed nucleosynthesis by metal-free stars, relying on the conclusions of *ab initio* stellar evolution and supernova models. There has been some good work on the evolution and nucleosynthesis by low- and intermediate-mass stars at zero metallicity (Marigo et al. 2001; Siess, Livio, & Lattanzio 2002), but as these stars are predicted not to form in the first generation (though they are hard to exclude observationally) and are not the focus of this meeting anyway, I do not discuss them in detail.

4.1. *Massive stars—supernovae or hypernovae?*

Although we think we know a great deal about the evolution of massive stars into core-collapse SNe, we have very little if any empirical information about the metallicity dependence of this phenomenon. The core mass, explosion energy, and yields may vary, and these factors will need to be understood before we can hope to uncover the unique signatures of metal-free stars in the EMP data. Yet there is already progress to report. A surprising link exists between the first stars and the low-redshift universe.

The trends in Fe-peak elements in the mainstream EMP stars cannot be easily understood in terms of standard core-collapse SNe models (e.g., Woosley & Weaver 1995). In an attempt to match these unusual ratios and their trends, the Tokyo group led by Ken'ichi Nomoto has taken their inspiration from the diversity in light curves of SNe in the local universe. Followup observations of low-z SNe and gamma-ray bursts (GRBs) indicate that core-collapse events can vary widely in their explosion energy and nickel production, from $E_{51} = E/(10^{51}$ erg$) = 0.5$–50 and $M(^{56}\mathrm{Ni}) = 0.002$–$0.5\,M_\odot$ (as estimated from optical light curves; Zampieri et al. 2003; Mazzali et al. 2003; Nomoto et al. 2003). Following this guidance, Nomoto et al. have asked: If these different types of supernovae also exist at low and/or zero metallicity, what would their nucleosynthetic signature be, and would it resemble any of the strange trends and outliers seen in the EMP abundance data? The answers to these questions hint at a powerful connection between low-z explosions and primordial stars.

First, Umeda & Nomoto (2005) and Tominaga et al. (2007) have argued that energetic hypernovae (HNe) with $E_{51} \gtrsim 10$ can best explain the observed trends in Fe-peak abundances below [Fe/H] = −3. The physical cause of the increase in Co and Zn (above Fe) and decrease in Cr and Mn (below Fe) is that in more energetic SNe, the complete Si-burning region is larger, and the elements above Fe are enhanced at the expense of the elements below. To match the observed trends, these authors must also include "SN-triggered" star formation in the second generation. In this simple picture, the mass of pristine gas into which the ejecta of the first stars is mixed is directly proportional to the explosion energy E_{51}, such that more energetic explosions yield lower [Fe/H] in the "second stars." The variation in [Cr, Mn, Co, Zn/Fe] caused by varying E_{51} is spread out to low [Fe/H], as it is in the data. Thus, the puzzling trends in the data have a simple physical explanation in terms of physical variations in supernovae also seen in the nearby universe.

The diversity of low-z SNe also provides a possible explanation for the surprising abundance patterns of the two HMP stars, again proposed by the Tokyo group. Inspired by the apparent evidence of $E_{51} < 1$ explosions from SN1997D and SN1999bw, they have suggested that the Fe-poor, CNO-enhanced patterns of HE 0107−5240 and HE 1327−2326 could be produced in a "faint SN" which has enough energy to eject virtually all the light-element-rich outer layers, but only a small portion of the Fe-rich inner core. Iwamoto et al. (2005) closely matched the observed patterns with mixing-fallback models of $E_{51} = 0.71$ and 0.74 for the two stars, respectively. The strong differences in the abundances of Na,

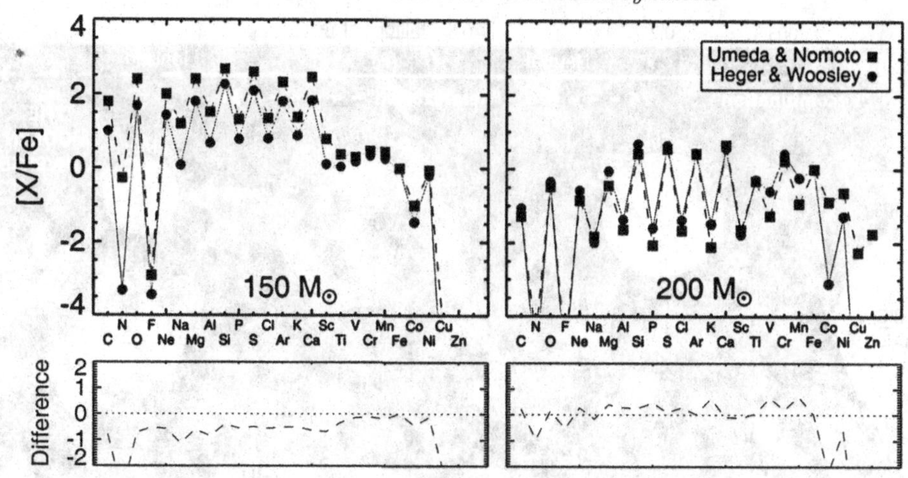

FIGURE 2. Comparison of HW02 and UN05 PISN yields.

Mg, and Al between the two stars is explained by a slight difference in the placement of the mass cut separating ejected from fallback material.

It has also been suggested that these peculiar stars acquired their light-element enhancements from the metal-enriched winds of a nearby massive-star companion, possibly a binary companion, that lost significant mass due to fast rotation (Meynet, Ekstrom, & Maeder 2006; Chiappini et al. 2006; Karlsson 2006). This critical issue of rotation is a key uncertainty in nucleosynthetic constraints on the first stars that will be discussed further below.

It is encouraging that the extraordinary trends in the lowest metallicity stars can be explained by such simple ideas. It may turn out that nucleosynthetic yields from massive stars depend more sensitively on E_{51} or rotation than on stellar mass, which would inhibit our efforts to constrain the first-stars IMF (see following sections). However, large-ranged explosion energy, mixing and fallback parameters, and rotation have not been fully explored, and the small difference in E_{51} that causes the large differences in the HMP abundance patterns suggests a fine-tuning problem. A full synthesis of the observed patterns, including stochastic effects in interstellar mixing, is needed to judge the true fraction of massive stars that explode as HNe in the first generation. Finally, I note that the apparent connection of HNe and GRBs at low redshift suggests a strong link between GRBs and the first stars that needs further investigation.

4.2. *Very massive stars?*

There are strong theoretical reasons for the first, metal-free stars to be skewed to higher masses than stars today. This is an old idea that has recently received close attention from investigators using both analytic calculations and numerical simulations to follow the formation of metal-free stars in pre-Galactic dark-matter haloes. A short summary of star formation from primordial gas is warranted to introduce the basic ideas that will later be tested with the techniques of chemical evolution.

Numerical simulations of primordial star-forming "minihaloes" have now settled on a robust prediction that the mass scale of the first stars is skewed toward high masses, with $M \sim 100\,M_{\odot}$ (Abel, Bryan, & Norman 2002; Bromm, Coppi, & Larson 2001). Recent extensions to this pioneering work have determined that star-to-star variations in environment or collapse redshift force variations in the protostellar accretion rate, which is expected to cause the final stellar mass to scatter perhaps a decade in either direction

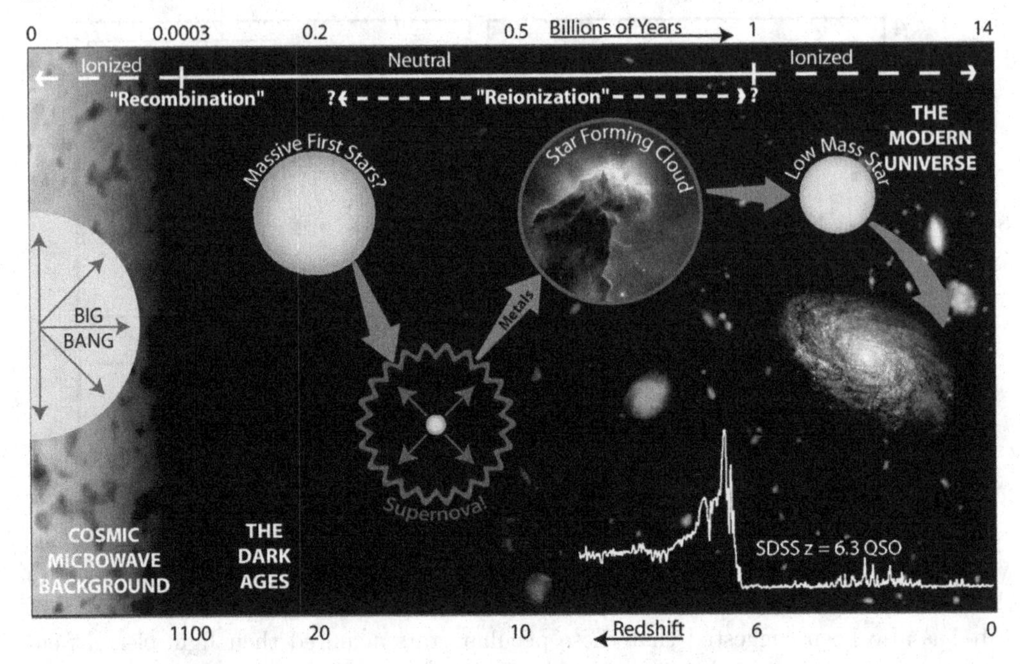

FIGURE 3. A pretty picture of the means we have of constraining the first stars with different types of data on chemical evolution and reionization.

from this characteristic mass scale (Yoshida et al. 2006; O'Shea & Norman 2007). This work makes two strong predictions—that we should find no low-mass metal-free stars in the Galactic halo (this is borne out—see below), and that the nucleosynthetic signature of EMP stars should reflect the expected yields of very massive, metal-free stars.

Both Heger & Woosley (2002; hereafter HW02) and Umeda & Nomoto (2005; hereafter UN05) have followed the evolution of very massive ($M \gg 100\,M_{\odot}$) stars of zero metallicity to the end of their evolution. Between 140 and 260 $M_{\odot}$, metal-free stars are believed to be disrupted completely by the pair-creation instability just after the end of core He burning. The contribution to this volume from Evan Scannapieco treats these objects in much greater detail, including models of their light curves, expected rate of incidence with redshift, and nucleosynthetic yields. For our purposes here, we are concerned mainly with their nucleosynthetic yields, which are compared in Figure 2 for 150 and 200 $M_{\odot}$ models from HW02 and UN05. The agreement between the two groups is encouragingly good, perhaps because calculating stellar evolution and nucleosynthesis is fairly straightforward for non-rotating massive stars. This comparison means we can compare the yields to the data in reasonable confidence that the models have converged within the limited range of their simplifying assumptions. However, the main caution is that fast rotation and/or mass loss by these VMS could substantially affect the calculated yields, but calculations for rotating stars are significantly more expensive and have not yet been carried out (A. Heger, private communication).

5. Chemical evolution in the hierarchical context

To interpret all these abundance data in the proper cosmological context of galaxy assembly, it is necessary to develop a formalism for connecting Galactic chemical evolution to the theory of galaxy formation, so that different observational data can be integrated

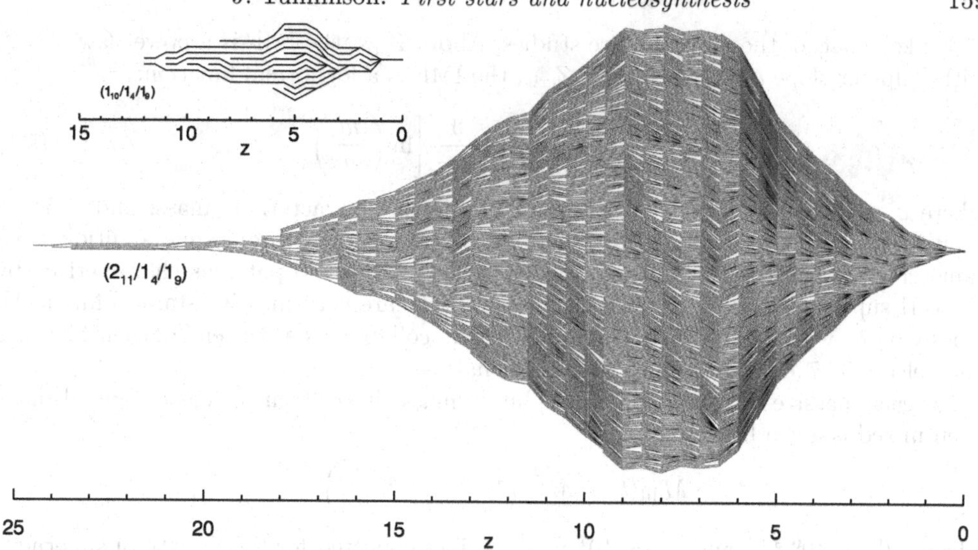

FIGURE 4. Examples of large and small halo merger trees. The single halo at $z = 25$ in the main panel is the first-born progenitor of the Galaxy's most massive predecessor.

together, as schematically illustrated in Figure 3. The goal is to derive empirical constraints on the first-stars IMF, as a complement to purely theoretical approaches that are being pursued by many groups (Bromm et al. 2001; Abel et al. 2002; Omukai et al. 2005; Tan & McKee 2004). To achieve these goals, Tumlinson (2006) developed a new stochastic, hierarchical treatment of chemical evolution that treats the problem in its proper structure-formation context. The structure-formation component first decomposes the present-day Galactic dark-matter halo ($M_h = 10^{12}\, M_\odot$) into its constituent sub-haloes, working backward in time using the extended Press-Schechter framework and the "N-fold merger with accretion" method of Somerville & Kollatt (1999). Full details on the calculation of the merger trees is presented in Tumlinson (2006).

The chemical evolution driver works forward in time, starting with the earliest "child" halo ($z = 28$ in Figure 4) and following the chemical enrichment history of all individual haloes; the $z = 0$ Galactic halo is the last to be merged. Between mergers, an isolated halo evolves according to equations describing its stellar populations, gas mass, and metallicity. Halo mergers are pairwise sums over all the child haloes, and so simply inherit all their stars and gas. Gas residing in bound haloes forms stars at a fixed mass efficiency, such that the total mass in stars formed in time interval Δt is

$$M_* = 2.0 \times 10^{-10} \times M_{\rm gas} \times \Delta t \quad ,$$

where the coefficient is set to match the total gas mass and stellar mass in the Milky Way for the final halo of a $10^{12}\, M_\odot$ tree. To match the basic peaked shape of the halo metallicity distribution function, a single timescale, $\tau_{\rm disk} = 0.8$–3.0×10^8 yr, controls the time after which a virialized parcel of gas in the dark-matter halo has settled into the disk.

Above a critical metallicity $Z \sim 10^{-5.5}$–$10^{-3.5}\, Z_\odot$, protostellar clouds are able to cool and fragment more efficiently, leading to a "normal" IMF (Schneider et al. 2002; Bromm & Loeb 2003; Santoro & Shull 2006). For simplicity, this model switches the IMF at a single [Fe/H] as a parameter of the model, $Z_{\rm crit} = 10^{-4}\, Z_\odot$.

Because it controls both the mass budget of metals released into the ISM and the number and mass distribution of low-mass stars that survive to $z = 0$, the IMF is the most important ingredient for modeling early chemical enrichment and, of course, one of the key goals of these abundance studies. Above $Z_{\rm crit}$, the IMF is a power law, $M^{-\alpha}$, with Salpeter slope $\alpha = 2.35$. Below $Z_{\rm crit}$, the IMF is a lognormal function:

$$\ln\left(\frac{dN}{d\ln m}\right) = A - \frac{1}{2\sigma^2}\left[\ln\left(\frac{m}{m_c}\right)\right]^2 \tag{5.1}$$

where σ is the width of the distribution, m_c is the characteristic mass, and A is an arbitrary normalization. This IMF offers flexible behavior with only one additional parameter. Stars with 8–140 $M_\odot$ are "massive stars" with yield patterns characterized by Type II supernovae, while stars with $M > 140\,M_\odot$ are very massive stars (VMSs). All massive stars with $M = 8$–$40\,M_\odot$ experience core-collapse SN at the end of their lifetimes and release $0.07\,M_\odot$ of iron into the parent halo.

For each massive star or VMS, the "dilution mass" into which the ejected metals have been mixed is given by

$$M_{\rm dil}(t) = M^0_{\rm dil}\left(1 - e^{(t-t_{\rm SN})/t^0_{\rm dil}}\right) \quad ,$$

where $M^0_{\rm dil} = 10^6\,M_\odot$ and $t^0_{\rm dil} = 10^7$ yr are values expected for the growth of supernova remnants in the Galactic ISM during the radiative phase and $t - t_{\rm SN}$ measures the time since the SN. New star formation is assigned a metallicity by stochastically sampling the available metal-enriched parcels, whose mass fraction relative to the total gas reservoir are truncated at unity under the assumption of closed boxes.

6. Observational constraints on the primordial IMF

From the MDF and Pop III Stars: The observed MDF shows a rise from [Fe/H] ≈ -3.8 to a peak at [Fe/H] ≈ -1.8, and a decline at higher metallicity, where the numbers are unreliable owing to disk/halo selection effects. Good fits to the peak of the halo MDF are obtained for $\tau_{\rm disk} = 0.8$–3×10^8 yr for $T_{\rm vir} = 10^3$ and 10^4 K. The best-fit MDF is shown in Panel A of Figure 7. This basic agreement confirms that the overall model is a fair description of early Galactic chemical evolution.

The absence of true Population III stars from the Galactic halo is a key constraint on the IMF of the first stars. The relevant quantity is F_0, the number of truly metal-free stars known in the Galactic halo divided by the total number at [Fe/H] $\leqslant -2.5$ (for the purposes of definition). The 373 stars reported by Beers, Preston, & Schechtman (1992), and the 146 Hamburg-ESO survey stars reported by Barklem et al. (2005) yield $F_0 \leqslant 0.0019$. If no $Z_{\rm crit}$ is employed, the IMF is the same at all metallicities and $F_0 \gtrsim 0.1$ for IMFs ranging over (1.0, 1.0) to (100, 2.5), and $F_0 = 0.25$ for a Miller-Scalo IMF (0.1, 1.57), even in the limit of instantaneous mixing. These values are clearly excluded by the data, and so there *must* be an evolution in the IMF that inhibits the formation of low-mass stars below some finite metallicity, in accord with the general prediction from *ab initio* simulations of first-star formation. Variations in F_0 with IMF overwhelm uncertainty in the other parameters (M_h, $T^{\rm min}_{\rm vir}$, etc.) and make F_0 a robust constraint on the IMF that should only improve with growing samples of metal-poor stars. Results for F_0 are displayed in panel A of Figure 5 over a large range of IMF parameters. F_0 provides a strong requirement that the initial IMF is top-heavy, with $m_c \gtrsim 6\,M_\odot$ and mean mass of $\langle M \rangle \gtrsim 8\,M_\odot$. The Miller-Scalo Galactic IMF has $m_c = 0.1$, $\sigma = 1.57$, and $\langle M \rangle \approx 0.64\,M_\odot$, roughly ten times smaller than the mean mass at zero metallicity.

From reionization: The reionization of the IGM at high redshift is a key indicator of cosmic star-formation activity at $z \gtrsim 6$. Our primary guide here is the unexpectedly high electron-scattering optical depth to the cosmic microwave background found by *WMAP*,

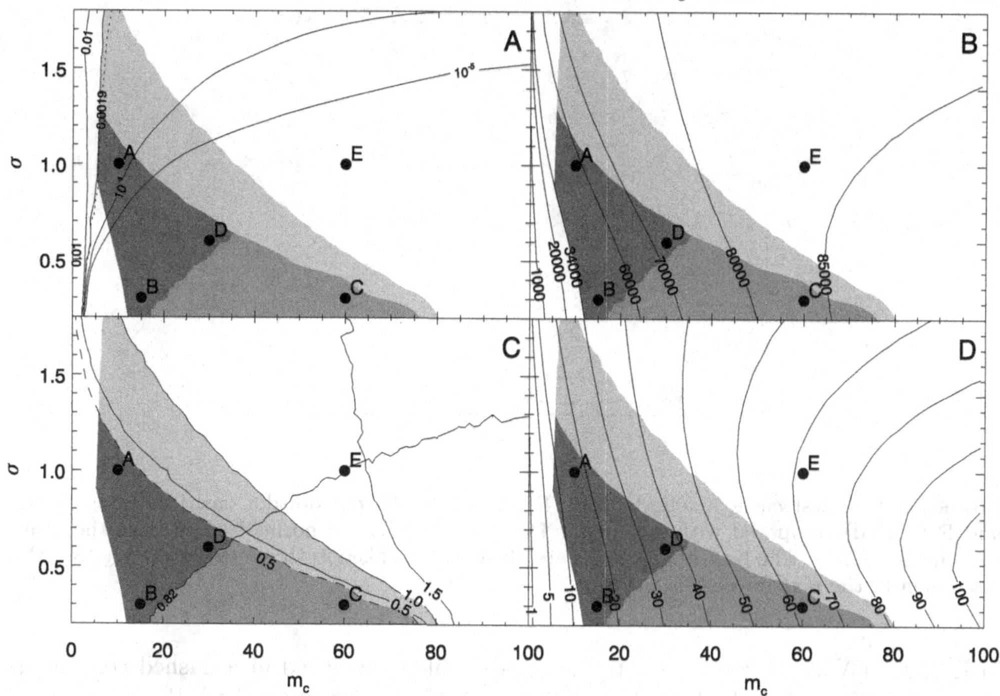

FIGURE 5. Observational constraints on the parameters for a primordial IMF. Each panel shows contours for an individual constraint, while the shaded regions that meet all constraints are the same in all panels. See text for discussion.

$\tau_{es} = 0.17^{+0.08}_{-0.07}$ (Spergel et al. 2003), which suggests that reionization began quite early and required an unusually high efficiency of ionizing photon production. The relevant quantity for IMF constraint is the time-integrated ionizing photons per baryon, γ_0. For a Salpeter IMF from 0.5–140 $M_\odot$, $\gamma_0 = 17,000$, insufficient to reproduce the $\tau_{es} \geqslant 0.10$. For this study we require that $\gamma_0 \geqslant 34,000$ to achieve the minimum τ_{es} consistent with the *WMAP* data at 1σ confidence, $\tau_{es} \geqslant 0.10$. Panel B of Figure 5 shows that $\gamma_0 \geqslant 34,000$ requires an IMF skewed to high mass. This model enables a full treatment of the detailed interrelationships between chemical evolution and reionization by construction; such a new study is underway now, and results in light of the third year *WMAP* data release are pending.

From detailed nucleosynthesis: The iron-peak elements (Cr–Zn) are produced in explosive events from massive stars and show consistent abundance patterns across the different studies (Cayrel et al. 2004; Cohen et al. 2004). PISNe yields match poorly with the stellar abundance data (Tumlinson, Venkatesan, & Shull 2004, hereafter TVS04); the most discrepant element is zinc, which is produced in very small quantities by α-rich freeze-out in PISNe. Thus high Zn yields must be produced from 8–40 $M_\odot$ to yield the observed $\langle[\mathrm{Zn/Fe}]\rangle = 0.3$ (Cayrel et al. 2004). The solid contours in panel C of Figure 5 mark the [Zn/Fe] excess needed, *uniformly* across the range 8–40 $M_\odot$, to compensate for the strong [Zn/Fe] deficit of PISNe and still give a mean $[\mathrm{Zn/Fe}] \approx 0.3$ for stars with $[\mathrm{Fe/H}] \leqslant -2$. IMFs in the upper right corner require very high Zn excess, $[\mathrm{Zn/Fe}] \gtrsim 1.0$. This [Zn/Fe] limit is conservative because (1) this Zn excess is much higher than that produced in typical published models of core-collapse supernova nucleosynthesis (Umeda & Nomoto 2005), and (2) $[\mathrm{Zn/Fe}] \leqslant 1.0$ for 8–40 $M_\odot$ generously leaves 90% of the Fe produced by PISNe, while 50% is still a poor match to the observed odd-even abundances

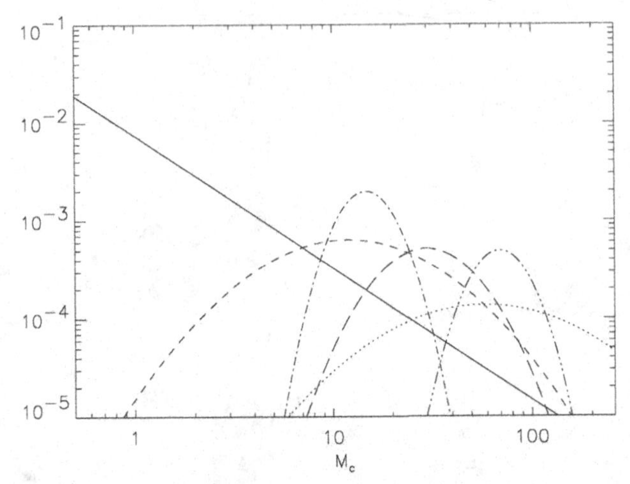

FIGURE 6. IMF test cases A (short dash), B (dot-dash), C (dot-dot-dot-dash), D (long dash), and E (dotted) compared with a Salpeter IMF, $\alpha = -2.35$, all normalized to have the same total mass $10^6\,M_\odot$. The best-fitting IMFs are all sharply peaked in the range 12–70 $M_\odot$. See the discussion in the text for details.

in EMPs (HW02; TVS04). This tighter constraint is displayed in a dashed contour in panel C of Figure 5 and adopted as a more realistic constraint on the IMF.

The r-process elements ($A \gtrsim 60$) are produced by rapid neutron captures in hot, dense, neutron-rich explosive events. The exact physical site is still uncertain, but the proposed mechanisms are all associated with massive stars in the range $M = 8$–$40\,M_\odot$ and may be restricted to 8–12 $M_\odot$ (Truran et al. 2002; Wanajo & Ishimaru 2006). Population II stars show r-process elements down to [Fe/H] ~ -3.5 (Barklem et al. 2005) with a mean [r/Fe] similar to the solar value at all [Fe/H] and scatter steadily increasing to 2 dex at [Fe/H] ~ -3. The presence of the r-process elements in the EMPs do not allow the early budget of iron to be provided exclusively by PISNe, which have no r-process (HW02; TVS04). Of the 146 HERES stars with [Fe/H] $\leqslant -2.5$, 119 (82%) show detected Ba at [Ba/Fe] > -2 (Barklem et al. 2005). This fraction is reproduced only if all SNe with 8–12 $M_\odot$ are assumed to produce Ba in the highest observed pure r-process ratio relative to Fe, [Ba/Fe] = 1.5 and the IMF lies to the right of the "0.82" contour in panel C of Figure 5. The parameter space excluded by this test is reproduced in the medium gray shade in the other panels. This r-process constraint is somewhat more speculative because it relies on an assumption that the r-process is uniquely associated with a narrow range of the IMF, between 8 and 12 $M_\odot$, but it is potentially more powerful for exactly that same reason.

The various constraints together suggest an IMF for the first stars that is restricted to the dark shaded area in Figure 5, with $m_c = 6$–$35\,M_\odot$, $\sigma \approx 0.3$–1.2 and mean mass $\langle M \rangle = 8$–$42\,M_\odot$. Five IMF test cases allow for further examination of their detailed properties and consequences: IMFs A (10, 1.0), B (15, 0.3), C (60, 0.3), D (30, 0.5), and E (60, 1.0). These cases are shown in Figure 6 compared against the power-law Salpeter IMF. IMFs A and B provide the overall best fit to the MDF. IMF B has a strong deficiency of VMP and EMP stars relative to the Ryan & Norris (1991) MDF, while IMFs C and E have an excess caused by the strong deficiency of stars that produce a nucleosynthetic signature (8–40 $M_\odot$) relative to those that do not (40–140 $M_\odot$). IMFs C and E are potentially excluded with the [Zn/Fe] and [r/Fe] results presented above, and also produce poor fits to the overall MDF. These MDF results are not decisive for

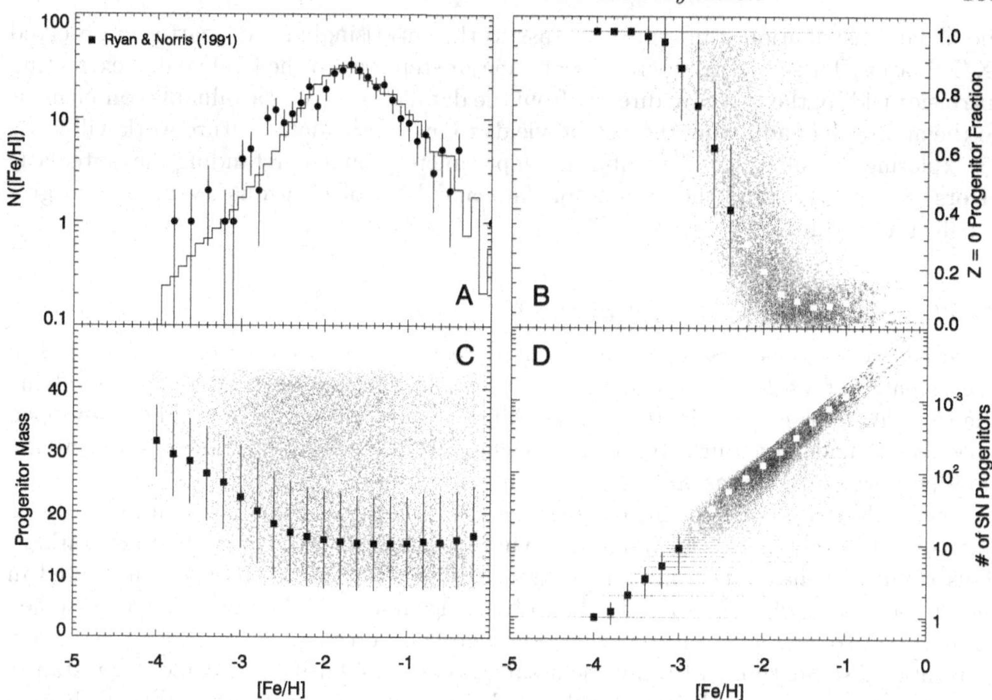

FIGURE 7. The MDF and detailed chemical evolution histories for model metal-poor stars. In panels B–D, squares and error bars mark the mean and 1σ dispersion in 0.2 dex bins. See text for discussion.

determining the IMF because of the poor statistics in the data below [Fe/H] ~ -2.5, which should be improved by future large surveys of Galactic halo metal-poor stars.

Different redshift-dependent star-formation histories for metal-free stars can also help discriminate the IMF test cases. An IMF that is more top-heavy—which places more mass into the 40–140 $M_\odot$ range of stellar mass that does not produce a nucleosynthetic signature—will take longer to produce a fixed overall enrichment and so should form metal-free stars for a longer time in each halo. This effect could potentially be detected at high redshift.

A given merger tree with chemical evolution provides detailed chemical histories of individual metal-poor stars in their proper context for the first time. This information is critical to properly interpreting the relative element abundances in metal-poor stars. Figure 7 shows detailed results for IMF A in a fiducial Milky Way tree. A number of important results appear in this figure. First, the model preserves a great deal of useful information about the chemical history of the model Galaxy, most importantly, the mass and metallicity distribution of SNe precursors for metal-poor stars (Panels B and C). Second, below [Fe/H] ~ -3, EMP stars have 1–10 SNe precursors, all of which are metal-free. In fact, the average [Fe/H] ~ -4 (UMP) star had exactly one metal-free precursor. This leaves only precursor mass as a variable, such that the variation of [Zn/Fe] with [Fe/H] below [Fe/H] $= -3$ (Cayrel et al. 2004), for example, may be explained simply as the more massive, short-lived precursors, $M \simeq$ 30–40 $M_\odot$, coming in first at [Fe/H] $\lesssim -4$, followed by longer-lived precursors of 8–10 $M_\odot$ only at [Fe/H] ~ -3. This is an alternative explanation to the triggered star-formation picture of Nomoto et al. (2003), although HNe would still be needed to give the increase in [Zn/Fe] with progenitor mass. Finally, the predicted number of precursors per metal-poor star, and their mass and metallicity distribution, can be used to assess the surprisingly small scatter in observed [X/Fe] ratios. These key results represent a major step toward the final goal of extracting intrinsic yields of the first SNe directly from the data, and serve as a valuable complement to the approach of adjusting theoretical yields to match the data. Future work will focus on exploring the full range of possibilities in parameters, on understanding the systematic errors, and on assessing the volume of data and level of data quality needed to give satisfactory results.

7. Unsolved problems and outlook

Recent work on the first stars has made a great deal of encouraging progress, and to the extent that we know anything empirical about the first stars, it is from studying their surviving signatures in the low-redshift universe. However, there are many open questions that deserve our attention. I conclude with the bright outlook offered by the rapidly growing databases on Galactic metal-poor stars.

First, it has yet to be conclusively shown that the metal-poor halo stars we take as immediate descendants of the first stars do in fact carry the ejecta of the first generations. This connection has been routinely assumed by generations of observers interested in metal-poor stars, who often justify their observing time on the grounds of this assumed connection. To me, it is an open question whether we can hope to prove this in any conventional sense, though I think the assumption can be falsified. It could, for instance, be shown that in a detailed and well-tested theory of Galaxy formation that there is essentially zero chance that the immediate descendants of the first stars end up within a few kpc of the Sun, where our most sensitive spectroscopic surveys are concentrated. Some limited attempts have been made to evaluate where in the galaxy the "second stars" end up, and to assess statistically the overlap between the oldest and most metal-poor Galactic populations, which are not necessarily identical populations (Diemand, Madau, & Moore 2005; Scannapieco in this volume). The early indications from new numerical experiments suggest that, although the immediate descendants of the earliest pre-Galactic halos are concentrated near the Galactic center and in the bulge, there are enough at the solar radius for chemical constraints from metal-poor stars to be relevant. For now, the identification of metal-poor stars with the "second stars" will persist until it is proven otherwise, and as such, this is still a largely unsolved problem for structure-formation theorists to solve.

Second, we are still unsure how much influence stellar rotation and mass loss have in determining the nucleosynthetic signature of metal-free stars. If these factors dominate over stellar mass as the decisive influence, then efforts to empirically derive the IMF are hindered. Intuition gives different answers on even so simple a question as: Are metal-free stars rotating faster or slower, on average, than their mass counterparts at higher metallicity? The consequences of fast rotation are everywhere, from the nucleosynthetic yields to the SNe and GRB rates from metal-free stars. This is the major unsolved problem for theorists of stellar evolution.

Finally, the stochastic treatment of chemical evolution presented here uses an abstract and highly parameterized prescription for dilution and mixing of supernova ejecta. A proper physical treatment of these processes is extremely complicated, and we have no complete theory at the moment. For the near term, our best hope lies in developing a realistic description of mixing that is statistically faithful to the distribution of metallicities, even if it is not rigorously physical. Mixing of metals in the first few generations is the core problem facing both theoretical formulations of the second stars and the proper interpretation of their living descendants.

The samples of metal-poor Galactic halo stars are expected to grow dramatically in the next decade owing to systematic surveys dedicated to studying large stellar populations in the outer Galaxy. The SEGUE extension to the Sloan Digital Sky Survey will discover up to 20,000 stars with [Fe/H] < -2.0 (and perhaps a few at [Fe/H] $\lesssim -5$), while the WFMOS project aims to obtain high-resolution spectra of 1 million stars with a 1000-fiber optical spectrograph on the Subaru Telescope. Next-generation large telescopes (GSMTs) can push these studies into nearby galaxies and probe the different chemical evolution histories of different types and masses. These surveys hope to achieve "chemical tagging," or the identification of distinct groups of kinematically and chemically related stars arising from a common formation history, an impressive achievement in "near-field cosmology" (Freeman & Bland-Hawthorn 2002). The results presented here show that these studies also have the potential to powerfully constrain the masses, feedback, and epoch of the first stars. They may also provide enough data for individual SNe yields to be extracted empirically. It may turn out that the history of cosmic star formation is written in the Galaxy.

REFERENCES

Abel, T., Bryan, G. L., & Norman, M. L. 2002 *Science* **295**, 93.
Aoki, W., Frebel, A., et al. 2006 *ApJ* **639**, 897.
Barklem, P., et al. 2005 *A&A* **439**, 129.
Beers, T. C. & Christlieb, N. 2005 *ARA&A* **43**, 531.
Beers, T. C., Preston, G. W., & Schechtman, S. A. 1992 *AJ* **103**, 1987.
Bromm, V., Coppi, P., & Larson, R. B. 2001 *MNRAS* **328**, 969.
Bromm, V. & Larson, R. B. 2004 *ARA&A* **42**, 79.
Bromm, V. & Loeb, A. 2003 *Nature* **425**, 812.
Cayrel, R., et al. 2004 *A&A* **416**, 1117.
Chiappini, C., et al. 2006 *A&A* **449**, 27.
Christlieb, N., et al. 2002 *Nature* **419**, 904.
Cohen, J. G., et al. 2004 *ApJ* **612**, 1107.
Diemand, J., Madau, P., & Moore, B. 2005 *MNRAS* **364**, 367.
Frebel, A., et al. 2005 *Nature* **434**, 871.
Freeman, K. C. & Bland-Hawthorn, J. 2002 *ARA&A* **40**, 487.
Heger, A. & Woosley, S. 2002 *ApJ* **567**, 532 (HW02).
Iwamoto, N., Umeda, H., Tominaga, N., Nomoto, K., & Maeda, K. 2005 *Science* **309**, 451.
Karlsson, T. 2006 *ApJ* **641**, L41.
Marigo, P., Girardi, L., Chiosi, C., & Wood, P. R. 2001 *A&A* **371**, 152.
Mazzali, P. A., et al. 2003 *ApJ* **599**, 95.
McWilliam, A., Preston, G. W., Sneden, C., & Searle, L. 1995 *AJ* **109**, 2757.
Meynet, G., Ekstrom, S., & Maeder, A. 2006 *A&A* **447**, 623.
Nomoto, K., et al. 2003 *Nuclear Physics A* **718**, 277.
Omukai, K., Tsuribe, T., Schneider, R., & Ferrara, A. 2005 *ApJ* **626**, 627.
O'Shea, B. W. & Norman, M. L. 2007 *ApJ* **654**, 66.
Ryan, S. G. & Norris, J. E. 1991 *AJ* **101**, 1865.
Santoro, F. & Shull, J. M. 2006 *ApJ* **643**, 26.
Schneider, R., Ferrara, A., Natarajan, P., & Omukai, K. 2002 *ApJ* **571**, 30.
Siess, L., Livio, M., & Lattanzio, J. 2002 *ApJ* **570**, 329.
Somerville, R. S. & Kolatt, T. S. 1999 *MNRAS* **305**, 1.
Spergel, D. N., et al. 2003 *ApJS* **148**, 175.
Tan, J. & McKee, C. M. 2004 *ApJ* **603**, 683.
Tominaga, N., Umeda, H., & Nomoto, K. 2007 *ApJ* **660**, 516.
Travaglio, C., et al. 2004 *ApJ* **601**, 864.

Truran, J. W., Cowan, J. J., Pilachowski, C. A., & Sneden, C. 2002 *PASP* **114**, 1293.
Tumlinson, J. 2006 *ApJ* **641**, 1.
Tumlinson, J., Venkatesan, A., & Shull, J. M. 2004 *ApJ* **612**, 602 (TVS04).
Umeda, H. & Nomoto, K. 2005 *ApJ* **619**, 427 (UN05).
Wanajo, S. & Ishimaru, Y. 2006 *Nuclear Physics A* **777**, 676.
Woosley, S. E., et al. 1994 *ApJ* **433**, 229.
Woosley, S. E. & Weaver, T. A. 1995 *ApJS* **101**, 181.
Yoshida, N., Abel, T., Hernquist, L., & Sugiyama, N. 2003 *ApJ* **592**, 645.
Yoshida, N., Omukai, K., Hernquist, L., & Abel, T. 2006 *ApJ* **652**, 6.
Zampieri, L., et al. 2003 *MNRAS* **338**, 711.

Optically observable zero-age main-sequence O stars

By NOLAN R. WALBORN

Space Telescope Science Institute, 3700 San Martin Drive, Baltimore, Maryland 21218, USA

A list of 50 optically observable O stars that are likely on or very near the ZAMS is presented. They have been selected on the basis of five distinct criteria, although some of them exhibit more than one. Three of the criteria are spectroscopic (He II λ4686 absorption stronger than in normal luminosity class V spectra, abnormally broad or strong Balmer lines, weak UV wind profiles for their spectral types), one is environmental (association with dense, dusty nebular knots), and one is photometric (derived absolute magnitudes fainter than class V). Very few of these stars have been physically analyzed, and they have not been considered in the current framework of early massive stellar evolution. In particular, they may indicate that the earliest, embedded phases are not as large a fraction of the main-sequence lifetimes as is currently believed. Detailed analyses of these objects will likely prove essential to a complete understanding of the early evolution of massive stars.

---

1. Introduction

It is often stated that zero-age main-sequence (ZAMS) O stars should not be and are not observed. This view arises from at least three sources: star-formation theory, which suggests that the embedded accretion (merger?) phases constitute a significant fraction of the main-sequence lifetimes of massive stars (2.5 Myr for the most massive); statistical studies of UCHII and IR objects relative to optically observed ones; and detailed physical analyses of optical O-star samples that find very few on the ZAMS. For instance, Repolust, Puls, & Herrero (2004) analyzed 24 relatively bright O stars and found only one, HD 93128 in the Carina Nebula compact cluster Trumpler 14, on the ZAMS. However, selection effects may be contributing to this view. If the optically observable near-ZAMS phase of massive stars is relatively brief, it must be sought in very young regions, which may be distant and/or extincted. Also, it is possible that some IR objects are no longer embedded, but rather viewed along unfavorable sightlines in galactic disks or through local, peripheral remanent dust clouds. For example, if we did not have such fortunate lines of sight toward NGC 3603 and 30 Doradus, we might be quite confused about their evolutionary status (Walborn 2002). Such must be the case for at least some objects.

Over the past 35 years, this author and others have encountered numerous optical O stars that appear to be very young for various morphological reasons. These results are scattered throughout the literature and have not been generally recognized by star-formation and evolutionary specialists, or even by quantitative spectroscopists, whose analyses are essential for the former. It is hoped that this summary presentation will provide a useful stimulus toward rectifying the omission. It will be seen that many of the objects in question are in the Magellanic Clouds. But also, HD 93128 will reappear in the plot.

2. Categories of candidate ZAMS O stars

The current sample of 50 optically observable, likely ZAMS O stars is listed in Table 1, along with some normal standard stars for comparison. The ZAMS candidate list was

ID	Sp Type	V	$B-V$	$E(B-V)$	V_0-M_V	M_V	Comment	Reference
					Vz			
HD 64315	O6 Vnz	9.24	0.25	0.57	13.2	−5.7		V. Niemela priv. com (Fig. 3)
HD 64568	O3 V((f*))	9.39	0.11	0.43	13.2	−5.1	also sublum	" (Fig. 3)
HD 92206B	O6 V((f))	9.16	0.17	0.49	12.2	...		N. Morrell priv. com
HD 93128	O3.5 V((f+))	8.77	0.24	0.56	12.2	−5.1	also sublum	Walborn 1973b, 1982, 1995
HD 93129B	O3.5 V((f+))	8.9	0.22	0.54	12.2	−4.9	also sublum	"
CPD−58°2611	O6 V((f))	9.63	0.28	0.60	12.2	−4.4	also sublum	"
CPD−58°2620	O6.5 V((f))	9.27	0.18	0.50	12.2	−4.4	also sublum	"
HDE 303311	O5 Vz	9.05	0.13	0.45	12.2	−4.5	also sublum	V. Niemela priv. com (Fig. 3)
FO 15	O5.5 Vz	12.05	...	1.21	12.2	−4.2	also sublum, $R = 4.15$	Niemela et al. 2006 (Fig. 3)
HD 150135	O6.5 Vz((f))	6.89	0.17	0.49	...	...		Niemela & Gamen 2005 (Fig. 3)
HD 152590	O7.5 V	8.42	0.14	0.46	11.5	−4.5	also sublum	Martins et al. 2005
LH2−96	O7.5 Vz	14.95	−0.17	0.15	18.6	−4.1	also sublum	Parker et al. 2001
LH9−1486	O6.5 Vz	14.20	−0.21	0.11	18.6	−4.7	also sublum	Parker et al. 1992
LH10−3073	O6.5 Vz	14.71	−0.10	0.22	18.6	−4.6	also sublum	" (Fig. 2)
LH10−3102	O7 Vz	13.55	−0.10	0.22	18.6	−5.7		" (Fig. 2)
LH10−3126	O6.5 Vz	14.32	0.00	0.32	18.6	−5.2	also Vb	" (Fig. 2)
LH10−3204	O6−7 Vz	14.02	−0.17	0.15	18.6	−5.0	also sublum	"
30 Dor−171	O6−8 Vz	15.67	0.26	0.58	18.6	−4.7		Walborn & Blades 1997
30 Dor−341	O8−9 Vz	14.40	−0.03	0.28	18.6	−5.0		"
30 Dor−803	O3−5 Vz	15.61	0.33	0.65	18.6	−4.9	also sublum	"
30 Dor−1340	O7 Vz	14.94	0.01	0.33	18.6	−4.6		"
30 Dor−1643	O3−5 Vz	15.51	0.15	0.47	18.6	−4.5	also sublum	"
30 Dor−1892	O8.5 Vz	15.63	0.23	0.54	18.6	−4.6		"
30 Dor−2270	O7 Vz	15.31	0.09	0.41	18.6	−4.5	also sublum	Parker 1993
NGC 346−113	OC6 Vz	14.93	−0.22	0.10	19.1	−4.5	also sublum	Walborn et al. 2000
					Vb			
θ^1 Ori C	O6 Vp var	5.13	0.00	0.32	8.3	−4.5	also wk wind, sublum, $R = 5.5?$	Morgan & Keenan 1973
θ^2 Ori A	O9.5 Vp	5.08	−0.11	0.19	8.3	−4.0	$R = 5.5?$	
HD 92206C	O8.5 Vp	9.05	0.15	0.46	12.2	−4.5		Walborn 1982
Herschel 36	O7.5 V(n)	10.30	...	0.85	10.9	−5.2	also knot, $R = 5.39$	Arias et al. 2006
					Weak Winds			
HD 5005A	O6.5 V((f))	7.76	...	0.42	...	...	from $b-y$	Walborn et al. 1985
HD 42088	O6.5 V	7.55	0.06	0.38	...	...	also Vz, "knot"	Martins et al. 2005
HD 54662	O6.5 V	6.21	0.03	0.35	...	...		Walborn et al. 1985

TABLE 1. O-Type ZAMS Candidates

ID	Sp Type	V	$B-V$	$E(B-V)$	V_0-M_V	M_V	Comment	Reference
					Knots			
N11A−7	O3–6 V	14.69	...	0.19	18.6	−4.5	y, also Vz, sublum	Heydari-Malayeri et al. 2001 (Fig. 2)
30 Dor−409A	O8.5 V	17.05	...	0.56	18.6	−3.3	WFPC2, also sublum	Walborn et al. 2002a, CHORIZOS
30 Dor−409B	O9 V	17.08	...	0.49	18.6	−3.0	WFPC2, also sublum	”
30 Dor−1201	O9.5 V	15.83	...	0.37	18.6	−3.9	WFPC2	”
30 Dor−1222	O9 V(n)p	15.11	...	0.39	18.6	−4.7	WFPC2	”
30 Dor−1429A	O3-4 V	15.88	...	0.55	18.6	−4.4	WFPC2, also sublum	”
N81−1	O6–8:	14.38	...	0.07	19.1	−4.9	y, also wk wind	Heydari-Malayeri et al. 2002
N81−2	O6–8:	14.87	...	0.06	19.1	−4.4	y, also wk wind, sublum	”
N81−3	O6–8:	16.10	...	0.10	19.1	−3.3	y, also wk wind, sublum	”
N81−11	O6–8:	15.74	...	0.07	19.1	−3.6	y, also wk wind, sublum	”
					Subluminous (or $R > 3$?)			
30 Dor−83	O9–9.5 V	15.44	−0.02	0.29	18.6	−4.0		Walborn & Blades 1997
30 Dor−324	O7–8 V	15.18	0.04	0.36	18.6	−4.5		”
30 Dor−466	O9 V	15.55	−0.11	0.20	18.6	−3.6		”
30 Dor−661	O3–6 V	15.03	0.14	0.46	18.6	−5.0		”
30 Dor−713	O3–6 V	14.61	0.03	0.35	18.6	−5.0		”
30 Dor−791	O3–5 V	15.84	0.31	0.63	18.6	−4.6		”
30 Dor−1035	O3–6 V	14.73	−0.05	0.27	18.6	−4.7		”
30 Dor−1170	O3–6 V	15.94	0.14	0.46	18.6	−4.0		”
					Normal V Comparisons			
HD 46149	O8.5 V	7.56	0.17	0.48	10.5	−4.9	$R=4$	Walborn & Fitzpatrick 1990
HD 46150	O5 V((f))	6.72v	0.13	0.45	10.5	−5.6	$R=4$	Walborn et al. 2002b
HD 46202	O9 V	8.17	0.17	0.48	10.5	−4.2	$R=4$	Martins et al. 2005
HD 46223	O4 V((f+))	7.25	0.22	0.54	10.5	−5.4	$R=4$	”
15 Mon	O7 V((f))	4.65	−0.25	0.07	9.4	−5.0		Walborn & Fitzpatrick 1990
HD 93027	O9.5 V	8.72	−0.02	0.28	12.2	−4.3		”
HD 93028	O9 V	8.36	−0.06	0.25	12.2	−4.6		Martins et al. 2005
HD 93204	O5 V((f))	8.42	0.10	0.42	12.2	−5.0		”
HDE 303308	O4 V((f+))	8.17	0.13	0.45	12.2	−5.4		Walborn et al. 2002b

Notes:

$(B-V)_0$: O2–O7, −0.32; O8–O9, −0.31; O9.5, −0.30

$(b-y)_0 = -0.15$; $E(B-V) = 1.49E(b-y)$ (Heydari-Malayeri et al. 2001, 2002)

$R = 3$ unless otherwise noted

TABLE 1. *Continued*

complete to the author's knowledge as of May 2006, although further candidates are being discovered as of this writing, e.g., in the SMC cluster NGC 346 by Evans et al. (2006), and in a new LMC fiber survey by I. Howarth (private communication). The sample is divided into five categories according to the distinct, principal discovery criteria, although many of them actually display more than one, as noted in the Comments. Three of these criteria are spectroscopic, one is environmental, and one is photometric. Discovery and/or data references are included in the Table. The five categories will now be discussed in turn.

2.1. *Vz spectra*

A luminosity classification for stars earlier than spectral type O9 was introduced by Walborn (1971, 1972, 1973a). It is based upon the selective emission effects (Walborn 2001) in He II λ4686 and the N III triplet $\lambda\lambda$4634-4640-4642, i.e., the Of phenomenon. These same lines display a *negative* effect in absorption with increasing luminosity in the MK O9–B0 classification, which was hypothesized to be caused by filling in of the absorptions by the same emission effects producing the Of phenomenon, thus providing the basis for a luminosity classification at the earlier types. A luminosity sequence at spectral type O6.5 in modern digital data is shown in Figure 1. As can be seen there, this He II line is a strong absorption feature in class V, which then weakens, neutralizes, and finally comes into emission above the continuum in the Ia supergiant. (Correlatively, the N III, already weakly in emission at class V, increases in strength with increasing luminosity.)

Walborn (1973b) first noted that in the Trumpler 14 O-dwarf spectra, the λ4686 absorption appeared stronger relative to the other He lines than in typical class V spectra. This very compact cluster in the Carina Nebula appears to be very young; it contains the O2 If* prototype, likely pre-WN object HD 93129A (Walborn et al. 2002b). Penny et al. (1993) found that this cluster is approximately 550,000 years old, while Repolust et al. (2004) derived an age of 150,000 years for HD 93128. Subsequently, even more extreme examples were found in the LMC giant H II regions 30 Dor and N11 (Parker et al. 1992; Walborn & Parker 1992; Parker 1993; Walborn & Blades 1997). Some of the N11 spectra are reproduced in Figure 2, and some new Galactic examples kindly provided for this presentation by V. Niemela and N. Morrell are shown in Figure 3.

A fairly obvious hypothesis is that the stronger λ4686 absorption in these probably very young objects is an "inverse" Of effect, or more precisely, that typical class V spectra already have some emission filling in that line while these objects have less, and hence may be less luminous and less evolved, i.e., nearer to the ZAMS. To denote that hypothesis, the luminosity class notation Vz has been introduced. Some analytical support has been provided by the work of Venero, Cidale, & Ringuelet (2002), Martins et al. (2005), and Mokiem et al. (2006), but a homogeneous analysis of the full sample in high-resolution data with state-of-the-art photospheric/wind models is essential to investigate whether the Vz stars as a class have systematically higher gravities and lower luminosities than class V. In Table 1, the three blocks of Vz stars correspond to the Galaxy, LMC, and SMC.

A misunderstanding of the Vz definition in late-O spectra should be clarified here. At early O types, the He II λ4686 absorption should be stronger than He II λ4541. At type O7, He II λ4541 is equal to He I λ4471, so that λ4686 is stronger than both in a Vz spectrum. At later O types, however, λ4541 weakens more rapidly with advancing type than λ4686 in normal class V spectra; thus, a late-O Vz spectrum must have λ4686 *stronger than He I λ4471.*

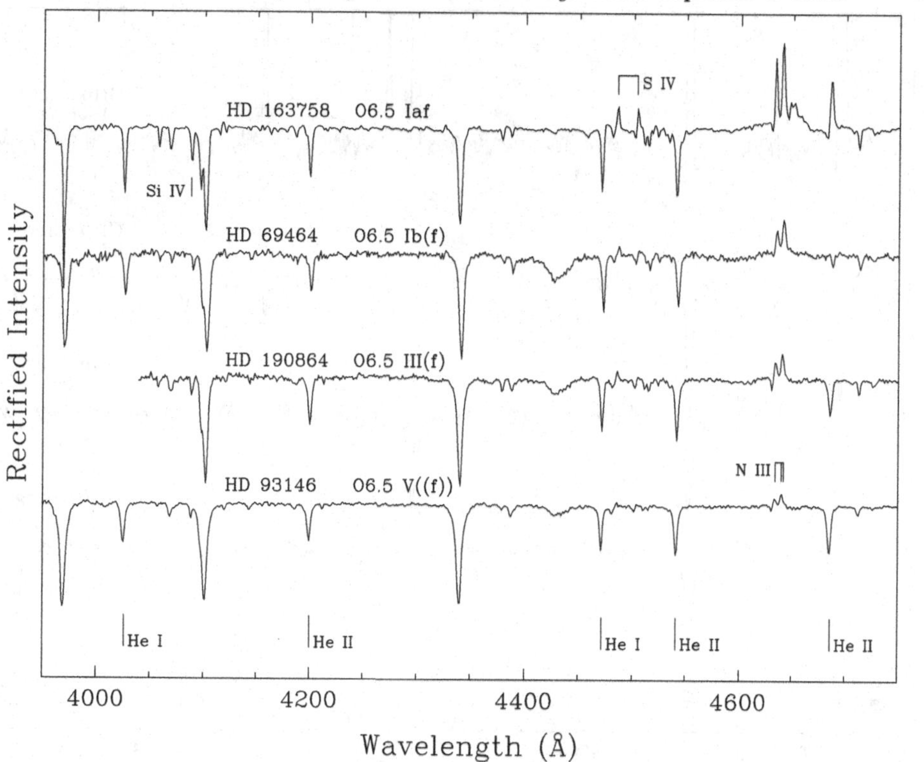

FIGURE 1. A luminosity sequence at spectral type O6.5. The rectified spectrograms are separated by 0.4 continuum units. The spectral lines identified below are He I λλ4026, 4471 and He II λλ4200, 4541, 4686. N III λλ4634-40-42 emission is marked above HD 93146, likewise Si IV λ4089 and S IV λλ4486-4504 (Werner & Rauch 2001) in HD 163758. Note the comparable strengths of the λλ4541, 4686 absorptions in the class V spectrum, and the weakening, then transition to emission of the latter with increasing luminosity, while the N III emission increases smoothly. The Si IV absorption has a positive luminosity effect, which is more sensitive at later types. Courtesy of Ian Howarth.

2.2. *Vb spectra*

W. Morgan frequently remarked on a peculiarity of the Orion Nebula Cluster OB stars, namely broader Balmer lines than in normal class V spectra, with profiles that did not appear to be rotational; Morgan & Keenan (1973) reproduced photographic spectrograms of θ^1 Ori C and the O7 V standard 15 Mon to illustrate the effect. At some point, he introduced the notation Vb to denote such spectra, by analogy with Ib for less luminous supergiants and IIIb in Keenan's subdivision of late-type giants. (Abt 1979 cites a photographic atlas prepared with Morgan that appeared in 1978 as the source of the new notation, but it does not appear there.) Abt (1979) and Levato & Malaroda (1981, 1982) presented further examples among B and A dwarf spectra in very young clusters.

As for the Vz category, the question arises whether the Vb phenomenon might be caused by higher gravity and lower luminosity than in class V, since it is well known that the Balmer lines weaken with increasing luminosity in OB spectra due to the decreasing Stark effect. A careful analysis of the five OB components of θ^1 and θ^2 Ori has been presented by Simón-Díaz et al. (2006), including a thorough investigation of line-broadening mechanisms and comparison with standard objects (including 15 Mon) analyzed with the same techniques. Interestingly, they find that H and He lines in the

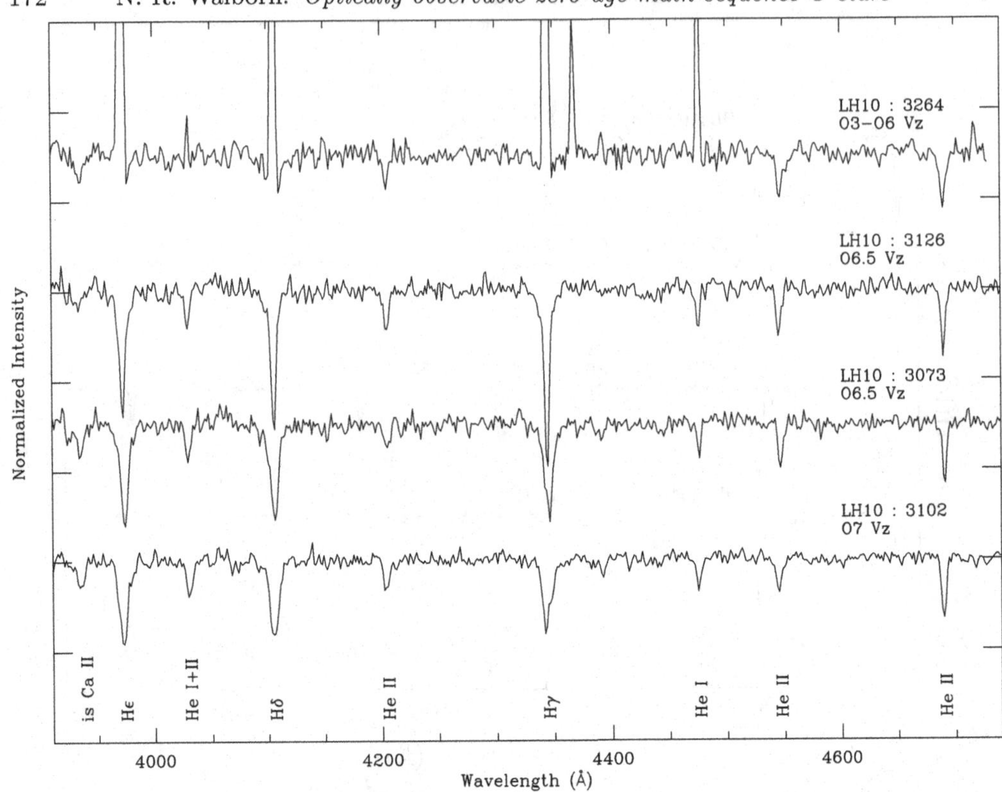

FIGURE 2. Vz spectra in Lucke-Hodge 10/Henize N11 in the LMC (Walborn & Parker 1992). The ordinate ticks are separated by 0.25 continuum units. See the Figure 1 caption for identifications of the He lines. Note the increased strength of the He II λ4686 absorption relative to the other He lines in these spectra, and also the great strength of the Balmer lines in LH10−3126. LH10−3264 is in the dense nebular knot N11A, which produces the very strong nebular emission lines. Courtesy of Joel Parker.

Orion stars tend to be broader than in the best-fitting models, most systematically in the B0.5 V spectrum of θ^1 Ori D. However, they derive similar gravities to those of the comparison stars and find that the Orion stars are somewhat off the ZAMS, although uncertainties in the latter result remain because of the extinction law and, as they point out, the effect of initial rotational velocities on the location of the ZAMS. In Table 1, a value of $R = 4.25$ (average of 3.0 and 5.5) has been used to calculate the absolute magnitudes of the Trapezium stars, because the actual value in the Orion Nebula remains uncertain (Robberto et al. 2004); thus their subluminosity is uncertain as well.

θ^1 Ori C is now known as the first O-type magnetic oblique rotator (Donati et al. 2002; Smith & Fullerton 2005; Gagné et al. 2005; Wade et al. 2006) and an extreme spectrum variable, including its abnormally weak UV wind profiles (Walborn & Nichols 1994; Stahl et al. 1996). Simón-Díaz et al. (2006) provide a detailed discussion of the effects of the variability on the quantitative analysis. Moreover, there is now evidence that the spectrum of 15 Mon may be variable as well; Simón-Díaz et al. also discuss the effects of its spectroscopic companion in that connection. Further quantitative analyses incorporating these complications are indicated to ascertain the nature of the Vb Balmer profiles and the physical origin of the peculiarity.

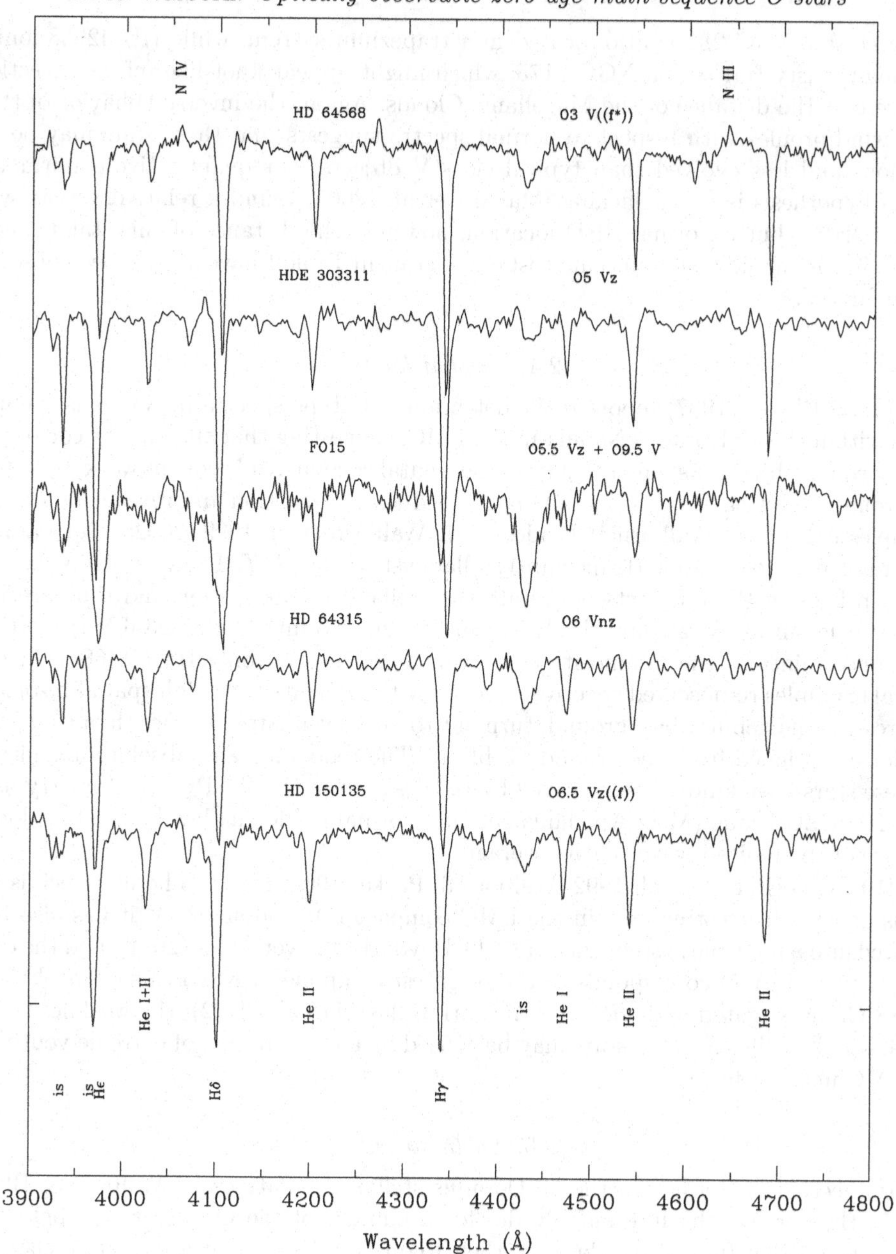

FIGURE 3. New Galactic Vz spectra, rectified and separated by 0.2 continuum units. See the Figure 1 caption for identifications of the stellar lines; in addition, N IV λ4058 emission is marked in HD 64568 here. Courtesy of Virpi Niemela and Nidia Morrell.

2.3. *Weak winds*

It is now well known that the ultraviolet stellar-wind profiles in O-type spectra display strong correlations with the optical spectral types, including an increase in strength with increasing luminosity (Walborn, Nichols-Bohlin, & Panek 1985). That study also detected four main-sequence stars with abnormally weak wind profiles for their spectral types, including θ^1 Ori C already discussed in the previous section. HD 5005A in the young cluster NGC 281 is also located in a trapezium system, while HD 42088 ionizes the small, dusty H II region NGC 2175, which might appear knot-like (cf. next section) if it were at the distance of the Magellanic Clouds. Again, the inverse behavior of these weak wind profiles with respect to normal spectra suggests that these stars may be less luminous and less evolved than typical class V objects, but quantitative confirmation of this hypothesis is so far lacking. Martins et al. (2005) found a relatively weak wind for HD 42088, but a normal HRD location; however, the distance of this star is highly uncertain. (It should be noted that most of the stars in Table 1 have not yet been observed in the far UV.)

2.4. *Nebular knots*

Walborn & Blades (1987) reported the detection of O-type spectra in two dense nebular knots within the 30 Doradus Nebula in the LMC, suggesting that they might correspond to very young objects just emerging from their natal cocoons. Subsequent work, both from the ground and with *HST*, has amply confirmed that suggestion and revealed additional examples in 30 Dor (Walborn & Blades 1997, Walborn et al. 1999, 2002a); several have been resolved into multiple (trapezium) stellar systems by *HST*. The very strong nebular emission lines in these objects obliterate the stellar He I absorption and thus preclude accurate classification, as only the He II can be seen, leading to the O3-6 V type (there is no uncertainty in the luminosity class, which depends on only He II λ4686); in fact, several examples re-observed spectroscopically with *HST*, its very high spatial resolution suppressing the nebular background, turn out to have even later O types than that range previously assigned from the ground (Table 1). The reddening and absolute magnitudes of these stars have kindly been derived by L. Úbeda from the WFPC2 photometry using the CHORIZOS code (Maíz-Apellániz 2004); a normal reddening law had to be adopted because of the limited wavelength coverage.

LH10−3264 (Parker et al. 1992, Walborn & Parker 1992; Figure 3 here), which is also Vz, is another interesting case in the LMC compact H II region N11A; it has also been resolved into a multiple system with *HST* by Heydari-Malayeri et al. (2001), and the entry in Table 1 (N11A-7) corresponds to the brightest component. N81 is a similar object in the SMC, investigated with *HST* by Heydari-Malayeri et al. (2002); the weakness of the stellar-wind profiles in these stars may be caused by a combination of extreme youth and the SMC metal deficiency.

2.5. *Subluminosity*

In the spectroscopic study of the 30 Doradus stellar populations by Walborn & Blades (1997), the derived absolute magnitudes for a number of the O V stars fall below the calibration of that luminosity class (cf. their Figure 3). Those that are not also classified as Vz are listed separately in Table 1. It is reasonable to hypothesize that they may be nearer to or on the ZAMS. However, a basic uncertainty, which also applies to some subluminous cases discussed in the previous sections, must be recognized: because of the lack of further information, the absolute magnitudes have to be derived with a normal value of $R = 3$. For a fixed distance modulus and variable extinction, a larger value of R would yield brighter absolute magnitudes. To resolve this issue, photometric observations covering a wider wavelength range (ideally, UV through IR) must be obtained to support the derivation of reddening laws toward each individual star. Indeed, there is evidence that the reddening law may vary among different lines of sight within a complex H II region environment, because of spatially diverse mixtures of dust grain properties. In any event, large values of R are associated with large dust grains and very young objects, so

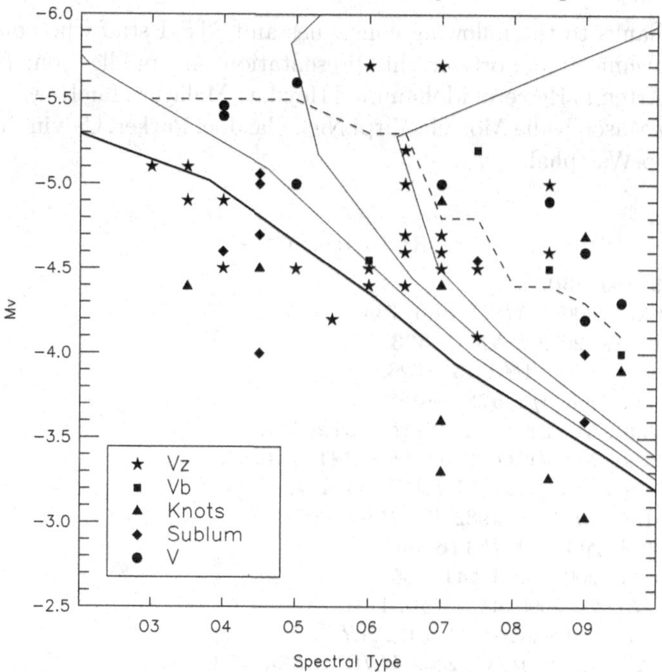

FIGURE 4. Derived absolute visual magnitudes for candidate ZAMS and normal O stars plotted against their spectral types. The bold solid line is the ZAMS from Schaller et al. (1992), while the lighter solid lines are the corresponding isochrones for 1, 2, 3 Myr. The dashed line is the luminosity class V calibration from Walborn (1973a). The normal stars fall near the latter, while most of the ZAMS candidates lie below it. Courtesy of Leonardo Úbeda.

these stars may be near the ZAMS in either case, but again, further observations and quantitative analyses are required to definitively establish their physical status.

3. Discussion

All of the absolute magnitudes from Table 1 are plotted against the spectral types in Figure 4, where they are also compared with the Schaller et al. (1992) ZAMS and isochrones, as well as the luminosity class V calibration (Walborn 1973). The different categories of ZAMS candidates are distinguished with different symbols. Subject to the uncertainties already discussed, the plot provides preliminary support for the present hypothesis: most of them fall below the class V calibration and near the ZAMS. Note that HD 93128 lies adjacent to the ZAMS, in agreement with the detailed analysis of Repolust et al. (2004). The normal class V stars cluster about the calibration, as well they should since they contributed to its derivation. Interestingly, the figure shows that a typical O3 V star is 1 Myr old, O5 V 2 Myr, O6–7 V 3 Myr, and O8–9 V perhaps 4–5 Myr. Less luminous stars at a given type must be younger. It is also clear that unresolved multiple systems move points upward in the diagram (recall that $0\farcs1$ corresponds to 5000 AU at the LMC), while underestimated values of R move them downward.

The present sample of candidate ZAMS stars is optimum for followup with high-resolution spectroscopy and state-of-the-art atmospheric/wind analysis. Beyond the basic question of their ZAMS status or otherwise, it may be hoped that such study will elucidate any sequential relationships among the different categories, thereby advancing our detailed understanding of early massive stellar evolution.

My sincere thanks to the following colleagues and STScI staff who contributed data, plots, and/or technical support for this presentation and publication: Rodolfo Barbá, Howard Bond, Artemio Herrero, Mohammad Heydari-Malayeri, Ian Howarth, Jesús Maíz-Apellániz, Greg Masci, Nidia Morrell, Virpi Niemela, Joel Parker, Calvin Tullos, Leonardo Úbeda, and Skip Westphal.

REFERENCES

ABT, H. A. 1979 *ApJ* **230**, 485.
ARIAS, J. I., ET AL. 2006 *MNRAS* **366**, 739.
DONATI, J.-F., ET AL. 2002 *MNRAS* **333**, 55.
EVANS, C. J., ET AL. 2006 *A&A* **456**, 623.
GAGNÉ, M., ET AL. 2005 *ApJ* **628**, 986.
HEYDARI-MALAYERI, M., ET AL. 2001 *A&A* **372**, 527.
HEYDARI-MALAYERI, M., ET AL. 2002 *A&A* **381**, 951.
LEVATO, H. & MALARODA, S. 1981 *PASP* **93**, 714.
LEVATO, H. & MALARODA, S. 1982 *PASP* **94**, 807.
MAÍZ-APELLÁNIZ, J. 2004 *PASP* **116**, 859.
MARTINS, F., ET AL. 2005 *A&A* **441**, 735.
MOKIEM, M. R., ET AL. 2006 *A&A* **456**, 1131.
MORGAN, W. W. & KEENAN, P. C. 1973 *ARA&A* **11**, 29.
NIEMELA, V. S. & GAMEN, R. C. 2005 *MNRAS* **356**, 974.
NIEMELA, V. S., ET AL. 2006 *MNRAS* **367**, 1450.
PARKER, J. WM. 1993 *AJ* **106**, 560.
PARKER, J. WM., ET AL. 1992 *AJ* **103**, 1205.
PARKER, J. WM., ET AL. 2001 *AJ* **121**, 891.
PENNY, L. R., ET AL. 1993 *PASP* **105**, 588.
REPOLUST, T., PULS, J., & HERRERO, A. 2004 *A&A* **415**, 349.
ROBBERTO, M., ET AL. 2004 *ApJ* **606**, 952.
SCHALLER, G., ET AL. 1992 *A&AS* **96**, 269.
SIMÓN-DÍAZ, S., ET AL. 2006 *A&A* **448**, 351.
SMITH, M. A. & FULLERTON, A. W. 2005 *PASP* **117**, 13.
STAHL, O., ET AL. 1996 *A&A* **312**, 539.
VENERO, R. O. J., CIDALE, L. S., & RINGUELET, A. E. 2002 *ApJ* **578**, 450.
WADE, G. A., ET AL. 2006 *A&A* **451**, 195.
WALBORN, N. R. 1971 *ApJS* **23**, 257.
WALBORN, N. R. 1972 *AJ* **77**, 312.
WALBORN, N. R. 1973a *AJ* **78**, 1067.
WALBORN, N. R. 1973b *ApJ* **179**, 517.
WALBORN, N. R. 1995 *RevMexAA (Ser. Conf.)* **2**, 51.
WALBORN, N. R. 1982 *AJ* **87**, 1300.
WALBORN, N. R. 2001. In *Eta Carinae and Other Mysterious Stars* (eds. T. Gull, S. Johansson, & K. Davidson). ASP Conf. Ser. 242, p. 217. ASP.
WALBORN, N. R. 2002. In *Hot Star Workshop III: The Earliest Stages of Massive Star Birth* (ed. P. A. Crowther). ASP Conf. Ser. 267, p. 111. ASP.
WALBORN, N. R. & BLADES, J. C. 1987 *ApJ* **323**, L65.
WALBORN, N. R. & BLADES, J. C. 1997 *ApJS* **112**, 457.
WALBORN, N. R. & FITZPATRICK, E. L. 1990 *PASP* **102**, 379.
WALBORN, N. R. & NICHOLS, J. S. 1994 *ApJ* **425**, L29.
WALBORN, N. R., NICHOLS-BOHLIN, J., & PANEK, R. J. 1985 *International Ultraviolet Explorer Atlas of O-Type Spectra from 1200 to 1900 Å*, NASA RP 1155.
WALBORN, N. R. & PARKER, J. WM. 1992 *ApJ* **399**, L87.
WALBORN, N. R., ET AL. 1999 *AJ* **117** 225.
WALBORN, N. R., ET AL. 2000 *PASP* **112**, 1243.

Walborn, N. R., et al. 2002a *AJ* **124**, 1601.
Walborn, N. R., et al. 2002b *AJ* **123**, 2754.
Werner, K. & Rauch, T. 2001. In *Eta Carinae and Other Mysterious Stars* (eds. T. Gull, S. Johansson, & K. Davidson). ASP Conf. Ser. 242, p. 229. ASP.

Metallicity-dependent Wolf-Rayet winds

By PAUL A. CROWTHER

Department of Physics & Astronomy, University of Sheffield, Hicks Building, Hounsfield Road, Sheffield, S3 7RH, UK

Observational and theoretical evidence in support of metallicity-dependent winds for Wolf-Rayet stars is considered. Well-known differences in Wolf-Rayet subtype distributions in the Milky Way, LMC and SMC may be attributed to the sensitivity of subtypes to wind density. Implications for Wolf-Rayet stars at low metallicity include a hardening of ionizing flux distributions, an increased WR population due to reduced optical line fluxes, plus support for the role of single WR stars as gamma-ray burst progenitors.

1. Introduction

Wolf-Rayet (WR) stars represent the final phase in the evolution of very massive stars prior to core collapse, in which the H-rich envelope has been stripped away via either stellar winds or close binary evolution, revealing products of H-burning (WN sequence) or He-burning (WC sequence) at their surfaces, i.e., He, N or C, O (Crowther 2007).

WR stellar winds are significantly denser than O stars, as illustrated in Figure 1, so their visual spectra are dominated by broad emission lines, notably He II λ4686 (WN stars) and C III λ4647–51, C III λ5696, C IV λ5801–12 (WC stars). The spectroscopic signature of WR stars may be seen individually in Local Group galaxies (e.g., Massey & Johnson 1998), within knots in local star-forming galaxies (e.g., Hadfield & Crowther 2006) and in the average rest frame UV spectrum of Lyman Break galaxies (Shapley et al. 2003).

In the case of a single massive star, the strength of stellar winds during the main sequence and blue supergiant phase scales with the metallicity (Vink et al. 2001). Consequently, one expects a higher threshold for the formation of WR stars at lower metallicity, and indeed the SMC shows a decreased number of WR to O stars than in the solar neighborhood. Alternatively, the H-rich envelope may be removed during the Roche lobe overflow phase of close binary evolution, a process which is not expected to depend upon metallicity.

WR stars represent the prime candidates for Type Ib/c core-collapse supernovae and long, soft gamma-ray bursts (GRBs). This is due to their immediate progenitors being associated with young massive stellar populations, compact in nature and deficient in either hydrogen (Type Ib) or both hydrogen and helium (Type Ic). For the case of GRBs, a number of which have been associated with Type Ic hypernovae (Galama et al. 1998; Hjorth et al. 2003), a rapidly rotating core is a requirement for the collapsar scenario in which the newly formed black hole accretes via an accretion disk (MacFadyen & Woosley 1999). Indeed, WR populations have been observed within local GRB host galaxies (Hammer et al. 2006).

At solar metallicity, single-star models predict that the core is spun down either during the red supergiant (via a magnetic dynamo) or Wolf-Rayet (via mass-loss) phases. The tendency of GRBs to originate from metal-poor environments (e.g., Stanek et al. 2006) suggests that stellar winds from single stars play a role in their origin since Roche lobe overflow in a close binary evolution would not be expected to show a strong metallicity dependence.

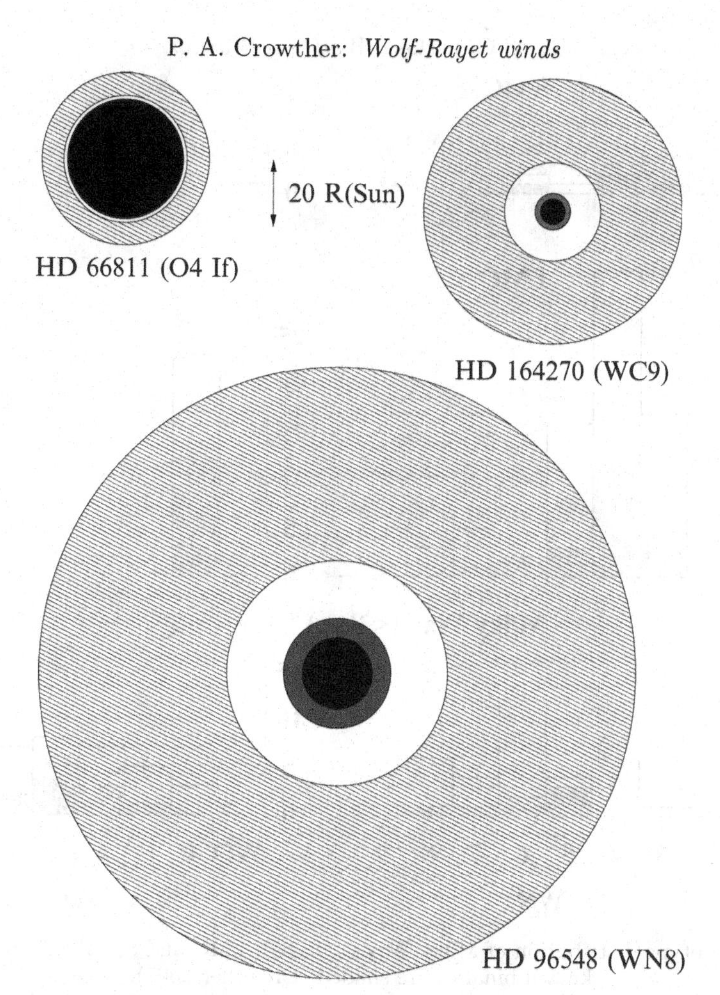

FIGURE 1. Comparisons between stellar radii at Rosseland optical depths of 20 (= R_*, black) and 2/3 (= $R_{2/3}$, grey) for HD 66811 (O4 If), HD 96548 (WN8) and HD 164270 (WC9), shown to scale, together with the wind region corresponding to the primary optical wind line-forming region, $10^{11} \leqslant n_e \leqslant 10^{12}$ cm^{-3} (hatched) in each case, illustrating the highly extended winds of WR stars with respect to O stars (Crowther 2007).

In this article, evidence in favor of a metallicity dependence for WR stars is presented, of application to the observed WR-subtype distribution in Local Group galaxies, plus properties of WR stars at low metallicity including their role as GRB progenitors.

2. WR subtype distribution

Historically, the wind properties of WR stars have been assumed to be metallicity independent (Langer 1989), yet there is a well-known observational trend to earlier, higher ionization, WN and WC subtypes at low metallicity as illustrated in Figure 2, whose origin is yet to be established.

Mass-loss rates for WN stars in the Milky Way and LMC show a very large scatter. The presence of hydrogen in some WN stars further complicates the picture, since WR winds are denser if H is absent (Nugis & Lamers 2000). This is illustrated in Figure 3, which reveals that the wind strengths of (H-rich) WN winds in the SMC are lower than corresponding H-rich stars in the LMC and Milky Way (Crowther 2006). Figure 4 shows

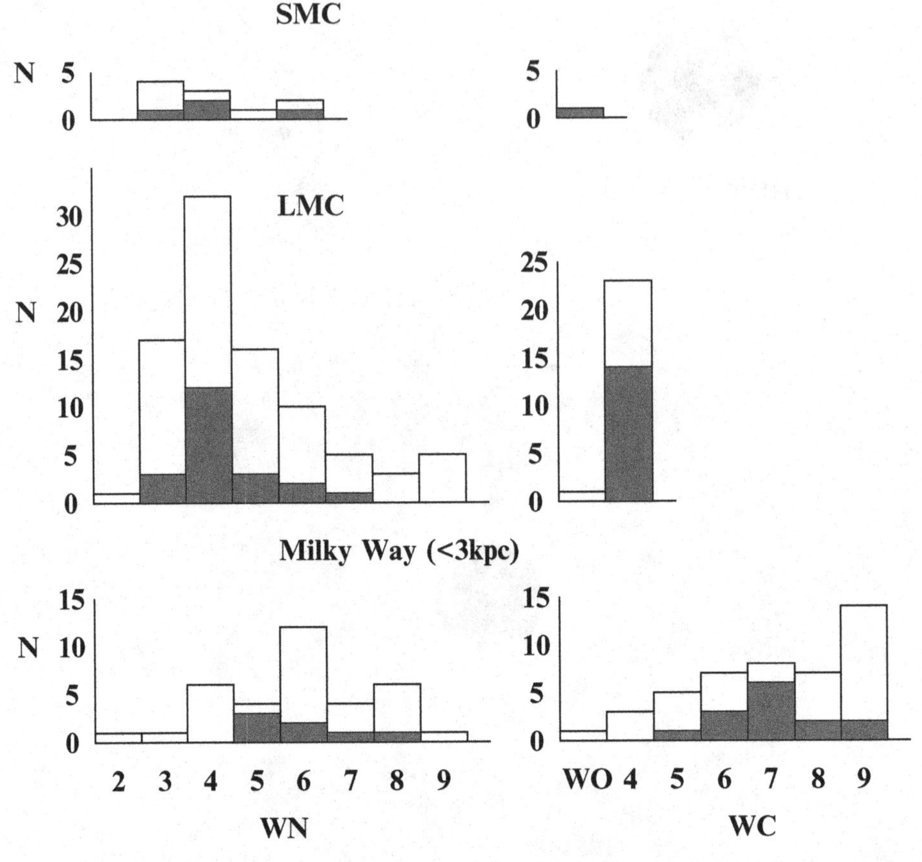

FIGURE 2. Subtype distribution of Milky Way (<3 kpc), LMC and SMC WR stars, in which known binaries are shaded (Crowther 2007).

that the situation is rather clearer for WC stars, for which LMC stars reveal ~0.2 dex lower mass-loss rates than Milky Way counterparts (Crowther et al. 2002).

The observed trend to earlier subtypes in the LMC (Figure 2) was believed to originate from a difference in carbon abundances relative to Galactic WC stars (Smith & Maeder 1991), yet quantitative analysis reveals similar carbon abundances (Koesterke & Hamann 1995; Crowther et al. 2002).

Theoretically, Nugis & Lamers (2002) argued that the iron opacity peak was the origin of the wind driving in WR stars, which Gräfener & Hamann (2005) supported via an hydrodynamic model for an early-type WC star in which lines of Fe IX–XVII deep in the atmosphere provided the necessary radiative driving. Vink & de Koter (2005) applied a Monte Carlo approach to investigate the metallicity dependence for cool WN and WC stars revealing $\dot{M} \propto Z^{\alpha}$, where $\alpha = 0.86$ for WN stars and $\alpha = 0.66$ for WC stars for $0.1 \leqslant Z \leqslant 1 Z_{\odot}$. The weaker WC dependence originates from a decreasing Fe content and constant C and O content at low metallicity. Empirical results for the solar neighborhood, LMC and SMC presented in Figures 3–4 are broadly consistent with theoretical predictions, although detailed studies of individual WR stars within galaxies' broader range in metallicity would provide stronger constraints. Theoretical wind models also predict smaller wind velocities at lower metallicity, as is observed for WO stars, which are presented in Figure 5 (Crowther & Hadfield 2006).

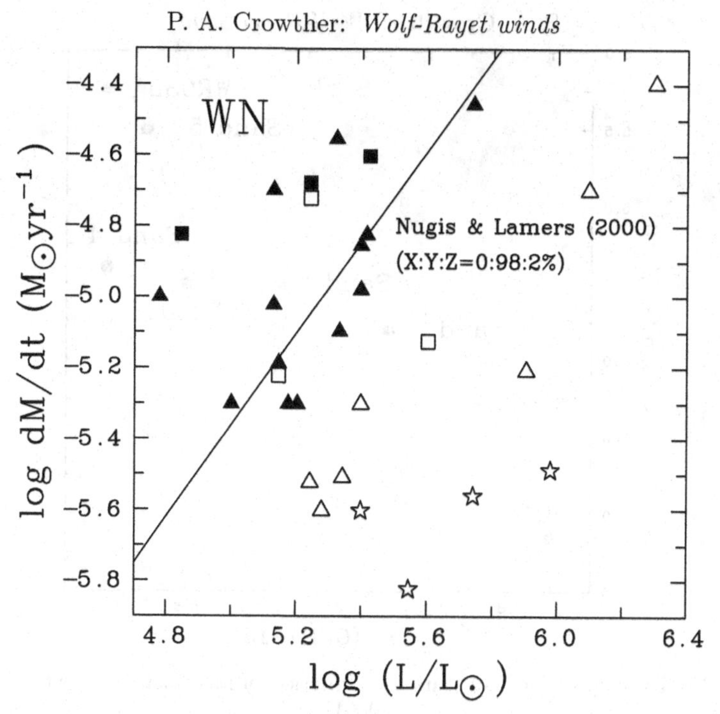

FIGURE 3. Mass-loss rates for WN stars in the Galaxy (squares), LMC (triangles) and SMC (stars) revealing a wide spread in wind densities for WN stars, for which stars without hydrogen (filled symbols) possess stronger winds. The solid line is from eqn. 22 of Nugis & Lamers (2000).

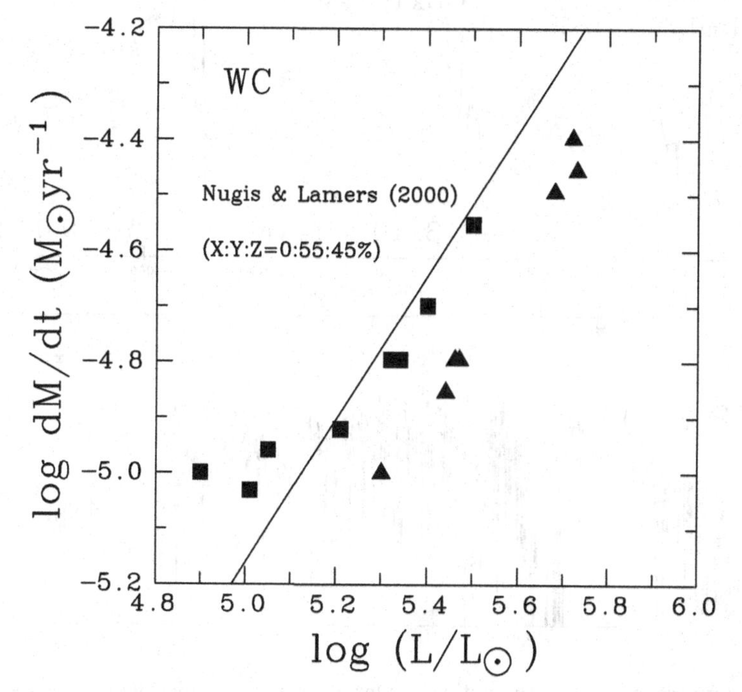

FIGURE 4. Mass-loss rates for WC and WO stars in in the Galaxy (squares) and LMC (triangles). The solid line is from eqn. 22 of Nugis & Lamers (2000).

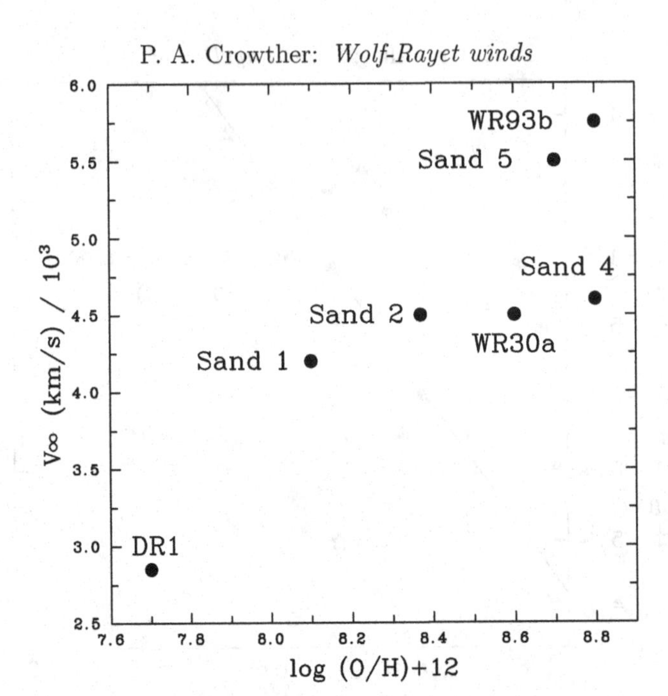

FIGURE 5. Wind velocities for WO stars as a function of metallicity (Crowther & Hadfield 2006).

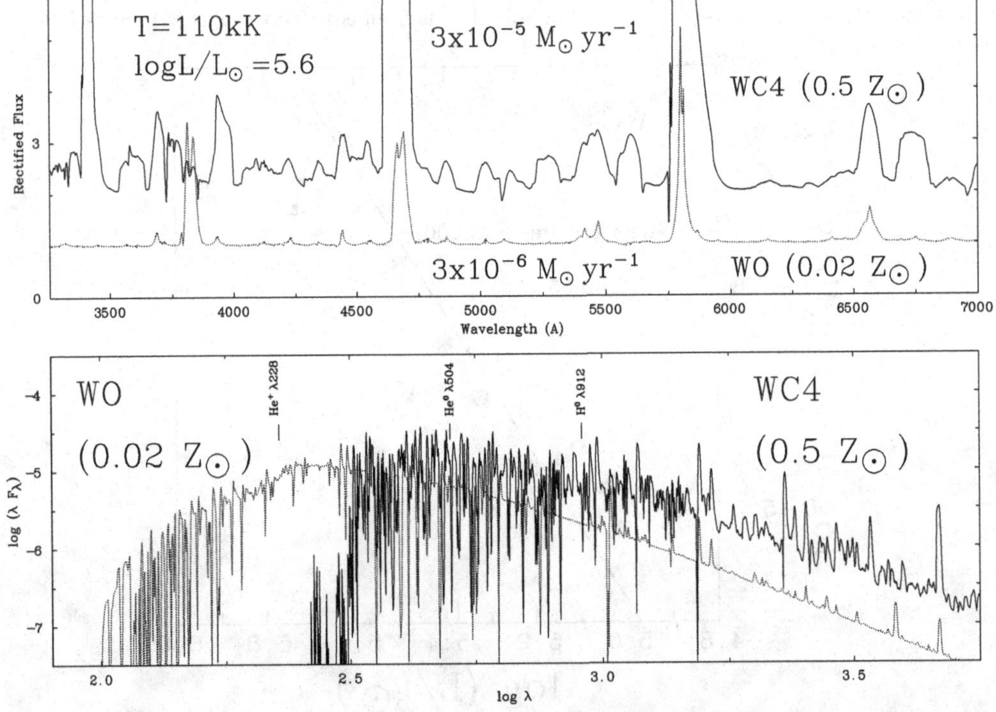

FIGURE 6. Comparison between theoretical stellar atmosphere models which differ solely in wind density (factor of 10), revealing an earlier spectral type (WC4 → WO) and harder ionizing flux distribution at low metallicity, adapted from Crowther & Hadfield (2006).

	LMC		SMC	
	$\log L$ (erg s^{-1})	N	$\log L$ (erg s^{-1})	N
WN2–4	35.92	36	35.23	8
WN5–6	36.24	15	35.63	2
WN7–9	35.86	9		

TABLE 1. Mean He II λ4686 line luminosities for Magellanic Cloud WN stars including known binaries (Crowther & Hadfield 2006)

The impact of a metallicity dependence for WR winds upon spectral types is as follows. At high metallicity, recombination from high to low ions (early to late subtypes) is very effective in very dense winds, while the opposite is true for low-metallicity, low-density winds. The situation is illustrated in the upper panel of Figure 6, where we present synthetic WC spectra obtained from identical models, except that their wind densities differ by a factor of 10, and the weak wind model is assumed to be extremely Fe-poor (adapted from Crowther & Hadfield 2006). The high wind density case has a WC4 spectral type while the low wind density case has an earlier WO subtype. Crowther et al. (2002) noted that a further increase in wind density by a factor of two predicts a WC7 subtype. Stellar temperatures further complicate this picture, such that the spectral type of a WR star results from a subtle combination of ionization and wind density, in contrast with normal stars.

3. WR populations at low metallicity

The effect of reduced WR wind densities at low metallicity on WR populations is as follows. WR optical recombination lines will (i) decrease in equivalent width, since their strength scales with the square of the density, and (ii) decrease in line flux, since the lower wind strength will reduce the line blanketing, resulting in an increased extreme UV continuum strength at the expense of the UV and optical. The equivalent widths of optical emission lines in SMC WN stars are well known to be lower than Milky Way and LMC counterparts (Conti et al. 1989). To date, the standard approach for the determination of unresolved WR populations in external galaxies has been to assume metallicity-independent WR line fluxes—obtained for Milky Way and LMC stars (Schaerer & Vacca 1998)—regardless of whether the host galaxy is metal-rich (Mrk 309; Schaerer et al. 2000) or metal-poor (I Zw 18, Izotov et al. 1997).

Ideally, one would wish to use WR template stars appropriate to the metallicity of the galaxy under consideration. Unfortunately, this is only feasible for the LMC, SMC and solar neighborhood, since it is challenging to isolate individual WR stars from ground-based observations in more distant galaxies, which span a larger spread in metallicity. Line luminosities for optical emission lines in LMC and SMC WR stars are compared in Table 1, illustrating significantly lower (factor of 5–6) luminosities for the lower metallicity of the SMC.

Reduced WR line fluxes are also predicted for WR atmospheric models at low metallicity if one follows the metallicity dependence from Vink & de Koter (2005), such that WR populations inferred from Schaerer & Vacca (1998) at low metallicity may underestimate actual populations by an order of magnitude. This is potentially problematic for single-star evolutionary models at very low metallicities ($\sim$1/50 $Z_\odot$), since the WR populations inferred for I Zw 18 and SBS0335−052E using Milky Way line fluxes compare well with evolutionary models (e.g., Izotov et al. 1997; Papaderos et al. 2006).

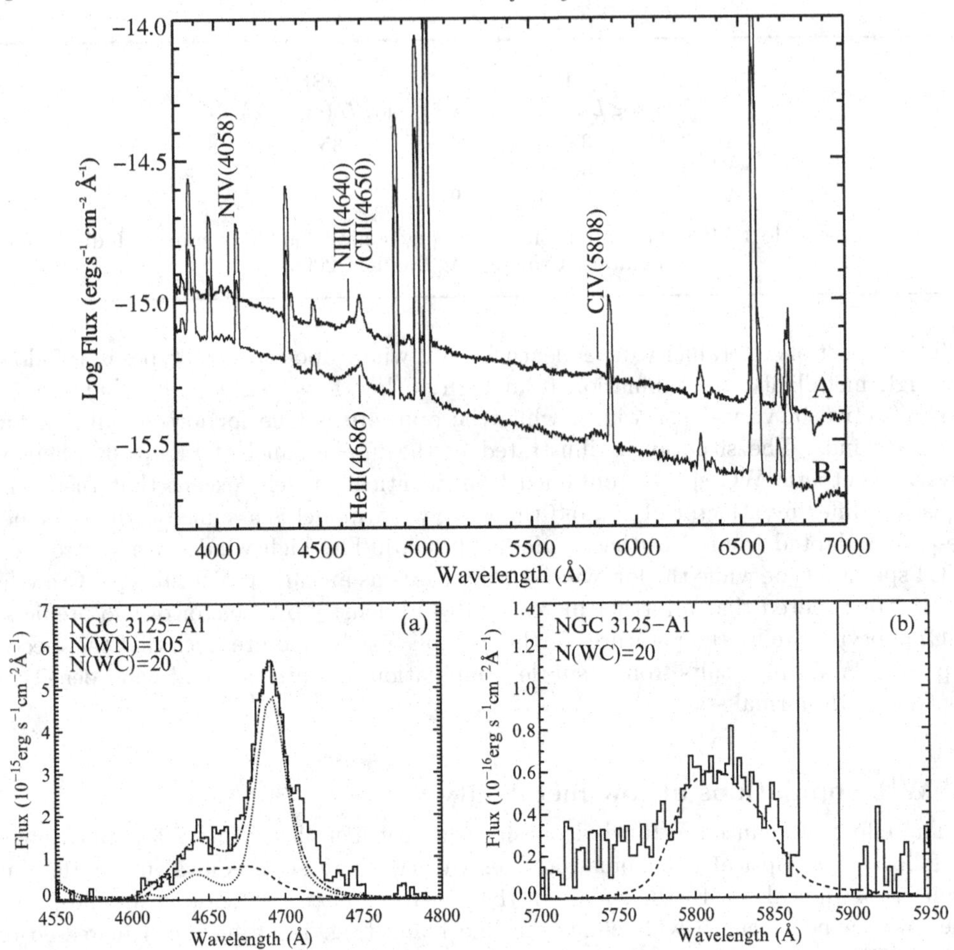

FIGURE 7. Top panel: Optical spectroscopy of knots A and B within the LMC metallicity starburst galaxy NGC 3125, indicating the C III 4650/He II 4686 (blue) and C IV 5801−12 (yellow) WR bumps. Lower panels: Fit to the blue and yellow bumps for cluster A1 using LMC template WN (dotted lines) and WC (dashed lines) stars (Hadfield & Crowther 2006).

If WR populations are in fact a factor of ∼10 larger, similar to that of the SMC, close binary evolution would represent the most likely origin for such large WR populations.

For Magellanic Cloud metallicity starburst galaxies, one may employ appropriate template spectra (e.g., Crowther & Hadfield 2006) to reproduce WR features, as shown in the upper panel of Figure 7 for the starburst galaxy NGC 3125. Indeed, consistent fits to the blue and yellow WR bumps may be achieved for the A1 cluster within NGC 3125 using LMC template WR stars, as shown in the lower panels of Figure 7 (Hadfield & Crowther 2006).

4. Ionizing fluxes and GRB progenitors

Schmutz et al. (1992) demonstrated that the ionizing fluxes from WR stars soften as wind density increases. Consequently, a metallicity dependence for WR wind strengths implies that WR ionizing-flux distributions soften at increased metallicity, as demonstrated by Smith et al. (2002). Indeed, relatively soft ionizing fluxes are observed in the super-solar metallicity WR starburst galaxy NGC 3049 (Gonzalez Delgado et al. 2002).

At low metallicities, one anticipates a combination of weak UV and optical spectral lines from WR stars (i.e., weak stellar He II λ4686) but very strong H and He Lyman continua (i.e., strong nebular He II λ4686), as is indicated in the lower panel of Figure 6. Indeed, low-metallicity star-forming galaxies display strong nebular He II λ4686, although shocks from supernovae remnants may also contribute to nebular emission.

The typical environment of nearby ($z < 0.25$) long-duration GRBs is unusually metal poor, as emphasized by Stanek et al. (2006) with respect to star-forming galaxies from the Sloan Digital Sky Survey. Reduced WR mass-loss rates at low metallicity will lead to reduced densities in the immediate environment of GRBs with respect to typical WR stars, as is observed (Chevalier et al. 2004). In addition, massive single stars undergoing homogeneous evolution in which WR mass-loss rates are low may maintain their rapidly spinning cores through to core collapse (Yoon & Langer 2005; Langer & Norman 2006).

5. Summary

Observational and theoretical evidence supports reduced wind densities and velocities for low-metallicity WR stars, which addresses the relative WR subtype distribution in the Milky Way and Magellanic Clouds, plus the reduced WR line strengths in the SMC with regard to the Galaxy and LMC. The primary impact at low metallicity is as follows; (a) an increased WR population due to lower line fluxes from individual stars, of particular relevance to I Zw 18 and SBS0335−052E; (b) harder ionizing fluxes from WR stars, potentially responsible for the strong nebular He II λ4686 seen in low-metallicity H II galaxies; (c) responsible for the reduced density of GRB environments with respect to solar metallicity WR counterparts.

Finally, a metallicity dependence for WR winds may help to reconcile the relative number of WN to WC stars observed in surveys (e.g., Massey & Johnson 1998) with evolutionary predictions. Evolutionary models for which rotational mixing is included, yet metallicity-dependent WR winds are not (Meynet & Maeder 2005), fail to predict the high N(WC)/N(WN) ratio observed at high metallicities (Hadfield et al. 2005), while models which account for the Vink & de Koter (2005) WR wind dependence compare much more favorably with observations (Eldridge & Vink 2006), in spite of the neglect of rotational mixing.

Many thanks to Lucy Hadfield, with whom the majority of the results presented here were obtained. PAC acknowledges financial support from the Royal Society.

REFERENCES

Chevalier, R. A., Li, Z.-Y., & Fransson, C. 2004 *ApJ* **606**, 369.

Conti, P. S., Garmany, C. D., & Massey P., 1989 *ApJ* **341**, 113.

Crowther, P. A. 2006. In *Stellar Evolution at Low Metallicity: Mass-Loss, Explosions, Cosmology* (eds. H. Lamers, N. Langer, & T. Nugis). ASP Conf. Ser. 353, p. 157. ASP.

Crowther, P. A. 2007 *ARA&A* **45**, 177.

Crowther, P. A. & Hadfield, L. J. 2006 *A&A* **449**, 711.

Crowther, P. A., Dessart, L., Hillier, D. J., Abbott, J. B., & Fullerton, A. W. 2002 *A&A* **392**, 653.

Eldridge, J. J. & Vink, J. S. 2006 *A&A* **452**, 295.

Galama, T. J., et al. 1998 *Nature* **395**, 670.

Gonzalez Delgado, R. M., Leitherer, C., Stasinska, G., & Heckman, T. M. 2002 *ApJ* **580**, 824.

Gräfener, G. & Hamann, W.-R. 2005 *A&A* **432**, 633.

Hadfield, L. J. & Crowther, P. A. 2006 *MNRAS* **368**, 1822.

Hadfield, L. J., Crowther, P. A., Schild, H., & Schmutz, W. 2005 *A&A* **439**, 265.

Hammer, F., Flores, H., Schaerer, D., Dessauges-Zavadsky, M., Le Floc'h, E., & Puech, M. 2006 *A&A*, **454**, 103.

Hjorth, J., Sollerman, J., Moller, P., et al. 2003 *Nature* **423**, 847.

Izotov, Y. I., Foltz, C. B., Green, R. F., Guseva, N. G., & Thuan, T. X. 1997 *ApJ* **487**, L37.

Koesterke, L. & Hamann, W.-R. 1995 *A&A* **299**, 503.

Langer, N., 1989 *A&A* **220**, 135.

Langer, N. & Norman, C. 2006 *ApJ* **638**, L63.

MacFadyen, A. I. & Woosley, S. E. 1999 *ApJ* **524**, 262.

Massey, P. & Johnson, O. 1998 *ApJ* **505**, 793.

Meynet, G. & Maeder A. 2005 *A&A* **429** 581.

Nugis, T. & Lamers, H. J. G. L. M., 2000 *A&A* **360**, 227.

Nugis, T. & Lamers, H. J. G. L. M. 2002 *A&A* **389**, 162.

Papaderos, P., Izotov, Y. I., Guseva, N. G., Thuan T. X., & Fricke K. J. 2006 *A&A* **454**, 119.

Schaerer, D., & Vacca, W. D. 1998 *ApJ* **497**, 618.

Schaerer, D., Guseva, N. G., Izotov, Y. I., & Thuan, T. X. 2000 *A&A* **362**, 53.

Schmutz, W., Leitherer, C., & Gruenwald, R. 1992 *PASP* **104**, 1164.

Shapley, A. E., Steidel, C. S., Pettini, M., & Adelberger, K. L. 2003 *ApJ* **588**, 65.

Smith, L. F. & Maeder, A. 1991 *A&A* **241**, 77.

Smith, L. J., Norris, R. P. F., & Crowther, P. A. 2002 *MNRAS* **337**, 1309.

Stanek, K. Z., Gnedin, O. Y., Beacom, J. F., et al. 2006 *Acta Astronomica* **56**, 333.

Vink, J. S. & de Koter, A. 2005 *A&A* **442**, 587.

Vink, J. S., de Koter, A., & Lamers, H. J. G. L. M. 2001 *A&A*, **369**, 574.

Yoon, S.-C. & Langer, N. 2005 *A&A* **443**, 643.

Eruptive mass loss in very massive stars and Population III stars

By NATHAN SMITH

Center for Astrophysics and Space Astronomy, University of Colorado, 389 UCB, Boulder, CO 80309, USA; nathans@astro.berkeley.edu

I discuss the role played by short-duration eruptive mass loss in the evolution of very massive stars. Giant eruptions of Luminous Blue Variables (LBVs) like the 19th century event of η Carinae can remove large quantities of mass almost instantaneously, making them significant in stellar evolution. They can potentially remove much more mass from the star than line-driven winds, especially if stellar winds are highly clumped such that previous estimates of O-star mass-loss rates need to be revised downward. When seen in other galaxies as "supernova impostors," these LBV eruptions typically last for less than a decade, and they can remove of order 10 $M_{\odot}$ as indicated by massive nebulae around LBVs. Such extreme mass-loss rates cannot be driven by radiation pressure on spectral lines, because the lines will completely saturate during the events. Instead, these outbursts must either be continuum-driven super-Eddington winds or outright hydrodynamic explosions, both of which are insensitive to metallicity. As such, this eruptive mode of mass loss could also play a pivotal role in the evolution and ultimate fate of massive metal-poor stars in the early universe. If they occur in these Population III stars, such eruptions would also profoundly affect the chemical yield and types of remnants from early supernovae and hypernovae thought to be the origin of long gamma-ray bursts.

1. Introduction

Mass loss is a critical factor in the evolution of a massive star. In addition to the direct reduction of a star's mass, it profoundly affects the size of its convective core, its core temperature, its angular momentum evolution, its luminosity as a function of time, and hence its evolutionary track on the HR diagram and its main-sequence (MS) lifetime (e.g., Chiosi & Maeder 1986). Wolf-Rayet (WR) stars are thought to be the descendants of massive stars as a consequence of mass loss in the preceding H-burning phases, during which the star sheds its H envelope (Abbott & Conti 1987; Crowther 2007). While the maximum initial mass of stars is thought to be $\sim$150 $M_{\odot}$ (Figer 2005; Kroupa 2005), WR stars do not have masses much in excess of 20 $M_{\odot}$ (Crowther 2007).† Thus, very massive stars have the immense burden of removing 30–130 $M_{\odot}$ during their lifetime before the WR phase, unless they explode first. Stellar evolution calculations prescribe $\dot{M}(t)$ based on semiempirical values, so we need to know when this mass loss occurs.

The main question I wish to address here is whether the majority of mass lost during the lifetime of the most massive stars occurs primarily via steady line-driven stellar winds, or instead through violent, short-duration eruptions or explosions. The two extremes are shown graphically in Figure 1. This question is critical for understanding how mass loss scales with metallicity, back to the time of the massive stars in the early universe. In this contribution I would like to draw attention to the specific role of LBV eruptions, advocating for their importance. The essential points of the argument are the following:

- Recent studies of hot star winds indicate that mass-loss rates on the MS are probably much lower than previously thought due to the effects of clumping in the wind. If so,

† By "WR stars" we mean H-deficient WR stars (core-He burning phases or later), and not the luminous H-rich WNL stars (Crowther et al. 1995), which are probably still core-H burning.

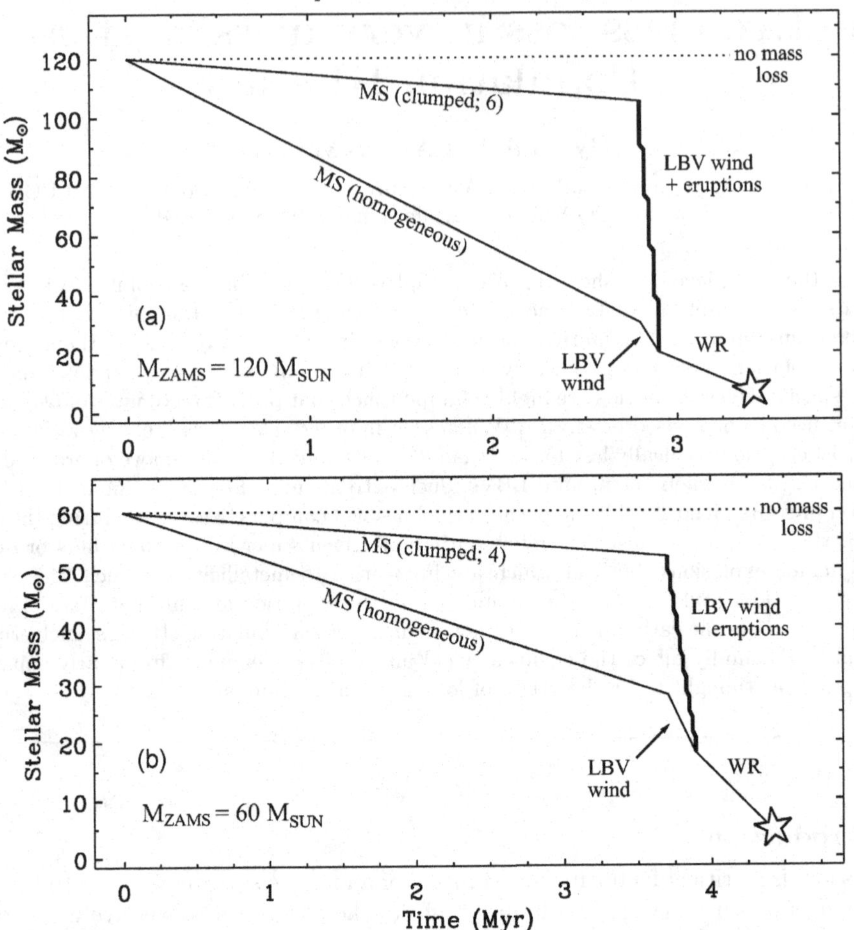

FIGURE 1. These plots are schematic representations of a star's mass as a function of time. Two extreme scenarios are shown: One has higher conventional O-star mass-loss rates assuming homogeneous winds on the main sequence (MS) with no clumping. This is followed by a brief LBV wind phase and a longer WR wind phase before finally exploding as a supernova; this is the type of scenario usually adopted in stellar evolution calculations. The second has much reduced mass-loss rates on the main sequence (assuming clumping factors of 4–6), followed by an LBV phase that includes severe mass loss in several brief eruptions, plus a steady wind in the time between them; this is the type of scenario discussed by Smith & Owocki (2006). Panel (a) shows the case for an initial stellar mass of 120 $M_{\odot}$ (appropriate for a luminous LBV like AG Carinae), and Panel (b) shows an initial mass of 60 $M_{\odot}$ (appropriate for a somewhat less luminous object like P Cygni, perhaps). The more numerous, more frequent, and less-extreme eruptions of the 60 $M_{\odot}$ scenario assume that the mass lost in an LBV burst may scale with proximity to the Eddington limit. Note that the clumping factors of 4–6 shown here are still fairly modest compared to some estimates of >10 for O-star winds.

then for the most massive stars, the revised mass-loss rates are inadequate to reduce the star's mass enough to reach the WR phase, where the H-rich envelope has been removed and the He-rich core material is exposed. Something other than the O star's line-driven wind must account for this mass loss.

- Observations of nebulae around LBVs and LBV candidates have revealed very high ejecta masses—of order 10 $M_{\odot}$. In some objects, there is evidence for multiple shell ejections on timescales of 10^3 years. Cumulatively, these sequential eruptions could, in principle, remove a large fraction of the total mass of the star. Thus, a few short-duration outbursts like the 19th century eruption of η Car could dominate mass lost during the lives of the most massive stars, and would be critical for the envelope shedding needed to form WR stars at any metallicity.

• The extreme mass-loss rates of these LBV bursts imply that line opacity is too saturated to drive them, so they must instead be either continuum-driven super-Eddington winds (see Owocki et al. 2004) or outright hydrodynamic explosions. Unlike steady winds driven by lines, the driving in these eruptions may be largely independent of metallicity, and might play a role in the mass loss of massive metal-poor stars (Population III stars).

These points have already been explained and justified in more detail by Smith & Owocki (2006), where more complete references can be found. That original paper was deliberately provocative, and some caveats had to be left out for the sake of brevity. Rather than repeat that discussion, I will briefly elaborate on a few of these issues, and will spend most of the time discussing alternatives and further implications.

2. The problem: Line-driven winds provide insufficient mass loss

In order to shed a massive star's envelope and reach the WR stage, models must prescribe semiempirical mass-loss rates, which can be scaled by a star's metallicity (e.g., Chiosi & Maeder 1986; Maeder & Meynet 1994; Meynet et al. 1994; Langer et al. 1994; Langer 1997; Heger et al. 2003). Often-adopted "standard" mass-loss rates are given by de Jager et al. (1988), and Nieuwenhuijzen & de Jager (1990). In order for stellar evolution models to match observed properties at the end of H burning, such as WR masses and luminosities, and the relative numbers of WR and OB stars, these mass-loss rates need to be enhanced by factors of $\sim$2 (Maeder & Meynet 1994; Meynet et al. 1994).

However, such enhanced mass-loss rates contradict observations. Recent studies suggest that mass-loss rates are in fact 3–10 or more times *lower* than the "standard" mass-loss rates, not higher. This is due to the influence of clumping in the winds (Fullerton et al. 2006; Bouret et al. 2005; Puls et al. 2006; Crowther et al. 2002; Hillier et al. 2003; Massa et al. 2003; Evans et al. 2004; Kudritzki & Urbaneja in these proceedings), such that mass-loss rates based on density-squared diagnostics like Hα and free-free radio continuum emission have led to overestimates if the wind is strongly clumped. The consequent reduced mass-loss rates mean that steady winds are simply inadequate for the envelope shedding needed to form a WR star. This is not such a problem for stars below $10^{5.8}\,L_\odot$, where the red supergiant (RSG) wind may be sufficient. However, above $10^{5.8}\,L_\odot$ (initial mass above 40–50 $M_\odot$) stars do not become RSGs (Humphreys & Davidson 1979), posing a severe problem if these stars depend upon line-driven winds for mass loss.

For example, consider the fate of a star with initial mass of 120 $M_\odot$. The most extreme O2 If* supergiant HD 93129A has a mass-loss rate derived assuming a homogeneous wind of roughly $2\times10^{-5}\,M_\odot\ \mathrm{yr}^{-1}$ (Repolust et al. 2004). If the true mass-loss rate is lower by a factor of 3–10 or more as indicated by clumping in the wind, then during a $\sim$2.5-Myr MS lifetime (Maeder & Meynet 1994), the star will only shed about 5–20 $M_\odot$, leaving it with $M > 100\,M_\odot$, and an additional 80 $M_\odot$ deficit to shake off before becoming a WR star. After this, the stellar wind mass-loss rates are higher during post-MS phases, but they are still insufficient to form a WR star. They therefore cannot make up for the lower $\dot{M}$ values on the MS. For a typical LBV lifetime of a few 10^4–10^5 yr (Bohannan 1997) and a typical $\dot{M}$ of $\sim10^{-4}\,M_\odot\ \mathrm{yr}^{-1}$ for most LBVs, the LBV phase will only shed a few additional solar masses through its line-driven wind. Thus, some mechanism other than just a steady wind is needed to reduce the star's total mass by several dozen $M_\odot$.

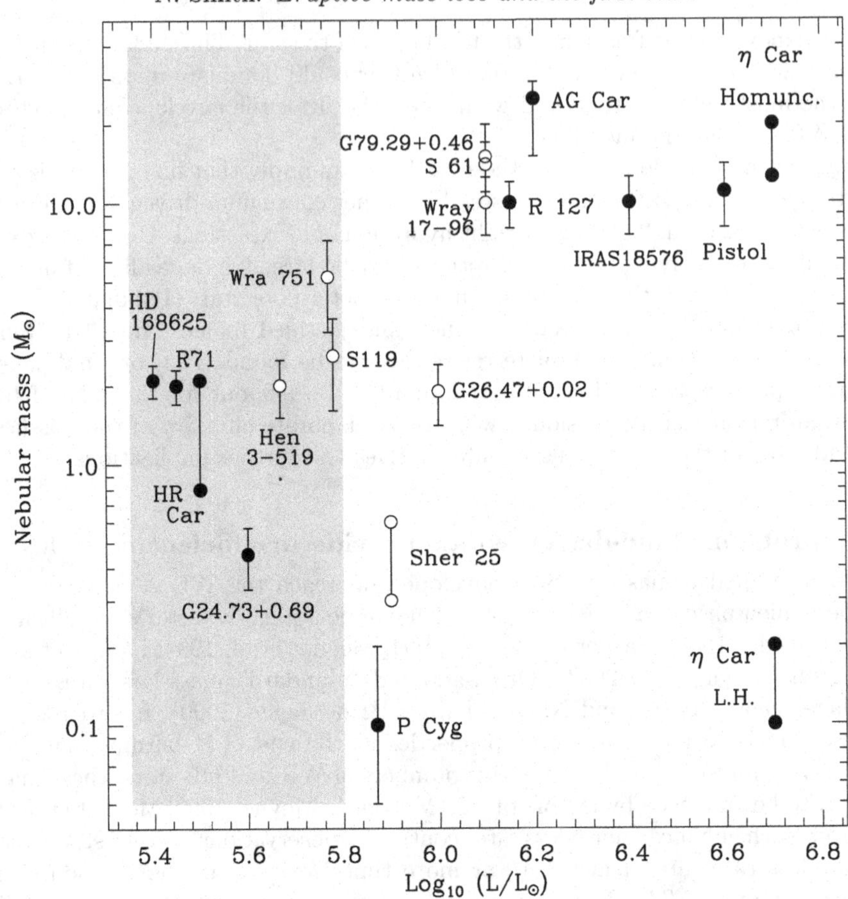

FIGURE 2. Masses of ejecta nebulae around LBVs (filled dots) and LBV candidates (unfilled) as a function of the central star's bolometric luminosity. Luminosities are taken from Smith, Vink, & de Koter (2004), while sources for the masses are given in Smith & Owocki (2006).

3. Balancing the budget: LBV eruptions

The most likely mechanism to rectify this hefty mass deficit is giant eruptions of LBVs (e.g., Davidson 1989; Humphreys & Davidson 1994; Humphreys, Davidson, & Smith 1999; Smith & Owocki 2006), where the mass-loss rate and bolometric luminosity of the star increase substantially. While we do not yet fully understand what causes these giant LBV outbursts, we know empirically that they do indeed occur, and that they drive substantial mass off the star. Deduced masses of LBV and LBV-candidate nebulae from the literature are plotted in Figure 2 as a function of the central star's luminosity. We see that for stars with $\log(L/L_{\odot}) > 6$, nebular masses of $10\,M_{\odot}$ are quite reasonable, *perhaps suggesting that this is a typical mass ejected in a giant LBV eruption.*

If such large masses are typical for LBV outbursts, then only a few such eruptions occurring sequentially during the LBV phase are needed to remove a large fraction of the star's total mass. This is shown schematically in Figure 1 for stars with initial masses of 120 and 60 $M_{\odot}$. Notice that in the 60 $M_{\odot}$ example, the LBV eruptions are more numerous and each one is less massive than in the 120 $M_{\odot}$ case; this is entirely hypothetical, but is based on the presumption that a more massive and more luminous star will have more violent mass ejections because of its closer proximity to the Eddington limit. For example, we might expect that η Car currently has an Eddington parameter of $\Gamma = 0.9$ or higher, whereas a less-luminous LBV like P Cygni probably has $\Gamma = 0.5$ or so. Further investigation of the amount of mass ejected in each burst, and their frequency and total number is probably the most important observational pursuit associated with LBVs and their role in stellar evolution.

Our best example of this phenomenon is the 19th century "Great Eruption" of η Carinae. The event was observed visually, the mass of the resulting nebula has been measured (12–20 $M_\odot$ or more; Smith et al. 2003), and proper-motion measurements of the expanding nebula indicate that it was ejected in the 19th century event (e.g., Morse et al. 2001). The other example for which this is true is the AD 1600 eruption of P Cygni, although its shell nebula has a much lower mass (Smith & Hartigan 2006). Both η Car and P Cyg are surrounded by multiple, nested shells indicating previous outbursts (e.g., Walborn 1976; Meaburn 2001). While the shell of P Cyg is less massive than η Car's nebula, it is still evident that P Cyg shed more mass in such bursts than via its stellar wind in the time between them (Smith & Hartigan 2006). This difference between P Cyg and η Car hints that LBV outbursts do indeed become progressively more extreme near the Eddington limit. However, the Homunculus and the Little Homunculus around η Car also caution that any one star can eject very different amounts of mass in each of its subsequent eruptions, with a corresponding wide range of luminous and kinetic energy.

Although LBV eruptions are rare, a number of extragalactic η Car analogs or "supernova impostors" have been observed, such as SN 1954J in NGC 2403 and SN 1961V in NGC 1058 (Humphreys et al. 1999; Smith et al. 2001; Van Dyk et al. 2002, 2005), V1 in NGC 2363 (Drissen et al. 1997), and several recent events seen as type IIn supernovae, like SN 1997bs, SN 2000ch, SN 2002kg, and SN 2003gm (Van Dyk et al. 2000, 2006; Wagner et al. 2004; Maund et al. 2006). Furthermore, massive circumstellar shells have also been inferred to exist around supernovae and gamma-ray bursters (GRBs). Some examples are the radio-bright SN 1988Z with a nebula as massive as 15 $M_\odot$ (Aretxaga et al. 1999; Van Dyk et al. 1993; Chugai & Danziger 1994), as well as similar dense shells around SN 2001em (Chugai & Chevalier 2006), SN 1994W (Chugai et al. 2004), SN 1998S (Gerardy et al. 2002), SN 2001ig (Ryder et al. 2004), GRB 021004 (Mirabal et al. 2003), and GRB 050505 (Berger et al. 2006).

These outbursts and the existence of massive circumstellar nebulae indicate that the 19th century eruption of η Car is not an isolated, freakish event, but instead may represent a common rite of passage in the late evolution of the most massive stars. A massive ejection event may even initiate the LBV phase, by lowering the star's mass, raising its L/M ratio, and drawing it closer to instability associated with an opacity-modified Eddington limit (Appenzeller 1986; Davidson 1989; Lamers & Fitzpatrick 1988; Humphreys & Davidson 1994). Mass loss in these giant eruptions may play a role in massive-star evolution analogous to thermal pulses of asymptotic giant-branch stars. In any case, meager mass-loss rates through stellar winds, followed by huge bursts of mass loss in violent eruptions at the end of core-H burning (see Figure 1) may significantly alter stellar evolution models.

4. Alternative scenarios

The scenario where LBV eruptions dominate the mass loss of the most massive stars, as shown in Figure 1, would represent a dramatic change in our understanding of mass loss in stellar evolution. Therefore, it certainly deserves close scrutiny, and it is worth considering some possible alternatives or modifications.

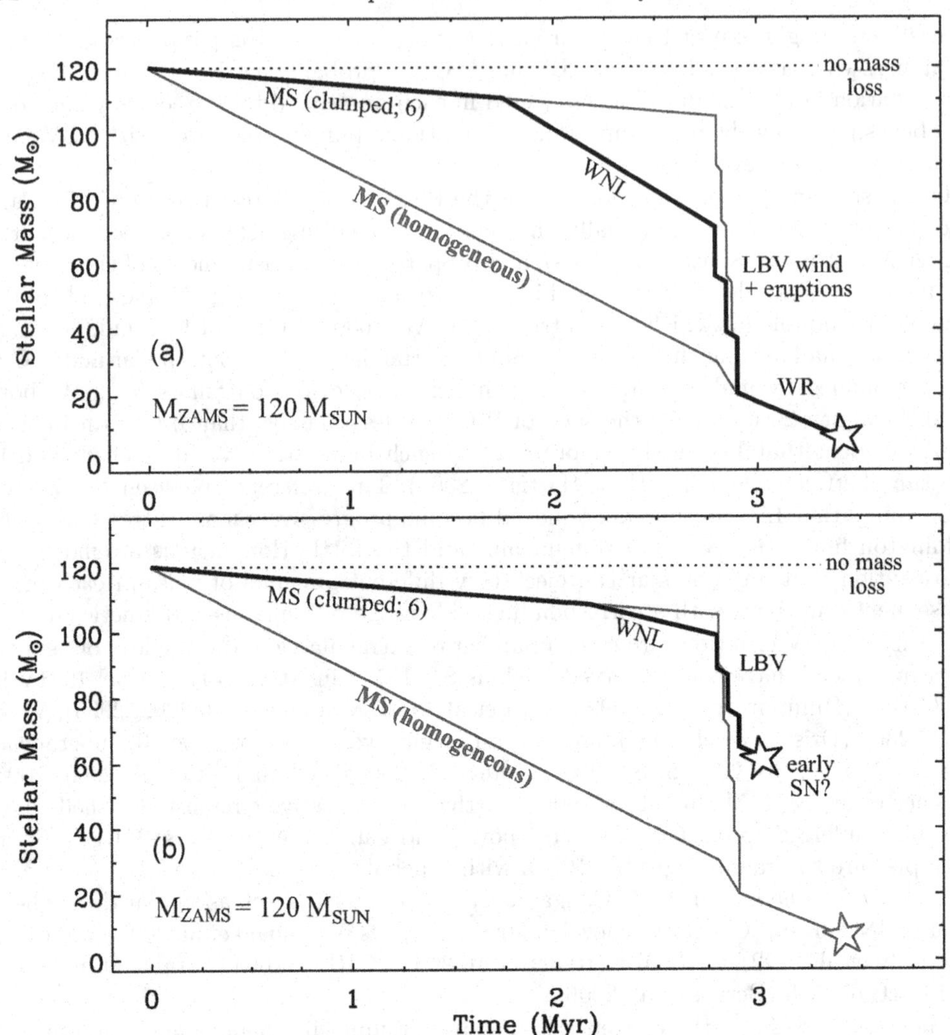

FIGURE 3. The same type of schematic plot as shown in Figure 1, but this figure shows some more complicated hypothetical alternatives in between the two extremes of Figure 1 (only one case of $M_{\rm ZAMS} = 120\ M_{\odot}$ is shown here). The first, Panel (a), allows for substantial mass loss via a wind during an extended late-type WN phase that lasts for almost half the MS lifetime. The total mass lost by the WNL wind is almost as much as through LBV eruptions, but one could easily adjust these depending on the duration of the WNL phase. The second alternative, in Panel (b), has a weaker WNL phase and relaxes the assumption that bona fide WR stars are the descendants of the most massive stars—i.e., it allows for the possibility that the most-massive stars might explode before reaching the WR phase, thereby reducing the amount of mass that needs to be shed through LBV eruptions.

First, however, I would like to emphasize that *the scenario in which homogeneous line-driven winds of O stars dominate the mass lost during the life of a star is almost certainly wrong.* There are two independent reasons to think so.

This need for recognizing the role of LBV eruptions in mass loss is partly motivated by recent studies of the mass-loss rates of O stars, where clumping in the winds suggests rather drastic reductions in the MS mass-loss rates. To be fair, the required amount of reduction in mass-loss rates is not yet settled; some indications favor reduction of more than an order of magnitude, while other estimates are more moderate, indicating factors of only a few. While this is debated, it is worth remembering that even if the mass-loss rate reduction is only a factor of three, *it would still indicate that LBV eruptions may dominate the total mass lost during the lifetime of a very massive star.* Note that the plots in Figure 2 adopt fairly modest mass-loss rate-reduction factors of only 4–6.

However, clumping in O-star winds is only part of the story. The other part is the observational reality that LBV eruptions like η Car's massive 19th century outburst do indeed occur, and we have evidence that they occur more than once, ejecting a mass of order $10\,M_\odot$ each time. A star's mass budget needs to allow for that. However, if we require several 10s of solar masses in LBV eruptions, plus enhanced mass loss during a WNL phase (see below), we run into a *serious* problem—homogeneous winds simply do not allow enough room for additional mass loss through WNL phases and LBV eruptions! Thus, the mass-loss rates implied by the assumption of homogeneous winds are not viable. I would then suggest that the existence of WNL and LBV mass loss is an independent argument that O star winds *must* be clumped, reducing their mass-loss rates by at least a factor of two to three.

Nevertheless, the provocative new scenario in Figure 1 may still seem a bit extreme, placing a huge and possibly unrealistic burden on LBV eruptions. The truth may lie somewhere in between, so let's consider two likely alternatives.

4.1. *A long WNL phase?*

One alternative is that a very massive star spends a good fraction of its H-burning MS lifetime as a late-type WN star (WNL; see Crowther et al. 1995). Even if their winds are clumped, WNL stars have much higher mass-loss rates than their O-star counterparts. Thus, if it is possible that massive stars spend something like a third or half of their MS lifetime as a WNL star, they can take a substantial chunk out of the star's total mass. This could temper the burden placed upon LBVs. This scenario is sketched in Figure 3(a). While Figure 3(a) seems reasonable (even likely) to me, there are a few caveats to keep in mind.

First, Figure 3(a) with its rather long WNL phase could only apply to the most freakishly massive stars, with initial masses above roughly 90–100 $M_\odot$. The justification for this comment is that spectral type O3 and even O2 stars still exist in clusters within star-forming regions that are 2.5–3 Myr old (like Tr16 in the Carina Nebula). O3 stars probably have initial masses around 80–100 $M_\odot$ or so, and MS lifetimes around 3 Myr. Therefore, these stars cannot spend a substantial fraction of their H-burning lifetime as a WNL star, because they evidently live for about 3 Myr without yet reaching the WNL phase. Only for the most massive stars, which are even more extremely rare, might a relatively long WNL phase be possible. This makes me wonder if we have yet another dichotomy in stellar evolution, with very different evolutionary sequences above and below 100–120 $M_\odot$—much like the dichotomy above and below 45–50 $M_\odot$. One could certainly make the case that the most luminous evolved stars that are sometimes called LBVs or LBV candidates—stars like η Car, the Pistol star, HD 5980, and possibly LBV 1806−20—have followed a different path than the "normal" LBVs like AG Car and R127.

Second, I would suggest that while WNL stars may make some contribution to the mass loss at the highest luminosities, their influence must be limited. They cannot provide the majority of mass lost by these stars, so the LBV eruption mass loss must still dominate. The reasoning behind this comment has to do with the available mass budget of η Carinae; namely, that η Car is probably a post-WNL star, *but it still has retained most of its original mass.*

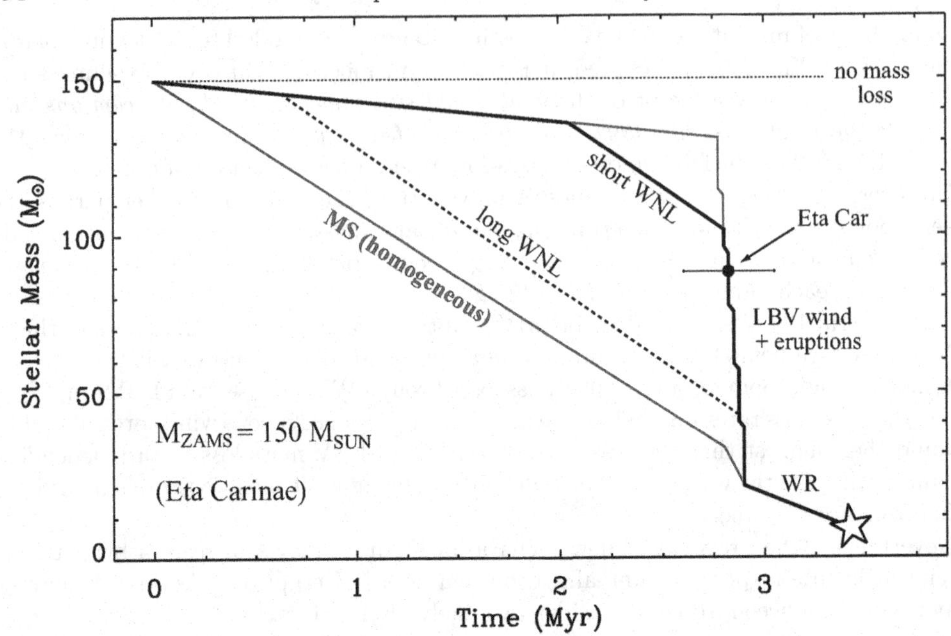

FIGURE 4. Same as Figures 1 and 3, but for a star with an initial mass at the upper mass limit of 150 $M_\odot$, perhaps appropriate for η Carinae. Here I show what the mass evolution might look like for a relatively short (small contribution) and a relatively long (dominant contribution) WNL phase, as well as the simpler extremes with a strong MS wind, as well as LBV eruptions with no WNL phase in gray. The dot shows the likely currently-observed locus of η Carinae (note that I am being quite generous here with the correction for η Car's companion star). Considering that we know η Car has already suffered 2–3 major LBV eruptions, which scenario is most consistent with its present mass?

Let's remember that η Car is the most luminous and most evolved member of a rich region containing over 65 O-type stars, as well as 3 WNL stars (see Smith 2006). It is fair to assume that the current LBV phase of η Car is not only a post-MS phase, but probably also a post-WNL phase, since its ejecta are more nitrogen rich than the WNL stars in Carina. It is also safe to assume that η Car has advanced further in its evolution sooner than the WNL stars of the same age in this region simply because it is more luminous and started with a higher initial mass. Now, η Car is seen today surviving as a very massive star of around 100 $M_\odot$ or more, and we measure a total of something like 20–35 $M_\odot$ in its circumstellar material ejected in only the last few thousand years (the Homunculus, plus more extended outer material; see Smith et al. 2003, 2005). That means η Car began its LBV phase—and ended its MS and/or WNL phase—with more than 120 $M_\odot$ still bound to the star!† If there really is an upper limit of about 150 $M_\odot$ to the mass of stars, then *this rules out the possibility that winds during the MS or WNL phases could dominate the mass lost by the star in its lifetime.* Consequently, it also requires that the MS and WNL winds were indeed highly clumped. If folks don't like relying on just η Car because it is an abomination, there's the Pistol star, which is also a post-MS object and has a present-day mass that probably exceeds 100 $M_\odot$.

This argument is made graphically in Figure 4, where options of "long" and "short" WNL phases are shown. Keeping three facts in mind—1) that we see more than 20 $M_\odot$ of

† Parameters could be chosen selectively to push this as low as perhaps 100–105 $M_\odot$, but not lower.

nebular material from recent LBV eruptions around η Car, 2) that η Car has a present-day mass around $100\,M_{\odot}$ if it is not violating the classical Eddington limit (I am being generous with the companion star's mass in Figure 4), and 3) that there is a likely upper-mass limit for stars of around $150\,M_{\odot}$—where would you place η Carinae on each track in Figure 4? What does that signify for the relative importance of the WNL phase?

4.2. *An early death at the end of the LBV phase?*

The main motivation for such huge amounts of mass loss in continuum-driven LBV eruptions is the assumption that even the most massive stars eventually reach the WR phase, requiring that their mass be reduced down to about $20\,M_{\odot}$ before that point (see Smith & Owocki 2006). If we can relax this constraint and say that the most massive stars above $100\,M_{\odot}$ perhaps *do not* make it to the WR phase, then we can alleviate the burden of removing so much mass through LBV explosions. This would be saying that the most massive stars might undergo core collapse at the end of the LBV phase, instead of entering the WR phase (Figure 3[b]).

That is easy to say and it would seem to fix the uncomfortable problem of depending on LBVs for such drastic mass shedding. However, we should be mindful that this alternative would require an *even more* radical paradigm shift in our understanding of stellar evolution than Figure 1(a). Namely, Figure 3(b) would require that not only are LBVs in a core He-burning phase,† but that LBVs even reach advanced stages like core O and Si burning. Current understanding implies that LBVs have not yet reached He burning. On the other hand, perhaps an early supernova explosion during the LBV phase is not crazy after all, since we really don't know what is going on deep inside the star. In fact, there are several reasons why an early explosion like in Figure 3(b) might be attractive:

- As noted earlier in Section 3, several observations of supernovae (especially Type IIn supernovae) and GRBs reveal that they have dense, massive circumstellar shells close to the star. In their talks at this meeting, H.-W. Chen and D. Fox noted additional examples of GRBs with dense ($\sim 10^6$ cm^{-3}) circumstellar shells seen in absorption spectra of afterglows. In some cases, these closely resemble the absorption features in the shell around η Car (T. Gull, these proceedings). Where did these compact and dense circumstellar shells come from if the WR phase has a sustained fast wind for a few 10^5 years? The answer may be that these shells did, in fact, originate in LBV-like outbursts that occurred within about 1000 years of the final death of the star. That would be astonishing and very important if true.
- So far, I don't know any example of a bona fide WR star that is surrounded by an extremely massive (like $\sim 50\,M_{\odot}$) group of nested shells left over from a previous LBV phase. Perhaps such objects would be rare anyway and don't last long in an observable phase like this, but it would be reassuring to see at least one example. If the most massive stars explode at the end of the LBV phase, then we wouldn't necessarily expect such massive shells around any WR star.
- Oxygen burning is unstable, and as noted by A. Heger in his talk, can lead to short pulsational bursts that may supply sufficient mechanical energy to power an eruption like η Car's 19th century event with $\sim 10^{50}$ ergs. The problem with this scenario, though, is that the duration of O burning is extremely short and could not account for the observational fact that η Car-like eruptions tend to recur on timescales of $\sim 10^3$ years. These O-burning blasts could only account for a last hurrah right before the star's final demise... but the possibility is interesting anyway.

† In fact, they must have reached it before the LBV phase, because the LBV phase is so short.

In any case, an explosion at the end of the LBV phase when the star is still very massive would almost certainly form a black hole, and this should happen in roughly 3 Myr. Are there any examples of massive black holes in massive-star clusters? If so, where are the expanding supernova remnants from these events? If this scenario were true, of course, it would mean that η Carinae and stars like it in other galaxies may explode as hypernovae at any moment. This would be good for my chances of getting future observing proposals accepted, but I assure the reader that this is not why I am mentioning the possibility.

4.3. *Binaries...?*

In addition to these two alternatives listed above, the potential role of close binaries—in particular, Roche Lobe Overflow (RLOF)—has been a glaring omission so far. In a wide variety of different scenarios depending on initial conditions, close binary evolution can modify a star's mass (see, for example, Vanbeveren et al. 1998). Given the fact that most stars are binaries, this should be considered as well, but I don't wish to get into this complex topic here. I would like to note, however, that like the continuum-driven mass loss in LBV eruptions, mass loss/transfer through RLOF will be relatively insensitive to metallicity compared to line-driven winds. Therefore, some of the comments in the next section apply to binary alternatives as well.

5. Potential implications for the first stars and their environments

The first stars, which should have been metal free, are generally thought to have been predominantly massive, exhibiting a flatter initial mass function than stars at the present epoch (e.g., Bromm & Larson 2004). With no metals, these stars should not have been able to launch line-driven winds, and thus, they are expected to have suffered no mass loss during their lifetimes. The lack of mass loss profoundly affects the star's evolution and the type of supernova it eventually produces (Heger et al. 2003), as well as the yield of chemical elements from the first supernovae and hypernovae that seeded the early interstellar medium of galaxies.

This view rests upon the assumption that mass loss in massive stars at the present time is dominated by line-driven winds, for which $\dot{M}$ can be scaled smoothly with metallicity—but this assumption may be problematic in view of recent observational constraints. As discussed above, massive shells around LBVs and the so-called "supernova impostors" in other galaxies indicate that short-duration eruptions contribute substantially—and may even dominate—the mass loss of very massive stars, while steady, line-driven winds on the MS contribute little to the total mass lost during their lifetime. Unlike line-driven winds, the driving mechanism for these outbursts is probably insensitive to metallicity, as explained in more detail by Smith & Owocki (2006).

Since the trigger of LBV eruptions is still unidentified, one of course cannot yet claim confidently that these eruptions will in fact occur in the first stars. However, the possibility that they offer a way for low-metallicity stars to shed large amounts of mass compels us to consider their potential influence for stellar evolution of the first stars and their surroundings. Some potential consequences are listed here:

- If the first stars were able to shed large amounts of mass through continuum-driven blasts at the end of MS evolution, then it could affect the type of explosion and the type of remnant the star leaves behind. Thus, the expected relative numbers of pair-instability explosions compared to supernovae that produce black holes or neutron stars as their remnants will change. Very massive stars that lose enough mass may fall below the threshold for pair-instability supernovae, allowing much of their core metals to remain trapped inside a black hole or neutron star. This change, in turn, will seriously alter expectations for the chemical yield returned to the ISM by early supernovae, and would affect the initial mass function (IMF) inferred from studies of very metal-poor stars (e.g., J. Tumlinson, these proceedings).

• The early ISM of galaxies was very different than it is today. In the early universe, all elements heavier than He came from massive stars and were recycled back into the ISM. This is partly because of the IMF skewed to higher masses, but mostly because at early times, intermediate-mass stars had not yet evolved off the MS to return C and O to the ISM as an AGB star. If the first stars were able to shed large amounts of mass *before* exploding as supernovae, the ISM would have been profoundly affected. Namely, the pollution of the early ISM could have a substantial contribution of nitrogen-rich CNO ashes, since these massive stars were likely mixed and self-enriched due to rotation. In other words, the early ISM may have been similar to the N-rich material in circumstellar LBV shells seen today. This could significantly affect the dust content of the early ISM as well, especially for the generation of stars immediately following the first stars.

• Continuum-driven blasts at the end of MS evolution might enable the first stars to reach and pass through a WR phase. In that case, the self-enrichment of CNO products in WR atmospheres would likely allow them to have line-driven winds, providing even further mass loss (Vink & de Koter 2005; Eldridge & Vink 2006). In addition to giving us further complications in determining the end product of stellar evolution for Population III stars (pair instability, BH, NS), the existence of a WR phase in the first stars would affect the mechanical energy of the surrounding ISM and would contribute additional N and C, not to mention affecting the immediate circumstellar environment into which a GRB shock expands.

In short, if mass loss of massive stars at the present epoch is dominated by mechanisms that are insensitive to metallicity, then we must question the prevalent notion that the first stars did not lose substantial mass prior to their final supernova event. If these outbursts can occur at low metallicity, it would profoundly alter our understanding of the evolution of the first stars and their role in early galaxies.

I thank Stan Owocki for many relevant discussions, and I thank Paul Crowther and Peter Conti for repeatedly reminding me of the potential importance of WNL stars. I was supported by NASA through grant HF-01166.01A from STScI, which is operated by the Association of Universities for Research in Astronomy, Inc., under NASA contract NAS5-26555.

REFERENCES

Abbott, D. C. & Conti, P. S. 1987 *ARA&A* **25**, 113.

Appenzeller, I. 1986. In *Luminous Stars and Associations in Galaxies* (eds. C. W. H. de Loore, A. J. Willis, & P. Laskarides). IAU Symp. 116, p. 139. D. Reidel Publishing Co.

Aretxaga, I., et al. 1999 *MNRAS* **309**, 343.

Berger, E., et al. 2006 *ApJ* **642**, 979.

Bohannan, B. 1997. In *Luminous Blue Variables: Massive Stars in Transition* (eds. A. Nota & H. Lamers). ASP Conf. Ser. 120, p. 3. Astronomical Society of the Pacific.

Bouret, J. C., Lanz, T., & Hillier, D. J. 2005 *A&A* **438**, 301.

Bromm, V. & Larson, R. B. 2004 *ARA&A* **42**, 79.

Chiosi, C. & Maeder, A. 1986 *ARA&A* **24**, 329.

Chugai, N. N. & Chevalier, R. A. 2006 *ApJ* **641**, 1051.

Chugai, N. N. & Danziger, I. J. 1994 *MNRAS* **268**, 173.

Chugai, N. N., et al. 2004 *MNRAS* **352**, 1213.

Crowther, P. A. 2007 *ARA&A* **45**, 177.

CROWTHER, P. A., ET AL. 2002 *ApJ* **579**, 774.

CROWTHER, P. A., ET AL. 1995 *A&A* **293**, 427.

DAVIDSON, K. 1989. In *Physics of Luminous Blue Variables* (eds. K. Davidson, A. F. J. Moffat, & H. J. G. L. M. Lamers). IAU Coll. 113, p. 101. Kluwer Academic.

DE JAGER, C., ET AL. 1988 *A&AS* **72**, 259.

DRISSEN, L., ROY, J. R., & ROBERT, C. 1997 *ApJ* **474**, L35.

ELDRIDGE, J. J. & VINK, J. S. 2006 *A&A* **452**, 295.

EVANS, C. J., ET AL. 2004 *ApJ* **610**, 1021.

FIGER, D. F. 2005 *Nature* **434**, 192.

FULLERTON, A. W., MASSA, D. L., & PRINJA, R. K. 2006 *ApJ* **637**, 1025.

GERARDY, C. L., ET AL. 2002 *ApJ* **575**, 1007.

HEGER, A., ET AL. 2003 *ApJ* **591**, 288.

HILLIER, D. J., LANZ, T., HEAP, S. R., ET AL. 2003 *ApJ* **588**, 1039.

HUMPHREYS, R. M. & DAVIDSON, K. 1979 *ApJ* **232**, 409.

HUMPHREYS, R. M. & DAVIDSON, K. 1994 *PASP* **106**, 1025.

HUMPHREYS, R. M., DAVIDSON, K., & SMITH, N. 1999 *PASP* **111**, 1124.

KROUPA, P. 2005 *Nature* **434**, 148.

LAMERS, H. J. G. L. M. & FITZPATRICK, E. 1988 *ApJ* **324**, 279.

LANGER, N. 1997 *A&A* **329**, 551.

LANGER, N., ET AL. 1994 *A&A* **290**, 819.

MAEDER, A. & MEYNET, G. 1994 *A&A* **287**, 803.

MASSA, D., ET AL. 2003 *ApJ* **586**, 996.

MAUND, J. R., ET AL. 2006 *MNRAS* **369**, 390.

MEABURN, J. 2001. In *P Cygni 2000: 400 Years of Progress* (eds. M. de Groot & C. Sterken). ASP Conf. Ser. 233, p. 253. Astronomical Society of the Pacific.

MEYNET, G., ET AL. 1994 *A&AS* **103**, 97.

MIRABAL, N., ET AL. 2003 *ApJ* **595**, 935.

MORSE, J. A., ET AL. 2001 *ApJ* **548**, L207.

NIEUWENHUIJZEN, H. & DE JAGER, C. 1990 *A&A* **231**, 134.

OWOCKI, S. P., GAYLEY, K. G., & SHAVIV, N. J. 2004 *ApJ* **616**, 525.

PULS, J., ET AL. 2006 *A&A* **454**, 625.

REPOLUST, T., PULS, J., & HERRERO, A. 2004 *A&A* **415**, 349.

RYDER, S. D., ET AL. 2004 *MNRAS* **349**, 1093.

SMITH, N. 2006 *MNRAS* **367**, 763.

SMITH, N. & HARTIGAN, P. 2006 *ApJ* **638**, 1045.

SMITH, N. & OWOCKI, S. P. 2006 *ApJ* **645**, L45.

SMITH, N., ET AL. 2003 *AJ* **125**, 1458.

SMITH, N., HUMPHREYS, R. M., & DAVIDSON, K. 2001 *PASP* **113**, 692.

SMITH, N., MORSE, J. A., & BALLY, J. 2005 *AJ* **130**, 1778.

SMITH, N., VINK, J., & DE KOTER, A. 2004 *ApJ* **615**, 475.

VANBEVEREN, D., VAN RENSBERGEN, W., & DE LOORE, C., EDS. 1998 *The Brightest Binaries.* Astrophys. Space Sci. Libr., Vol. 232. Kluwer Academic.

VAN DYK, S. D., ET AL. 1993 *ApJ* **419**, L69.

VAN DYK, S. D., ET AL. 2000 *PASP* **112**, 1532.

VAN DYK, S. D., FILIPPENKO, A. V., & LI, W. 2002 *PASP* **114**, 700.

VAN DYK, S. D., ET AL. 2005 *PASP* **117**, 553.

VAN DYK, S. D., ET AL. 2006 *PASP* **118**, 351.

VINK, J. S. & DE KOTER, A. 2005 *A&A* **442**, 587.

WAGNER, R. M., ET AL. 2004 *PASP* **116**, 326.

WALBORN, N. R. 1976 *ApJ* **204**, L17.

From progenitor to afterlife

By ROGER A. CHEVALIER

Department of Astronomy, University of Virginia, P.O. Box 400325,
Charlottesville, VA 22904, USA

The sequence of massive-star supernova types IIP (plateau light curve), IIL (linear light curve), IIb, IIn (narrow line), Ib, and Ic roughly represents a sequence of increasing mass loss during the stellar evolution. The mass loss affects the velocity distribution of the ejecta composition; in particular, only the IIP's typically end up with H moving at low velocity. Radio and x-ray observations of extragalactic supernovae show varying mass-loss properties that are in line with expectations for the progenitor stars. For young supernova remnants, pulsar wind nebulae and circumstellar interaction provide probes of the inner ejecta and higher velocity ejecta, respectively. Among the young remnants, there is evidence for supernovae over a range of types, including those that exploded with much of the H envelope present (Crab Nebula, 3C 58, 0540–69) and those that exploded after having lost most of their H envelope (Cas A, G292.0+1.8).

---

1. Introduction: Core-collapse supernovae

Core-collapse supernovae show considerable diversity among their properties. A basic observational division is into the SNe II (Type II supernovae), which have hydrogen in their spectra, and SNe Ib/c, which do not (or have weak hydrogen lines). The reason for the difference is that the progenitors of the SN Ib/c have lost their H envelopes, and perhaps more, during their evolution leading up to the supernova. The mass loss can occur either through the winds from a single star or can be aided by interaction with a binary companion.

The SNe II show strong diversity themselves. Their observational classification is based on a variety of factors, but it is clear that pre-supernova mass loss plays a significant role in determining the type. Two types are distinguished by their light curves: IIP (plateau) and IIL (linear). Models of Type IIP light curves have long showed that the likely progenitors of the SNe IIP are the red supergiants that end their lives with most of their H envelopes retained (Grasberg et al. 1971; Chevalier 1976). The plateau phase of the light curve is due to the internal energy deposited by the initial explosion. This progenitor hypothesis has been directly confirmed by observations of the progenitors of a number of SNe IIP (Hendry et al. 2006 and references therein). While the SNe IIP might explode with a hydrogen envelope of $\sim$10 $M_{\odot}$, the more rapid decline of the SNe IIL imply that they explode with an envelope of $\sim$1 $M_{\odot}$ (Blinnikov & Bartunov 1993). Because of higher rates of mass loss for more luminous stars, the reduced H envelope is expected to occur for single stars with initial masses of $\gtrsim 20$ $M_{\odot}$. Alternatively, mass loss in a binary system could play a role in the reduced envelope mass.

The prototype of the SNe IIb was SN 1993J, which made a transition from a Type II at early times to a Type Ib/c at late times, based on spectroscopic observations. The H-envelope mass required for SN 1993J was $\sim$0.2 $M_{\odot}$ (Woosley et al. 1994). For this to occur in a single star requires special timing, so a binary origin is preferred. A likely binary companion for SN 1993J has been directly observed (Maund et al. 2005).

SNe IIn have the spectroscopic feature of narrow emission lines (Schlegel 1990), typically Hα, which indicates that circumstellar interaction plays a role in the emission from early times. A supernova can be a Type IIn and another type; e.g., SN 1998S was both a IIn and IIL. Because of the strong circumstellar interaction, it can be difficult to determine the nature of the photospheric emission in an SN IIn. The H emission from circumstellar interaction implies strong mass loss before the supernova in an SN IIn, so the H envelope is likely to be depleted at the time of the supernova.

The most noteworthy peculiar SN II is the nearby SN 1987A, which was relatively compact at the time of the explosion, although it had a massive H envelope. The best explanation for the explosion as a blue supergiant star and the axisymmetric ring features around it is probably that it was in a binary system (Podsiadlowski 1992).

The Type Ib/c supernovae are believed to be H-poor Wolf-Rayet stars at the time of their explosion. There is some observational evidence that SNe Ib, which have He lines, can have a small amount of high-velocity H at the time of the explosion (Elmhamdi et al. 2006). Although the presence of an H envelope in a massive star typically leads to the formation of a red supergiant in the late evolutionary stages, a small amount of H mass ($\lesssim 0.01\ M_\odot$) is not expected to support an extended envelope.

These considerations show that a major factor in the determination of supernova type is the amount of H left in the envelope at the time of the supernova. If the H-envelope mass is greater than the core mass, then the core is effectively decelerated by the envelope and there is mixing between them by Rayleigh-Taylor instabilities. There is not only the outward mixing of heavy elements, but also the inward mixing of H to low velocities. This can be directly observed in the late spectra of SNe IIP; e.g., late spectra of SN 1999em showed H moving at several 100 km s^{-1} (Elmhamdi et al. 2003).

The relative numbers of the different kinds of supernovae is uncertain from an observational point of view. If all the supernovae came from single stars, the stellar mass function was the Salpeter function ($n(M) \propto M^{-2.35}$), and Type IIP supernovae came from 8–20 $M_\odot$ stars, Type IIL 20–25 $M_\odot$, and Type Ib/c $> 25\ M_\odot$, the relative fractions of IIP:IIL:Ib/c would be 0.71:0.08:0.21. Binary evolution could increase the relative number of IIL and Ib/c events (Nomoto et al. 1996). However, SNe IIP are likely to be an important component of the core-collapse supernovae.

After the explosion of the progenitor star, the event has an afterlife in two ways: through its interaction with the surrounding medium and through the possible activity of a central compact remnant (neutron star or black hole). I will discuss how the expectations for the afterlife phases depend on the supernova type and what can be learned about these phases from observations. The early circumstellar interaction observed in extragalactic supernovae is discussed in Section 2, the pulsar wind nebula expansion inside the supernova in Section 3, and circumstellar interaction in young remnants in Section 4. More details on the material in Sections 3 and 4 can be found in Chevalier (2005). Section 5 contains a discussion of future prospects.

2. Early circumstellar interaction

Circumstellar interaction begins soon after the supernova shock wave has emerged from the progenitor star. The radiation-dominated shock front accelerates out the outer edge of the star until the point where radiative losses halt the acceleration process. Because more compact stars have a larger density contrast between the average density and the photospheric density, the shock waves attain higher velocities in more compact stars. The shock break-out radiation accelerates the gas out ahead of the shock, so that the shock front in fact disappears. However, the velocity of the radiatively accelerated gas declines with radius ($\propto r^{-2}$), so that a viscous shock eventually forms. The shocked region is driven by the supernova gas, which has a steep power-law profile in the region where shock acceleration has occurred. The interaction region is bounded by a reverse shock—where the supernova gas is shocked—on the inside and a forward shock—where the circumstellar gas is shocked—on the outside. If both the supernova density profile and the surrounding wind density profile can be described as power laws in radius, the structure and evolution of the interaction region can be described by a self-similar solution (Chevalier 1982).

The early interaction in extragalactic supernovae can be observed in a number of ways: radio emission from shock accelerated electrons, x-ray emission from hot gas and nonthermal processes, optical emission from cooling shock waves and radiatively heated gas, and infrared emission from radiatively or shock heated dust grains. Radio is the best marker of interaction, because it has been observed from all the types of massive-star supernovae (Weiler et al. 2005). On the other hand, radio emission has never been detected from SNe Ia.

Although there is not a good understanding of particle acceleration and magnetic field generation associated with shocked regions, simple models that assume some fraction of the postshock energy density goes into relativistic electrons and magnetic fields do a reasonable job of reproducing the observed evolution of radio supernovae. Shock compression of a stellar wind magnetic field is typically not adequate (unless the magnetic field completely dominates the wind energy flux), so field amplification in the shocked region is required. Possible mechanisms for amplification are hydrodynamic instabilities in the shocked region or field amplification associated with cosmic-ray-driven turbulence in the shock wave (Bell 2004). An additional factor in radio light curves is early low-frequency absorption of the radio emission, as is typically observed. The expected mechanisms are synchrotron self-absorption and free-free absorption by unshocked stellar wind gas.

A basic aspect of a radio light curve is thus the peak luminosity and the time of the peak. Values are shown in Figure 1 for those supernovae that have light curves. The dashed lines in Figure 1 result from a synchrotron self-absorption interpretation of the rise at radio wavelengths; the assumption of equipartition of energy between relativistic electrons and magnetic fields gives the radius, and thus the velocity, of the radio-emitting region. Although equipartition is by no means guaranteed, the results are not sensitive to this assumption. If some process other than synchrotron self-absorption is the dominant absorption process, the radio turn-on is further delayed and the indicated velocity is lower than actually present in the supernova. This is the reason that some of the SNe II have very low apparent velocities; the turn-on is likely due to free-free absorption.

The supernovae divide themselves into three main regions with regard to velocity. The SNe II have the lowest velocities, although not actually as low as indicated by Figure 1. The SNe Ib/c probably are dominated by synchrotron self-absorption and their typical velocities are $\sim 30,000$ km s^{-1}. There are three reasons for the higher velocities in the SNe Ib/c vs. SNe II: shock acceleration during the supernova continues to higher velocities in the SNe Ib/c because of the more compact progenitors, the lower circumstellar densities around the SNe Ib/c give less deceleration of the interaction region, and the SNe Ib/c typically have lower ejecta masses than, but similar energies to, the SNe II so the ejecta have higher mean velocities.

Figure 1 also shows the relatively nearby gamma-ray burst (GRB)-SN Ic associations. They require relativistic or semi-relativistic velocities and thus distinguish themselves from the normal SNe Ib/c. They do not have especially low ejecta mass ($\sim 6\ M_\odot$ was deduced for SN 1998bw, Woosley et al. 1999), so the high velocity must be due either to an extraordinarily large supernova energy or to a different source for the emission, a central GRB engine.

Among the SNe II, the range in radio luminosity is probably due to a range in mass-loss density. At the low-luminosity end are the SNe IIP, which have been detected only in recent years. The mass-loss rates that are implied by the radio (and x-ray) observations are consistent with those suggested by stellar evolution calculations, which are deduced

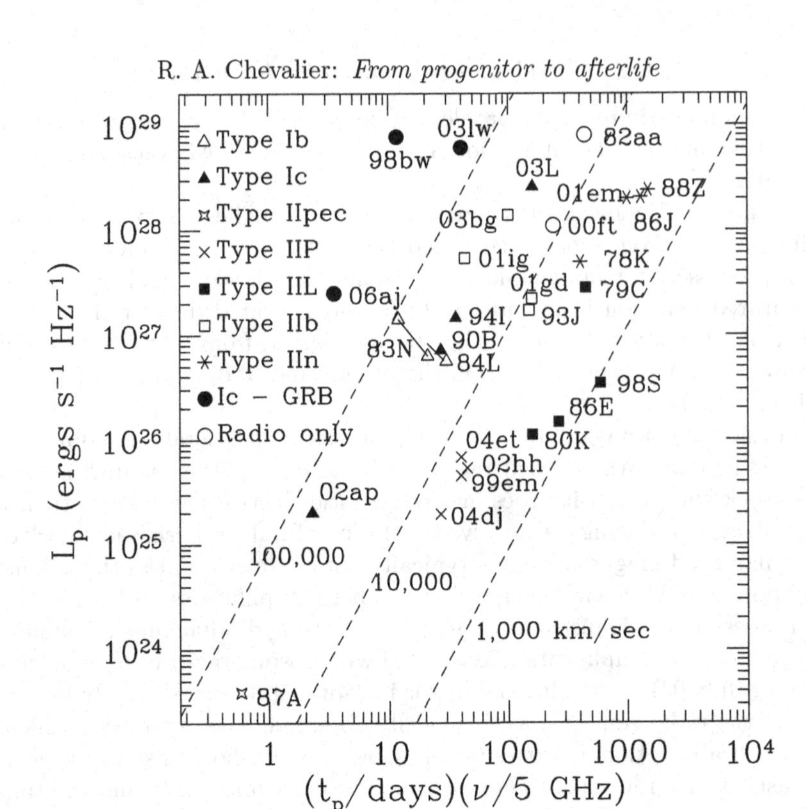

FIGURE 1. Peak radio spectral luminosity vs. the product of the time of peak and the frequency of the measurement. The observed supernovae are designated by the last two digits of the year and the letter, and the Types are indicated by the symbols. The dashed lines show the mean velocity of the radio shell if synchrotron self-absorption is responsible for the flux peak; a value of the energy index $p = 2.5$ is assumed. This is an update of Fig. 4 of Chevalier (1998).

from observations of Galactic stars (Chevalier, Fransson, & Nymark 2006). Even over the mass range 10–20 $M_\odot$, there is a considerable range of mass-loss rate. The existing data are consistent with the expected correlation between mass-loss rate and progenitor mass determined from direct observations, although there are not yet enough data to provide a good test.

The SNe IIL have a higher mass-loss rate—as expected if they came from single stars. If binaries play a role, the expected mass-loss rates are not so clear. At the high radio-luminosity end of the SNe II are the SNe IIn, which appear to have massive, clumpy circumstellar media. In order to obtain the high radio, x-ray, and optical luminosities, the progenitor star must have lost several $M_\odot$ within $\sim 10^3$ years of the supernova.

Results on mass-loss densities are summarized in Table 1, where $\dot{M}$ is the mass-loss rate and v_w is the wind velocity of the progenitor; supernova observations just give the ratio $\dot{M}/v_w$, so the value of v_w has been assumed. For the SNe II (except for SN 1987A), free-free absorption is the likely absorption process, so there is a fairly direct estimate of the circumstellar density, although uncertainties arise because of the dependence of the absorption on the circumstellar temperature.

For the SNe IIb, Table 1 just lists the well-studied SN 1993J, which had a red supergiant progenitor. Figure 1 shows that SN 2001gd was probably similar. However, the radio observations of SN 2001ig and SN 2003bg indicate higher velocity expansion and they probably had Wolf-Rayet star progenitors (Ryder et al. 2004; Soderberg et al. 2006).

SN	$\dot{M}$ ($M_\odot$ yr^{-1})	Assumed v_w (km s^{-1})
IIP	10^{-6}–10^{-5}	10
IIL	10^{-5}–10^{-4}	10
93J (IIb)	3×10^{-5}	10
IIn	$\lesssim 10^{-3}$	10
87A (IIpec)	4×10^{-8}	500
Ib/c	10^{-6}–10^{-4}	1000

TABLE 1. Estimates of $\dot{M}$ for the supernova progenitors

There is presumably a continuous distribution between the SNe IIb, in which the H lines are clearly visible in spectra, and the SNe Ib, in which the Hα line is weak.

The position of SN 1987A in Figure 1 is determined by the low-luminosity radio emission that was observed over the first 200 days after the explosion (Turtle et al. 1987). The initial rise of the radio emission is likely due to synchrotron self-absorption, so estimates of mass-loss density are uncertain, but, if the efficiency of synchrotron production is similar to that for the SNe Ib/c (see below), the density is remarkably low (Table 1). This low density is supported by the rapid expansion that the supernova shock wave made to the time that the first radio imaging observations were carried out (Gaensler et al. 1997).

Since 1990, the radio flux from SN 1987A has been rising because of its interaction with mass lost during a previous red-supergiant stage. This increase was anticipated by the observation of dense gas that had been radiatively illuminated and the ensuing interaction has been observed over a broad wavelength range (McCray 2005). The transition from red-supergiant to blue-supergiant explosion took $\sim 10^4$ years. The radio light curve of SN 1987A is unusual because of its previous red-supergiant phase, although few radio supernovae are followed past an age of three years. Another object that seems to have made a transition to dense gas interaction is SN 2001em, which was initially observed as an SN Ib/c and within three years made a transition to SN IIn, at which time it was a luminous radio and x-ray source (Chugai & Chevalier 2006 and references therein). In this case, there was apparently a phase of dense mass loss within $\sim 10^3$ years of the supernova explosion which ended before the supernova occurred.

Like the SNe II, the SNe Ib/c also have a considerable range in peak radio luminosity (Figure 1). The observed range in luminosity is roughly consistent with the observed range of mass-loss densities for Galactic Wolf-Rayet stars if the efficiency factors (fractions of postshock energy density in magnetic fields and relativistic electrons) do not vary greatly between objects and are ~ 0.1. In this picture, SN 2002ap and SN 2003L roughly represent the low and high extremes for the radio luminosities expected for SNe Ib/c. The low density around SN 2002ap is a factor in the high velocity of the radio region that is inferred for this source.

In addition to radio emission, x-ray emission has been detected from essentially all types of massive-star supernovae, except perhaps SNe Ib. However, there is typically only a small amount of data for any particular supernova, so that the x-ray data can provide a consistency check on deductions from the radio emission, but do not yield much additional information on the mass-loss properties of the progenitors. In the case of SNe II, the x-ray emission is likely to be thermal emission from the shocked ejecta gas. The interpretation of the emission generally depends on the density structure of the supernova. In the case of SNe Ib/c, the thermal interpretation generally does not produce sufficient luminosity, so nonthermal mechanisms are indicated (Chevalier & Fransson 2006). Near maximum optical light, inverse Compton emission can be important, but it cannot explain later emission. Chevalier & Fransson (2006) suggested that the late emission can be explained by synchrotron radiation in a scenario where the forward shock wave is cosmic-ray dominated so that the electron energy spectrum flattens at high energy. More detailed observations are needed to check on this hypothesis.

Optical emission from circumstellar interaction occurs if the interaction is sufficiently dense to produce a radiatively cooling reverse shock wave. It is in this case that a significant fraction of the interaction power can appear at optical wavelengths. Thus, optical emission from interaction is detected from IIn, IIL, and IIb supernovae, but not from IIP or Ib/c supernovae. The Hα line profiles observed for IIL and IIb supernovae typically have the boxy shape that is expected for emission from a fairly narrow region near the reverse shock front. The SNe IIn have narrow centrally peaked Hα emission that is likely to be from slow shock waves driven into circumstellar clumps, although a detailed theory for the formation of such lines is not yet available.

Overall, there is reasonable agreement between the circumstellar media inferred from supernova observations with what is expected around the progenitor. One area of uncertainty is still what the expectations are where binary interaction has been important for the progenitor. In addition, the supernova observations are sensitive to clumping in the circumstellar wind and may provide a method to investigate clumping in winds from late-type stars (e.g., Weiler et al. 2005).

3. Pulsar wind nebulae

Massive stars undergo core collapse at the end of their lives, leading to the formation of a neutron star or black hole. If the neutron star is an active pulsar, the magnetic field and relativistic particles generated by the pulsar create a bubble within the supernova (Figure 2). Because the supernova gas is already freely expanding, the swept-up shell around the bubble accelerates with time and is thus subject to Rayleigh-Taylor instabilities. This picture provides a reasonable account of the properties of the Crab Nebula (Chevalier 1977). A similar model applied to nine young pulsar nebulae is also in accord with the observations, provided that the relativistic particles and magnetic fields are not far from equipartition in the pulsar nebulae (Chevalier 2005). In these models for the pulsar nebulae, the initial rotation periods of the pulsars are in the range 10–100 ms.

The fact that the pulsar nebula interacts with the inner part of the supernova ejecta gives constraints on the supernova type. As discussed in Section 1, the basic supernova types are related to the amount of H envelope that is lost leading up to the explosion. In SNe IIP, with most of their H envelope intact, the core material is slowed by the envelope material and the reverse shock wave during the supernova drives H-rich material back towards the center of the supernova. Once the H envelope has a mass less than that of the core, it does not effectively decelerate the core material and ends up at a high velocity. While the H can give indications of the amount of mass loss, the heavy-element production is related to the initial mass of the star. Below $\sim$12 $M_\odot$, most of the heavy elements synthesized during the stellar evolution end up in the compact object; above this mass, there are increasing amounts of O (oxygen) and other heavy elements.

When the pulsar bubble expands into the supernova, there is a shock wave driven in the ejecta. During the early phases of evolution, the shock is a radiative shock, which can give optical emission. The shock emission declines either because the decreasing density and increasing shock velocity cause a transition to a nonradiative shock, or because the pulsar power declines strongly so that the shell expansion tends toward free expansion. An

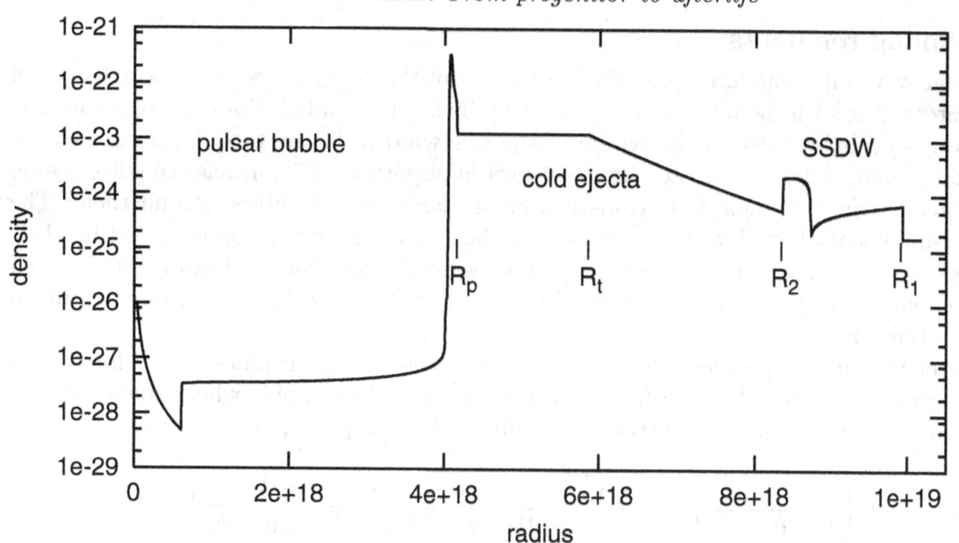

FIGURE 2. The density profile for the interaction of a pulsar nebula with the host supernova remnant. The supernova remnant is modeled as a self-similar driven wave (SSDW) bounded by a forward shock at R_1 and a reverse shock at R_2. The pulsar bubble has swept up a thin shell of ejecta at R_p. The freely expanding, cold ejecta have an inflection point in the density at R_t. The reverse shock wave has not yet reached the pulsar bubble. (from Blondin, Chevalier & Frierson 2001).

alternative source of optical emission is photoionization of the cool gas by the ultraviolet synchrotron radiation from the pulsar bubble.

There are three nebulae where the swept-up gas has apparently been observed: the Crab Nebula, 3C 58, and 0540–69 in the Large Magellanic Cloud. The Crab is especially well studied and shows filaments that are primarily composed of H and He, with not much in the way of heavier elements beyond those present in cosmic abundances. The small amount of heavy elements suggests a relatively low initial mass, perhaps 8–10 $M_\odot$ (Nomoto et al. 1982). The average velocity of the optical filaments is $\sim$1400 km s^{-1} and the gas was probably at a lower velocity before being accelerated by the pulsar bubble. The low velocity of the H is suggestive of an SN IIP in which the H has been mixed back toward the center; this type is also consistent with the estimated initial mass.

The optical filaments in 3C 58 are faint and more difficult to observe, but show lines of H and N, and velocities up to $\sim$1000 km s^{-1} (Fesen, Kirshner, & Becker 1988). These appear to be swept-up ejecta, as opposed to shocked circumstellar gas, because of their relatively high velocities and the fact that they appear in projection only over the pulsar nebula. As in the case of the Crab, the evidence points to a relatively low-mass progenitor and an SN IIP. Although 3C 58 has typically been identified with SN 1181, there are a number of lines of evidence that it is actually $\sim$2500 years old (Chevalier 2005).

The case of 0540–69 is different because strong lines of O and S are present. The presence of H in the filaments has been controversial and it was assumed not to be present in Chevalier (2005), but recent observations definitely show that it *is* in the filaments around the pulsar bubble (Serafimovich et al. 2005), which have velocities $\sim$1000 km s^{-1}. The composition suggests an initial mass $\gtrsim$15 $M_\odot$ and an SN IIP, which would place the supernova among the higher mass SNe IIP.

4. Young remnants

Nearby young remnants with ages up to several thousand years are expected to still be interacting with the mass-loss region set up by the progenitor. However, the mass-loss region is probably beyond the region of the free wind from the progenitor star, so the stellar evolution leading up to the supernova is important. When massive stars are on the main sequence, their fast winds can create large wind bubbles around them. The lower-mass stars have lower mass-loss rates, but this is partially compensated by their longer evolutionary lifetimes. The wind bubbles eventually slow to 10–20 km s^{-1}, which is comparable to the space velocities of the massive stars, so that they can catch up to the bubble on one side.

After the main-sequence phase, stars enter the red-supergiant phase, with slow, dense mass loss ($\sim$10 km s^{-1}). The free wind extends out to the point where the wind ram pressure equals the pressure in the surrounding medium, p, i.e., at

$$r_{RSG} = 5.0 \left(\frac{\dot{M}}{5 \times 10^{-5} \ M_{\odot} \ \mathrm{yr}^{-1}} \right)^{1/2} \left(\frac{v_w}{15 \ \mathrm{km \ s}^{-1}} \right)^{1/2} \left(\frac{p/k}{10^4 \ \mathrm{cm}^{-3} \ \mathrm{K}} \right)^{-1/2} \ \mathrm{pc} \ , \quad (4.1)$$

where k is Boltzmann's constant. This shows that the red-supergiant wind can extend out >5 pc from the supernova. The extended wind around SN 1987A was observed as it was illuminated by the radiation from the supernova, out to a radius $\sim$5 pc (Chevalier & Emmering 1989; Sugerman et al. 2005).

The supernova interaction with a red-supergiant wind can last for thousands of years if the wind is strong and extended, and gives rise to strong radio and x-ray emission. The best case of such interaction appears to be the 325-year-old remnant Cas A. The morphology, expansion rates, and masses are consistent with interaction with a freely expanding wind (Chevalier & Oishi 2003; Laming & Hwang 2003). The remnant contains slow-moving shocked circumstellar clumps, called the quasi-stationary flocculi, that are H and He rich. The fact that these give rise to narrow line emission means that Cas A can be regarded as a very old SN IIn. Whether it was an SN IIn in its early phases depends on how far back the wind extended to the progenitor star. If the progenitor made a transition to a Wolf-Rayet star before the supernova, it would have initially been an SN Ib/c. Chevalier & Oishi (2003) argued that the wind extended back to near the surface based on two points: there are some fast knots containing H, showing that the progenitor had some H at the time of the explosion and the formation of very fast cool knots might be aided by the presence of a dense surrounding wind. However, the knots with H in Cas A have velocities $\sim$10,000 km s^{-1} and SNe Ib can have H with velocities $\gtrsim$12,000 km s^{-1} (Elmhamdi et al. 2006).

A general expectation of strong interaction with a red-supergiant wind is that enough of the H envelope has been lost, so that no H in the ejecta is expected at low ($\lesssim$3000 km s^{-1}) velocity. This is the case for the fast ejecta knots in Cas A. A remnant with strong circumstellar interaction and a pulsar wind nebula is G292.0+1.8, which also has fast moving knots without H. Another remnant is 1E 0102.2–7219 in the Small Magellanic Cloud, which also has strong interaction and H-poor fast knots.

As discussed in the previous section, the suggested type of supernova for the Crab and 3C 58 is a low-mass SN IIP. However, in these cases, there has been no detection of interaction with the circumstellar medium; quite strong limits have been placed on x-ray emission from interaction around the Crab nebula (Seward, Gorenstein, & Smith 2006). A possible explanation for the low emissivity is that the supernova shock wave has passed through the red-supergiant wind, which in this case did not extend out far from the progenitor, and is currently moving in a low-density wind bubble left from the main-sequence phase. Some support for this picture comes from the observation of a faint x-ray shell around the pulsar wind nebula G21.5–0.9 (Matheson & Safi-Harb 2005; Bocchino et al. 2005). This remnant was regarded as a pure pulsar nebula, like the Crab, until long x-ray observations were undertaken with *Chandra.*

Chevalier (2005) suggested that the remnant 0540–69 came from an SN Ib/c based on the apparent rapid expansion of the outer ejecta, which implied a low circumstellar density in the region surrounding the progenitor. However, as discussed in Section 3, H is present in the slow-moving ejecta, which implies a IIP supernova. In this case, the interaction that is observed at a radius of 6–10 pc is probably with the interstellar medium; Hwang et al. (2001) estimate that mass of x-ray emitting gas is $\sim 40\ M_\odot$. The fact that the x-ray temperature is relatively low for the average shock velocity suggests that the remnant is interacting with clumps or clouds. The problem with the IIP designation is the rapid expansion despite the expected interaction with the slow wind from the red-supergiant progenitor. This issue requires more investigation.

5. Discussion and conclusions

There are excellent future prospects for developing a more complete picture of the massive-star evolution leading up to a supernova and the subsequent expansion of the supernova into the circumstellar medium. The increasing number of *Hubble Space Telescope* images of galaxies has improved the prospects for identifying the progenitor stars of nearby supernovae. Follow-up observations at radio and x-ray wavelengths can then reveal the mass-loss environment for that particular progenitor.

There is a growing number of young remnants that have observed pulsar wind nebulae and/or circumstellar interaction. Many of these have been well observed at x-ray wavelengths (owing to *Chandra* and *XMM*), but are less well observed at optical and infrared wavelengths. Infrared observations seem especially important because a number of the objects have high extinction. As the amount of information increases, there is the possibility of looking for correlations between the nature of the compact object in a remnant and the nature of the surrounding supernova. An initial examination of this point (Chevalier 2005) did not reveal any correlations.

Along with these endeavors, hydrodynamic modeling of the variety of supernova events, along with their interaction with mass loss, is needed. The result will be a better understanding of the final evolution of massive stars and the variety of possible outcomes.

This research was supported in part by NSF grant AST-0307366 and NASA grant NAG5-13272.

REFERENCES

Bell, A. R. 2004 *MNRAS* **353**, 550.
Blinnikov, S. I. & Bartunov, O. S. 1993 *A&A* **273**, 106.
Blondin, J. M., Chevalier, R. A., & Frierson, D. M. 2001 *ApJ* **563**, 806.
Bocchino, F., van der Swaluw, E., Chevalier, R., & Bandiera, R. 2005 *A&A* **442**, 539.
Chevalier, R. A. 1976 *ApJ* **207**, 872.
Chevalier, R. A. 1977. In *Supernovae* (ed. D. N. Schramm) p. 53. Reidel.
Chevalier, R. A. 1982 *ApJ* **258**, 790.
Chevalier, R. A. 1998 *ApJ* **499**, 810.
Chevalier, R. A. 2005 *ApJ* **619**, 839.
Chevalier, R. A. & Emmering, R. T. 1989 *ApJ* **342**, L75.
Chevalier, R. A. & Fransson, C. 2006 *ApJ* **651**, 381.

Chevalier, R. A., Fransson, C., & Nymark, T. 2006 *ApJ* **641**, 1029.

Chevalier, R. A. & Oishi, J. 2003 *ApJ* **593**, L23.

Chugai, N. N. & Chevalier, R. A. 2006 *ApJ* **641**, 1051.

Elmhamdi, A., et al. 2003 *MNRAS* **338**, 939.

Elmhamdi, A., Danziger, I. J., Branch, D., Leibundgut, B., Baron, E., & Kirshner, R. P. 2006 *A&A* **450**, 305.

Fesen, R. A., Kirshner, R. P., & Becker, R. H. 1988. In *Supernova Remnants and the Interstellar Medium* (eds. R. S. Roger & T. L. Landecker). p. 55. Cambridge University Press.

Gaensler, B. M., Manchester, R. N., Staveley-Smith, L., Tzioumis, A. K., Reynolds, J. E., & Kesteven, M. J. 1997 *ApJ* **479**, 845.

Grasberg, E. K., Imshenik, V. S., & Nadyozhin, D. K. 1971 *Ap. Sp. Sci.* **10**, 28.

Hendry, M. A., et al. 2006 *MNRAS* **369**, 1303.

Hwang, U., Petre, R., Holt, S. S., & Szymkowiak, A. E. 2001 *ApJ* **560**, 742.

Laming, J. M. & Hwang, U. 2003 *ApJ* **597**, 347.

Matheson, H. & Safi-Harb, S. 2005 *Adv. Sp. Res.* **35**, 1099.

Maund, J. R., Smartt, S. J., Kudritzki, R. P., Podsiadlowski, P., & Gilmore, G. F. 2005 *Nature* **427**, 129.

McCray, R. A. 2005. In *Cosmic Explosions, On the 10th Anniversary of SN1993J* (eds. J. M. Marcaide & K. W. Weiler). p. 77. Springer.

Nomoto, K., Iwamoto, K., Suzuki, T., Pols, O. R., Yamaoka, H., Hashimoto, M., Hoflich, P., & van den Heuvel, E. P. J. 1996. In *Compact Stars in Binaries* (eds. J. van Paradijs, E. P. J. van den Heuvel, & E. Kuulkers). p. 119. Kluwer.

Nomoto, K., Sugimoto, D., Sparks, W. M., Fesen, R. A., Gull, T. R., & Miyaji, S. 1982 *Nature* **299**, 803.

Podsiadlowski, E. M. 1992 *PASP* **104**, 717.

Ryder, S. D., Sadler, E. M., Subrahmanyan, R., Weiler, K. W., Panagia, N., & Stockdale, C. 2004 *MNRAS* **349**, 1093.

Schlegel, E. M. 1990 *MNRAS* **244**, 269.

Serafimovich, N. I., Lundqvist, P., Shibanov, Yu. A., & Sollerman, J. 2005 *Adv. Sp. Res.* **35**, 1106.

Seward, F. D., Gorenstein, P., & Smith, R. K. 2006 *ApJ* **636**, 873.

Soderberg, A. M., Chevalier, R. A., Kulkarni, S. R., & Frail, D. A. 2006 *ApJ* **651**, 1005.

Sugerman, B. E. K., Crotts, A. P. S., Kunkel, W. E., Heathcote, S. R., & Lawrence, S. S. 2005 *ApJS* **159**, 60.

Turtle, A. J., Campbell-Wilson, D., Bunton, J. D., Jauncey, D. L., & Kesteven, M. J. 1987 *Nature* **327**, 38.

Weiler, K. W., Dyk, S. D. V., Sramek, R. A., Panagia, N., Stockdale, C. J., & Montes, M. J. 2005. In *1604–2004: Supernovae as Cosmological Lighthouses* (eds. M. Turatto, S. Benetti, L. Zampieri, & W. Shea). p. 290. ASP.

Woosley, S. E., Eastman, R. G., & Schmidt, B. P. 1999 *ApJ* **516**, 788.

Woosley, S. E., Eastman, R. G., Weaver, T. A., & Pinto, P. A. 1994 *ApJ* **429**, 300.

Pair-production supernovae: Theory and observation

By EVAN SCANNAPIECO

Kavli Institute for Theoretical Physics, Kohn Hall, University of California–Santa Barbara, Santa Barbara, CA 93106, USA

Nonrotating stars that end their lives with masses $140\,M_\odot \leqslant M_\star \leqslant 260\,M_\odot$ should explode as pair-production supernovae (PPSNe). Here I review the physical properties of these objects, as well as the prospects for them to be observationally constrained.

In very massive stars, much of the pressure support comes from the radiation field, meaning that they are loosely bound, and that $(d\lg p/d\lg \rho)_{\rm adiabatic}$ near the center is close to the minimum value necessary for stability. Near the end of C/O burning, the central temperature increases to the point that photons begin to be converted into electron–positron pairs, softening the equation of state below this critical value. The result is a runaway collapse, followed by explosive burning that completely obliterates the loosely bound star. While these explosions can be up to 100 times more energetic than core collapse and Type Ia supernovae, their peak luminosities are only slightly greater. However, due both to copious Ni^{56} production and hydrogen recombination, they are brighter much longer, and remain observable for ≈1 year.

Since metal enrichment is a local process, PPSNe should occur in pockets of metal-free gas over a broad range of redshifts, greatly enhancing their detectability, and distributing their nucleosynthetic products about the Milky Way. This means that measurements of the abundances of metal-free stars should be thought of as directly constraining these objects. It also means that ongoing supernova searches, which limit the contribution of very massive stars to $\lesssim$1% of the total star-formation–rate density out to $z \approx 2$, already provide weak constraints for PPSN models. A survey with the NIRCam instrument on *JWST*, on the other hand, would be able to extend these limits to $z \approx 10$. Observing a 0.3 deg^2 patch of sky for ≈1 week per year for three consecutive years, such a program would either detect or rule out the existence of these remarkable objects.

1. Introduction

Pair-production supernovae (PPSNe) are the uniquely calculable result of nonrotating stars that end their lives in the 140–260 $M_\odot$ mass range (Heger & Woosley 2002, hereafter HW02). Their collapse and explosion result from an instability that generally occurs whenever the central temperature and density of star moves within a well-defined regime (Barkat, Rakavy, & Sack 1967). While this instability arises irrespective of the metallicity of the progenitor star, PPSNe are expected only in a primordial environment, and there are three main reasons for this association.

First, in the present metal-rich universe, it appears that stars this large are never assembled, as supported by a wide range of observations. Figer (2005) carried out a detailed study of the $Z \approx Z_\odot$ Arches cluster, which is large ($M_\star > 10^4\,M_\odot$), young (τ = 2.0–2.5 Myrs), and at a well-determined distance, making it ideal for such studies. No stars more massive than 130 $M_\odot$ were found, although more than 18 were expected. A similar ≈150 $M_\odot$ limit was found in the lower-metallicity cluster R136 in the Large Magellanic Cloud (Weidner & Kroupa 2004), and from a grab bag of clusters compiled by Oey & Clarke (2005).

However, there are good theoretical reasons to believe that the situation may have been very different under primordial conditions. In this case, the primary coolant at low temperatures is molecular hydrogen, which starts to be populated according to local thermodynamic equilibrium (LTE) at a typical density and temperature of $\approx 10^4$ cm^{-3} and 100 K. As the Jeans mass under these conditions is $\approx 10^3\, M_\odot$, the fragmentation of primordial molecular clouds may have been biased towards the formation of stars with very high masses (Nakamura & Umemura 1999; Abel, Bryan, & Norman 2002; Schneider et al. 2002; Tan & McKee 2004).

Finally, as very massive, radiatively supported stars are only loosely bound, they tend to drive large winds. However, these winds are primarily line-driven and scale with metallicity of $Z^{1/2}$ or faster (Kudritzki 2000; Vink et al. 2001; Kudritzki 2002). As long as another mechanism did not act to generate significant mass loss in primordial stars (e.g., Smith & Owocki 2006), this raises the real possibility that they may have not only been born, but have ended their lives in the mass range necessary to drive PPSNe.

This conference proceeding summarizes both the underlying physics and the prospects for observation of these most powerful of astrophysical explosions, and it is structured as follows: In Section 2, I describe the pair-production instability in detail and how it eventually leads to stellar disruption. Section 3 is a discussion of the post-explosion physics of these objects, and how it effects their luminosity and temperature evolution. In Section 4, I discuss the optical light curves of PPSNe, and contrast them with light curves of SNe Type II and Ia. Section 5 discusses the redshift evolution of metal-free stars, and its implications for PPSN environments. In Section 6, these estimates are used to determine the feasibility of present and future PPSN. In Sections 7 and 8, I discuss the likely distribution of the descendents of metal-free stars in the Galaxy, and I close with a short summary in Section 9.

2. Physics of pair-production supernovae

The instability that leads to the formation of PPSNe was first identified by Barkat, Rakavy, & Sack (1967), who carried out a detailed analysis of the relevant equation of state for very massive stars near the end of their lifetimes. These results are shown in Figure 1. In hydrostatic balance $p \propto M^{2/3}\rho^{4/3}$, which means that if the adiabatic coefficient softens to below $\gamma = 4/3$, the star will become unstable to the point of runaway collapse.

Figure 1 shows that such a collapse occurs for stars with central temperatures from 10^9 to 3×10^9 K, and central densities less than 5×10^5 g cm^{-1}. This region is bounded by three limits. Below the low-temperature boundary on the left, most of the central pressure is provided by the radiation field, such that $\gamma \approx 4/3$. As the temperature increases, photon energies rise to the point that electron–positron pairs begin to be made, removing energy from the radiation field and softening γ below the critical value. The high-density end of this region, on the other hand, is bounded by degeneracy pressure, which provides sufficient support to halt collapse at high densities. Finally, at the high-temperature boundary of the pair-instability regime, the energy consumed in creating the rest mass becomes less significant, and γ remains above 4/3.

As massive stars are very loosely bound, their evolution is somewhat more complicated than their lower-mass counterparts, and thus require more detailed numerical modeling. For this reason, Barkat, Rakavy, & Sack (1967; see also Fraley 1968) thought that a $M \geqslant 30\, M_\odot$ star would enter the PPSN regime, although we now know that this mass limit is much higher (Bond, Arnett, & Carr 1984; HW02).

As they collapse, the central temperature and density quickly increase in such stars, moving right through the unstable regime and starting explosive burning in the carbon-oxygen core. This explosive burning takes place in an environment in which there are very few excess neutrons, resulting in a large deficiency in the number of nuclei with an

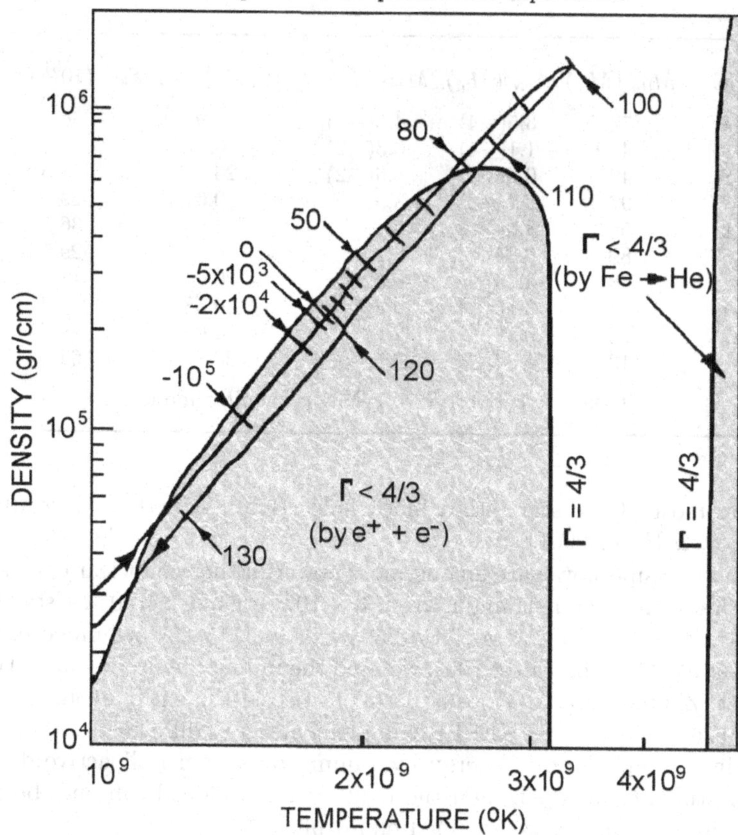

FIGURE 1. Original plot from Barkat, Rakavy, & Sack (1967) showing the range of central temperatures and densities at which the equation of state softens below the $\gamma = 4/3$ value required for stability. The solid lines give an early estimate of the evolution of a $40\,M_\odot$ star near the end of its life (which is now known not to lead to PPSNe). The region on the right shows the temperature and density region relevant in usual core-collapse supernovae.

odd charge above ^{14}N. Thus, elements such as Na, Al, and P are almost two orders of magnitude less abundant than neighboring elements with an even charge, such as Ne, Mg, and Si. This strong "odd-even effect" is a very general feature that is unavoidable in a model of the nucleosynthetic products of PPSNe (HW02).

Models of the kinematics of PPSNe are similarly robust. For stars with initial masses more than about $140\,M_\odot$, the energy released during explosive burning is sufficient to completely disrupt the star, resulting in a PPSN. This complete disruption means that while the evolution of these stars is somewhat more complex than lower-mass stars, the explosion mechanism driving the resulting supernovae is far simpler. There are no issues of fallback, mass-cut, or neutrino heating, and in the nonrotating case, the results are uniquely calculable.

A shock moves outward from the edge of the core, initiating the supernova outburst when it reaches the stellar surface. Just above the $140\,M_\odot$ limit, weak silicon burning occurs and only trace amounts of radioactive ^{56}Ni are produced. It is this ^{56}Ni that powers the late-time supernova light curves. The amount of ^{56}Ni produced increases in larger progenitors, and in $260\,M_\odot$ progenitors up to $50\,M_\odot$ may be synthesized, $\approx$100 times more than in a typical Type Ia supernova. For stars with masses above $260\,M_\odot$, however, the onset of photodisintegration in the center imposes an additional instability

Model	$M_{\rm He}$ ($M_\odot$)	$M_{\rm N}$ ($M_\odot$)	$M_{^{56}\rm Ni}$ ($M_\odot$)	R (10^{13} cm)	$\mathcal{E}_{\rm kin}$ (10^{51} ergs)
150-W	70	3.5(−4)	4.2(−2)	3.9	6.9
150-I	46	1.1(−4)	6.3(−2)	16	9.2
150-S	49	0.86	8.6(−2)	26	8.5
200-W	97	2.7(−6)	3.3	0.68	29.5
200-I	58	8.0(−6)	5.1	2.8	36.5
200-S	89	0.34	2.2	29	29.1
200-S2	78	4.75	0.82	20	18.7
250-W	123	3.1(−6)	6.2	0.58	47.2
250-I	126	9.1(−6)	32	4.0	76.7
250-S	113	1.34	24.5	26	64.6

TABLE 1. Properties of PPSN progenitor models

that collapses most of the star into a black hole (Bond, Arnett & Carr 1984; HW02; Fryer, Woosley, & Heger 2001).

Pair-production supernovae are among most powerful thermonuclear explosions in the universe, with a total energies ranging from 3×10^{51} ergs for a 140 $M_\odot$ star (64 helium $M_\odot$ core) to almost 100×10^{51} ergs for a 260 $M_\odot$ star (133 $M_\odot$ helium core; HW02). In Scannapieco et al. (2005, hereafter S05), we used the implicit hydrodynamical code KEPLER (Weaver, Zimmerman, & Woosley 1978) to model the entire evolution of the star and the resulting light curves. KEPLER implements gray diffusive radiation transport with approximate deposition of energy by gamma rays from radioactive decay of ^{56}Ni and ^{56}Co (Eastman et al. 1993), and the light curves obtained can only be followed as long as there is a reasonably well-defined photosphere.

A supernova can be bright either because it makes a lot of radioactive ^{56}Ni (as in Type Ia supernovae) or because it has a large low-density envelope and large radius (as in bright Type II supernovae). More radioactivity gives more energy at late times, while a larger initial radius results in a higher luminosity at early times. Here, the most important factors in determining the resulting light curves are the mass of the progenitor star and the efficiency of dredge-up of carbon from the core into the hydrogen envelope during or at the end of central helium burning. The specifics of the physical process encountered here are unique to primordial stars. Lacking initial metals, they have to produce the material for the CNO cycle themselves, through the synthesis of ^{12}C by the triple-alpha process. Just enough ^{12}C is produced to initiate the CNO cycle and bring it into equilibrium: a mass fraction of 10^{-9} when central hydrogen burning starts, and a mass fraction $\sim 10^{-7}$ during hydrogen-shell burning.

At these low values, the entropy in the hydrogen shell remains barely above that of the core, and the steep entropy gradient at the upper edge of the helium core that is typical for metal-enriched helium-burning stars is absent. This means that, during helium burning the central convection zone can get close, nip at, or even penetrate the hydrogen-rich layers. Once such mixing of high-temperature hydrogen and carbon occurs, the two components burn violently, and even without this rapid reaction, the hydrogen burning in the CNO cycle increases proportionately to the additional carbon. Thus mixing of material from the helium-burning core, which has a carbon abundance of order unity, is able to raise the energy generation rate in the hydrogen-burning shell by orders of magnitude over its intrinsic value.

This mixing has two major effects on the PPSN progenitor: first, it increases the opacity and energy generation in the envelope, leading to a red-giant structure for the pre-supernova star, in which the radius increases by over an order of magnitude. Second, it decreases the mass of the He core, consequently leading to a smaller mass of ^{56}Ni being synthesized and a smaller explosion energy. The former effect increases the luminosity of the supernova at early times, while the latter effect can weaken it. In S05 we accounted for these uncertainties by employing different values of convective overshooting.

A suite of representative models was chosen to address the expected range of pre-supernova models—from blue-supergiant progenitors with little or no mixing to well-mixed red hypergiants, which can have pre-SN radii of 20 AU or more. These models are summarized in Table 1, in which the names refer to the mass of the progenitor star (in units of $M_\odot$) and the weak (W), intermediate (I), or strong (S) level of convective overshoot. Here we show the final mass in helium, nitrogen, nickel, the radius just before the explosion, and the kinetic energy of the explosion in units of 10^{51} ergs.

3. Luminosity and temperature evolution

The KEPLER code can be used to compute approximate light curves and has been validated against much more complex and realistic codes such as EDDINGTON, and observations of a prototypical Type II-P supernova, SN 1969L (Weaver & Woosley 1980; Eastman et al. 1994), although its main deficiency is that it is a single-temperature code using flux-limited radiative diffusion. The evolution of the luminosities, effective temperatures, and photospheric radii for the models in Table 1 are shown in Figure 2. As the shock moves toward the low-density stellar surface, its energy is deposited into progressively smaller amounts of matter. This results in high velocities and temperatures when the shock reaches the stellar surface, causing a pulse of ultraviolet radiation with a characteristic timescale of a few minutes. This "breakout" phase is by far the most luminous and bluest phase of the PPSN burst, but its very short duration makes it difficult to use in observational searches. In fact, the analog of this phase in conventional SNe has so far only been indirectly detected in SN 1987A (e.g., Hamuy et al. 1988; Catchpole et al. 1988).

Following breakout, the star expands with $R_{\rm phot}$ initially proportional to time. Though a small fraction of the outer mass may move much faster, the characteristic velocity of the photosphere during this phase is a modest $v = (2KE/M)^{1/2} \sim (10^{53}\ {\rm ergs}/200\,M_\odot)^{1/2} \sim 5000\ {\rm km\ s^{-1}}$, because of the very large mass participating in the explosion. During the expansion, the radiation-dominated ejecta cool adiabatically, with T approximately proportional to R^{-1}, with an additional energy input from the decay of ^{56}Ni (if a significant mass was synthesized during the explosion) and hydrogen recombinations (when $T \approx 10^4$ K). As the scale radius for this cooling is the radius of the progenitor, the temperatures and luminosities are substantially larger throughout this phase in the cases with the strongest mixing.

After $\approx$50 days, the energy input from ^{56}Co decay becomes larger than the remaining thermal energy and the energy deposited by ^{56}Co in deeper layers that were enriched in ^{56}Ni can diffuse out. For stars that were compact to begin with, this can cause a delayed rise to the peak of the light curve. For stars with larger radii, the radioactivity just makes a bright tail following the long plateau in emission from the expanding envelope. Eventually, even the slow-moving inner layers recombine and there is no longer a well-defined photosphere. At this time, the assumption of LTE breaks down, and more detailed radiative transfer calculations are required, which are beyond the scope of our S05 modeling. The SN is fainter and redder during this phase, however, and thus is difficult to detect at cosmological distances in optical and near infrared (NIR) surveys.

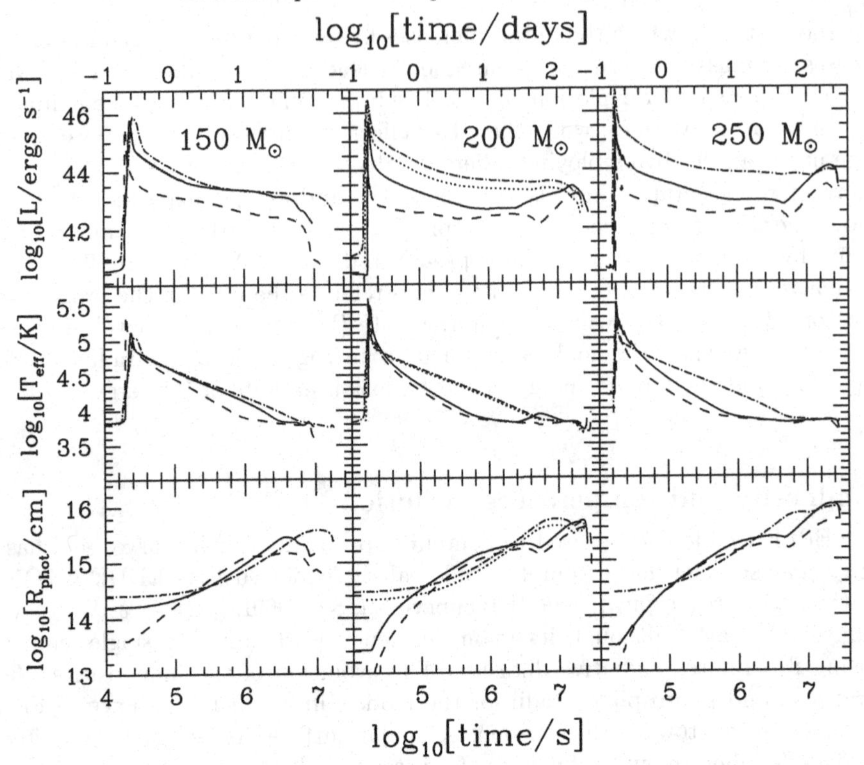

FIGURE 2. Luminosities (top row), effective temperatures (center row), and photospheric radii (bottom row) of PPSNe for ten different representative models. Each set of panels is labeled with the total mass of the progenitor star, and models with weak (dashed), intermediate (solid), and strong (dot-dashed) convective overshoot are shown. Finally, the dotted lines in the central panels correspond to model 200-S2. See text and Table 1 for details.

4. PPSNe light curves

In S05, we calculated approximate PPSNe light curves, assuming a blackbody distribution with the color temperature equal to the effective temperature. Recall that the peak frequency in this case occurs at a wavelength of $5100(T_{\rm eff/10,000\ K})^{-1}$ Å. The resulting light curves are shown in Figure 3, which gives the evolution of AB magnitudes at three representative wavelengths: 5500 Å, corresponding to the central wavelength of the V band; 4400 Å, corresponding to the B band; and 3650 Å, corresponding to the U band. We focus on blue wavelengths, as it is features in these bands that will be redshifted into the optical and NIR at cosmological distances. For comparison, we also include observed light curves for an SN Type Ia (1994D as measured by Patat et al. 1996 and Cappellaro et al. 1997), an SN Type II-P (1999em as observed by Elmhamdi et al. 2003), and the very bright Type II-L SN 1979C (de Vaucouleurs et al. 1981; Barbon et al. 1982).

The most striking feature from this comparison is that despite enormous kinetic energies of $\sim 50 \times 10^{51}$ ergs, the peak optical luminosities of PPSNe are similar to those of other SNe, even falling below the Ia and II curves in many cases. This is because the higher ejecta mass produces large optical depth, and most of the internal energy of the gas is converted into kinetic energy by adiabatic expansion. Furthermore, the colors of the PPSN curves are not unlike those of more usual cases. In fact, pair-production supernovae spend most of their lives in the same temperature range as other SNe. Clearly,

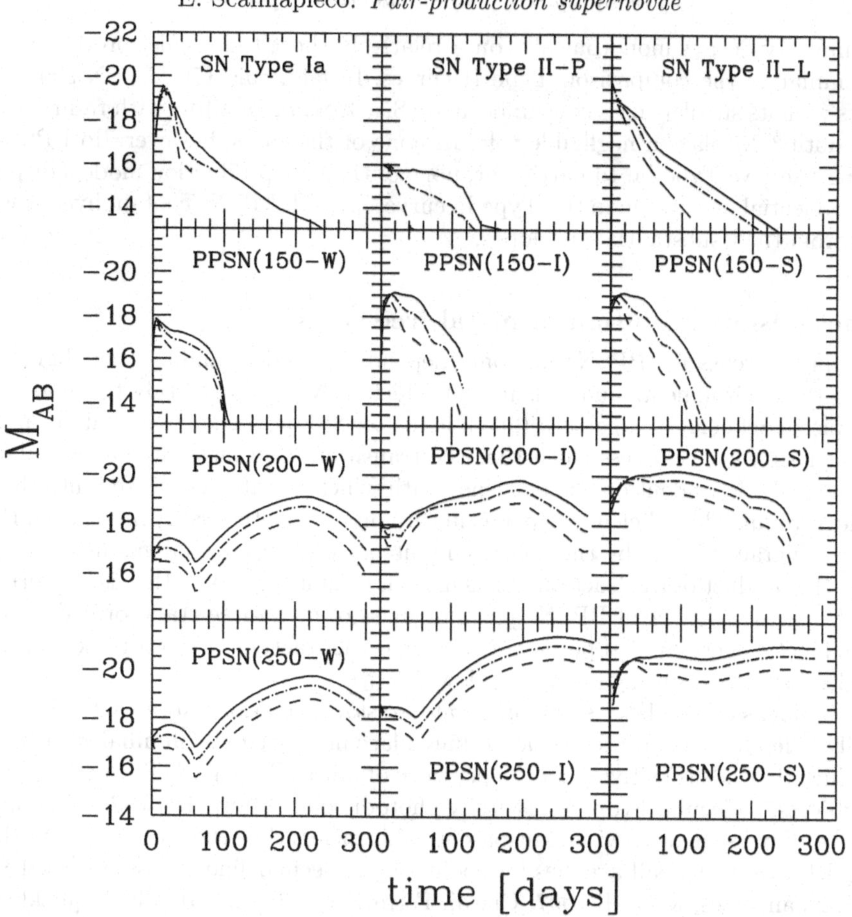

FIGURE 3. Comparison of light curves of an SN Type Ia, an SN Type II-P, a bright SN Type II-L, and PPSNe models with varying progenitor masses and levels of dredge-up. In all cases, the solid lines are absolute V-band AB magnitudes, the dot-dashed lines are the absolute B-band AB magnitudes, and the dashed lines are the absolute U-band AB magnitudes.

then, PPSNe will not be obviously distinguishable from their more usual counterparts "at first glance."

Rather, distinguishing PPSNe will from other SNe will require multiple observations that constrain the time evolution of these objects. In particular, there are two key features that are uniquely characteristic to PPSNe. The first of these is a dramatically extended intrinsic decay time, which is especially noticeable in the models with the strongest enrichment of CNO in the envelope. This is due to the long adiabatic cooling times of supergiant progenitors whose radii are $\sim$20 AU, but whose expansion velocities are similar, or even less than, those of other SNe. Second, PPSNe are the only objects that show an extremely late rise at times $\gtrsim$100 days. This is due to energy released by the decay of ^{56}Co, which, unlike in the Type Ia case, takes months to dominate over the internal energy imparted by the initial shock. In this case, the feature is strongest in models with the least mixing and envelope enrichment during helium burning, as these have the largest helium cores and consequently the largest ^{56}Ni masses.

Note, however, that neither of these features is generically present in all PPSNe, and both can be absent in smaller VMS that fail to expand to large sizes through dredge-up and do not synthesize appreciable amounts of ^{56}Ni. In the case of 150-W, for example, the luminosity decays monotonically on a relatively short time scale, producing a light curve similar to the comparison Type II curves. In fact, this $150\,M_\odot$ SN shares many features with its smaller-mass cousin: both are SNe from progenitors with radii $\sim 10^{13}$ cm, and in both ^{56}Ni plays a negligible role. In none of the cases, however, do PPSNe look anything like SNe Type Ia. In particular, none of the pair-production models display the long exponential decay seen in the Type Ia curves, and all PPSNe contain hydrogen lines, arising from their substantial envelopes.

5. The redshift evolution of metal-free stars

Planning searches for PPSNe not only depends on understanding their light curves, but also the environments and redshifts at which they are most likely to be located. In Scannapieco, Schneider, & Ferrara (2003; hereafter SSF03) we showed that cosmological enrichment is a local process, such that the transition from metal-free to Population II stars is heavily dependent on the efficiency with which metals were mixed into the intergalactic medium. This efficiency depends in turn on the kinetic energy input from PPSNe, which was parameterized by the "energy input per unit primordial gas mass" $\mathcal{E}_{\rm g}^{III}$, defined as the product of the fraction of gas in each primordial object that is converted into stars ($f_\star^{III}$), the number of PPSNe per unit mass of metal-free stars formed ($\mathcal{N}^\pm$), the average kinetic energy per supernova ($\mathcal{E}_{\rm kin}$), and the fraction of the total kinetic energy channeled into the resulting galaxy outflow ($f_{\rm wind}$).

Here we focus on the later stage of metal-free star formation. Note that this is fundamentally different than star formation taking place in very small "minihalos" at redshifts ≈ 25, which depends sensitively on the presence of initial H_2 (e.g., O'Shea et al. 2005). In small objects, molecular hydrogen is easily photodissociated by 11.2–13.6 eV photons, to which the universe is otherwise transparent. This means that the emission from the first stars quickly destroyed all avenues for cooling by molecular line emission (Dekel & Rees 1987; Haiman, Rees, & Loeb 1997; Ciardi, Ferrara, & Abel 2000), which quickly raised the minimum virial temperature necessary to cool effectively to approximately 10^4 K. Thus the majority of primordial star formation is likely to have occurred in objects above this limit, who form their own H_2 at high densities and are largely impervious to the photodissociating background (Oh & Haiman 2002).

Incorporating outflows into a detailed analytical model of such "primordial galaxies" leads to the approximate relation that, by mass, the fraction of the total star formation in metal-free stars at $z = 4$ is

$$F_\star^{III}(z=4) \sim 10^{-5}\left(\mathcal{E}_{\rm g}^{III}\right)^{-1}, \tag{5.1}$$

where, as above, $\mathcal{E}^\pm$ is in units of 10^{51} ergs per $M_\odot$ of gas (see Figure 3 of SSF03 for details). Extrapolating the results in SSF03 to $z = 0$ gives

$$F_\star^{III}(z=0) \sim 10^{-5.5}\left(\mathcal{E}_{\rm g}^{III}\right)^{-1}. \tag{5.2}$$

These fractions can be related to the underlying population of stars by adopting fiducial values of $f_\star^{III} = 0.1$ for the star-formation efficiency, which is consistent with the observed star-formation–rate density at intermediate and high redshifts (Scannapieco, Ferrara, & Madau 2002), and $f_w = 0.3$ for the wind efficiency, which is consistent with the dwarf galaxy outflow simulations of Mori, Ferrara, & Madau (2002). Finally, we assume that one pair-production SN occurs per $1000\,M_\odot$ of metal-free stars. This gives $F_\star^{III}$ values of $0.3(\mathcal{E}_{\rm kin})^{-1}$ at $z = 4$ and $0.1(\mathcal{E}_{\rm kin})^{-1}$ at $z = 0$, respectively. Or, in other words, for typical energies of 30×10^{51} ergs per PPSNe, $\sim 1\%$ of the star formation at $z = 4$ and $\sim 0.3\%$ of the star formation at $z = 0$ by mass should be in metal-free stars.

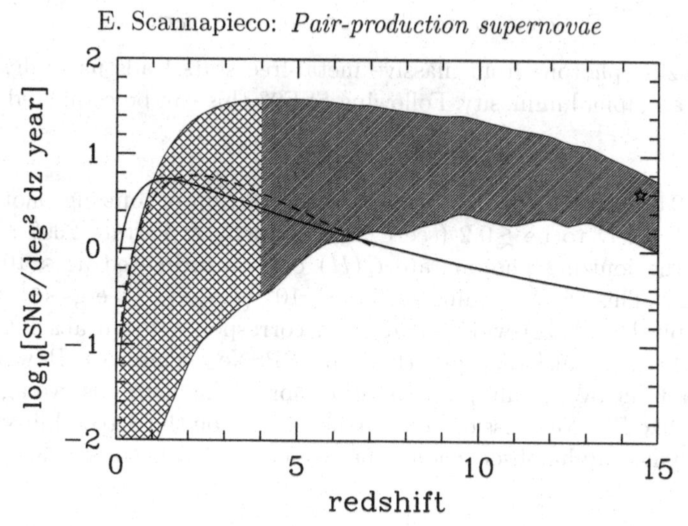

FIGURE 4. Number of PPSNe per square degree per unit redshift per year for a wide range of models. The solid and dashed curves assume Pop III star-formation–rate densities of $0.001\,M_\odot$ yr^{-1} Mpc^{-3} and 1% of the observed star-formation–rate density, respectively. The shaded region covers the range of metal-free star-formation–rate density models considered in SSF03, with the weakest feedback model ($\mathcal{E}_g^{III} = 10^{-4}$) defining the upper end, and the strongest feedback model ($\mathcal{E}_g^{III} = 10^{-2.5}$) defining the lower end. An extrapolation of these star-formation–rate densities to $z = 0$ leads to the crosshatched region. In all SSF03 models the highest rates occur at redshifts $\leqslant$10. Finally, the starred point is the $z = 15$ estimate by Weinmann & Lilly (2005).

In Figure 4 we show estimates of the number of SNe per deg^2 per *dz per year* over the wide range of models considered in SSF03, extrapolating to $z = 0$. In all cases we assume that one PPSN forms per $1000\,M_\odot$ metal-free stars, and for comparison we show two simple estimates, which we refer to further below. In the first simple model, we assume that metal-free star formation occurs at a constant rate density, which we take to be

$$\rho^{\pm}(z)0.001\,M_\odot\,\mathrm{yr}^{-1}\,\mathrm{Mpc}^{-3} \quad ,$$

independent of redshift. In the second case, we assume that at all redshifts metal-free stars form at 1% of the observed total star-formation–rate density, which we model as

$$\log_{10}\left[\rho_\star^{\mathrm{obs}}(z)/M_\odot\,\mathrm{yr}^{-1}\,\mathrm{Mpc}^{-3}\right] = -2.1 + 3.9\,\log_{10}(1+z) - 3.0\,\left[\log_{10}(1+z)\right]^2 \quad ,$$

a simple fit to the most recent measurements (Giavalisco et al. 2004; Bouwens et al. 2004). For both star-formation models we again assume that one pair-production SN occurs per $1000\,M_\odot$ of metal-free stars.

Note that for the full range of models in Figure 4, metal-free star formation naturally occurs in the smallest galaxies, which are just large enough to overcome the thermal pressure of the ionized IGM, but small enough not to be clustered near areas of previous star formation (SSF03). In our adopted cosmology, for a temperature of 10^4 K, the minimum virial mass is $3 \times 10^9(1+z)^{-3/2}\,M_\odot$ with a corresponding gas mass of $5 \times 10^8(1+z)^{-3/2}\,M_\odot$. This means the total stellar mass of primordial objects is likely to be around $M_\star \approx 10^8\,M_\odot(1+z)^{-3/2}$, many orders of magnitude below $L_\star$ galaxies. Thus in general blank-field surveys should be the best method for searching for PPSNe, as catalogs of likely host galaxies would be extremely difficult to construct.

Nevertheless, as VMS shine so brightly, a direct search for primordial host galaxies is not a hopeless endeavor. In particular, the lack of dust in these objects and the large number of ionizing photons from massive metal-free stars leads naturally to a greatly enhanced Lyman-alpha luminosity. Following SSF03 this can be estimated as

$$L_\alpha = c_L(1 - f_{\rm esc})Q(H)M_\star \ , \qquad (5.3)$$

where $c_L \equiv 1.04 \times 10^{-11}$ ergs, $f_{\rm esc}$ is the escape fraction of ionizing photons from the galaxy, which is likely to be $\lesssim$0.2 (see Ciardi, Bianchi, & Ferrara 2002 and references therein), and the ionizing photon rate $Q(H)$ can be estimated as $\approx 10^{48}\ {\rm s}^{-1}\, M_\odot{}^{-1}$ (Schaerer 2002). This gives a value of $L_\alpha \sim 10^{45}(1 + z)^{-3/2}$ ergs s^{-1} which, if observed in a typical ~1000-Å-wide broad band corresponds to an absolute AB mag $\sim -23 + 3.8 \log(1 + z)$, much brighter than the PPSNe themselves. However, this flux would be spread out over many pixels and be more difficult to observe against the sky than the point-like PPSNe emission. For further details on the detectability of metal-free stars though Lyman-alpha observations, the reader is referred to SSF03.

6. Pair-production supernovae in cosmological surveys

From the models developed above, it is relatively straightforward to relate the star-formation history of VMS to the resulting number of observable pair-production supernovae. In this section and below we adopt cosmological parameters of $h = 0.7$, $\Omega_m = 0.3$, $\Omega_\Lambda = 0.7$, and $\Omega_b = 0.045$, where h is the Hubble constant in units of 100 km s^{-1} and Ω_m, Ω_Λ, and Ω_b are the total matter, vacuum, and baryonic densities in units of the critical density (e.g., Spergel et al. 2003).

Here we focus on three PPSN light curves, which bracket the range of possibilities: the faintest of all our models, 150-W, in which there is neither significant dredge-up nor ^{56}Ni production; an intermediate model, 200-I, in which some dredge-up occurred, but 5.1 $M_\odot$ of ^{56}Ni were formed; and the model with the brightest light curves, 250-S, in which substantial dredge-up leads to an enormous initial radius of over 20 AU, and the production of 24.5 $M_\odot$ of ^{56}Ni causes an extended late-time period of high luminosity. Note that we do not address the possibility of extinction by dust, which amounts to assuming that pristine regions remain dust free throughout the lifetime of the very massive PPSN progenitor's stars.

For any given PPSN model, we can calculate $t(\lambda, F_\nu^{\rm min}, z)$, the total time the observed flux at the wavelength λ from an SN at the redshift z is greater than the magnitude limit associated with the specific flux $F_\nu^{\rm min}$. The total number of PPSNe shining at any given time with fluxes above $F_\nu^{\rm min}$, per square degree per unit redshift is then given by the product of the volume element, the (time-dilated) PPSN rate density, and the time a given PPSN is visible, that is

$$\frac{dN_{\rm deg^2}}{dz}\left(\lambda, F_\nu^{\rm min}, z\right) = [r(z)\ \sin(1\,{\rm deg})]^2 \frac{dr}{dz}\frac{\rho^\pm(z)}{1+z} t\left(\lambda, F_\nu^{\rm min}, z\right) \ , \qquad (6.1)$$

where the rate density $\rho^\pm(z)$ is the number of PPSNe per unit time, per comoving volume as a function of redshift.

The resulting observed PPSNe counts for these models are given in Figure 5 for two limiting magnitudes. In the upper panels, we take an $I_{\rm AB} = 26$ magnitude limit, appropriate for the Institute for Astronomy (IfA) Deep Survey (Barris et al. 2004), a ground-based survey that covered a total of 2.5 deg^2 from September 2001 to April 2002. As we are focused on lower-redshift observations, we only plot estimated counts for the two simple rate density models described in Section 5, rather than the more detailed (but higher redshift) SSF03 models.

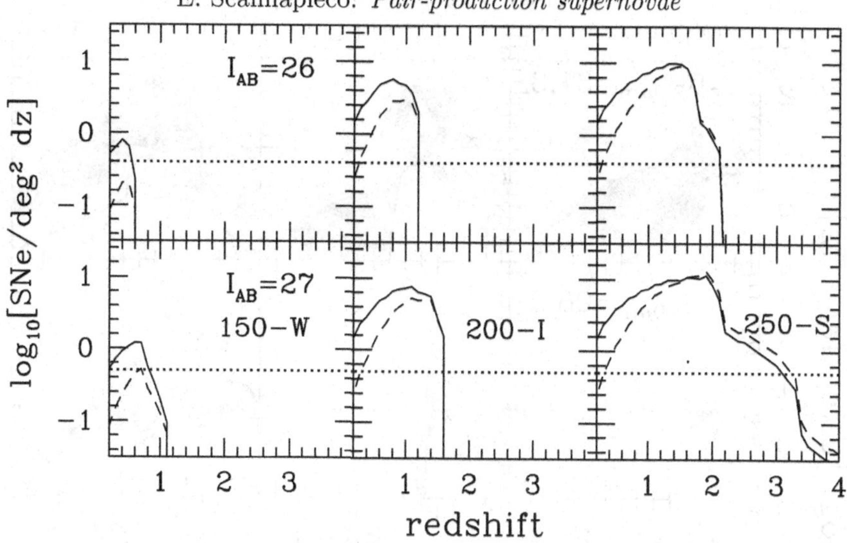

FIGURE 5. Number of PPSNe per square degree per unit redshift above a given I-band magnitude, assuming 0.001 $M_\odot$ yr^{-1} Mpc^{-3} (solid lines) or 1% of the observed star-formation–rate density (dashed lines). The $I_{\rm AB} = 26$ cut taken in the upper rows approximately corresponds to the magnitude limit of the Institute for Astronomy Deep Survey which covered 2.5 deg^2 (as shown by the dotted lines) from September 2001 to April 2002. The $I_{\rm AB} = 27$ limit in the bottom panel corresponds to that of the ongoing COSMOS survey, which will survey an area of 2 deg^2 (again indicated by the dotted lines). Note, however, that the COSMOS survey itself is primarily focused on large-scale structure issues and will not be able to find PPSNe, as each pointing is visited only once.

From this figure we see that existing datasets, if properly analyzed, are easily able to place useful constraints on PPSN formation at low redshifts. Given a typical model like 200-I, for example, the already realized IfA survey can be used to place a constraint of $\lesssim$1% of the total star-formation–rate density out to a redshift $\approx$1. Similarly, extreme models such as 250-S can be probed out to redshifts $\approx$2, all within the context of a recent SN search driven by completely different science goals. Note however that these limits are strongly dependent on significant mixing in the SN progenitor, or the production of ^{56}Ni, and thus models such as 150-W remain largely unconstrained by the IfA survey.

In the bottom panels of Figure 5, we consider a limiting magnitude of $I_{\rm AB} = 27$, appropriate for the COSMOS survey,† a project that covers 2 deg^2 using the Advanced Camera for Surveys on *HST*. Raising the limiting magnitude from $I_{\rm AB} = 26$ to $I_{\rm AB} = 27$ has the primary effect of extending the sensitivity out to slightly higher redshifts. This pushes the probed range from $z \lesssim 1$ to $z \lesssim 1.5$ in the 200-I case, and from $z \lesssim 2$ to $z \lesssim 3$ in the 250-S case. Again, this is all in the context of an ongoing survey. Even with this fainter limiting magnitude, however, low-luminosity PPSNe like 150-W are extremely difficult to find, and remain largely unconstrained.

This shortcoming is easily overcome by moving to NIR wavelengths. In Figure 6 we present for the first time similar limits computed for the *James Webb Space Telescope* (*JWST*), using both the simple models Pop III SFR density models shown in Figure 5, as well as the full range of more detailed models computed in SSF03. With *JWST*, $dN_{\rm deg^2}/dz$ is dramatically increased with respect to ground-based searches. This is due to the fact that for the majority of their lifetimes, the effective temperatures of PPSNe are just above the $\approx 10^{3.8}$ K recombination temperature of hydrogen, which corresponds

† See http://www.astro.caltech.edu/~cosmos/.

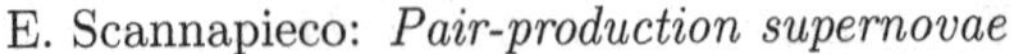

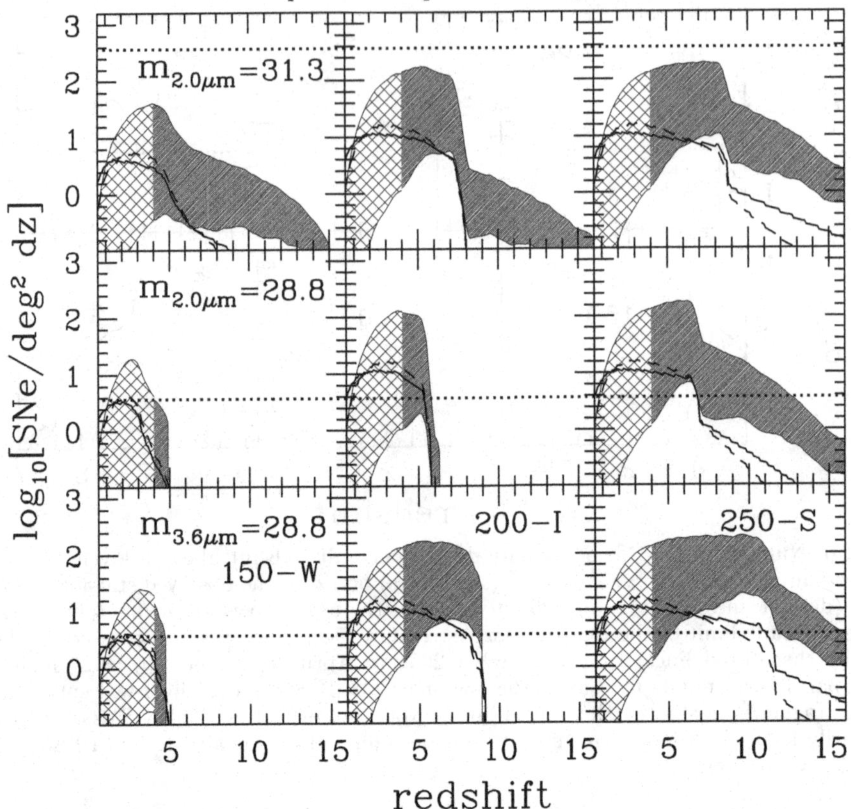

FIGURE 6. Number of PPSNe per square degree per unit redshift above a fixed AB magnitude, with limits appropriate for three possible surveys with the *James Webb Space Telescope.* Curves are as in Figure 5, while the shaded regions are predictions from the full range of SSF03 models described in Section 5. Top: The 31.1 AB magnitude limit taken in these panels corresponds to the 10 σ detection limit for a 10^6 sec integration at 2 μm with the NIRCam instrument. While this would be able to probe SN extremely deeply, it would only cover a 9.8 arcmin^2 patch of the sky, corresponding to the dotted lines. No detections would be expected in this small area. Center: The 28.8 AB magnitude 2 μm limit taken in these panels corresponds to the 10 σ detection limit for a 10^4 second integration with NIRCam. In 10^6 seconds, such measurements could be taken at roughly 100 pointings, covering a 0.3 deg^2 area, corresponding to the dotted lines. In this case ~10 PPSNe like 150-W or 200-I would be detectable out to $z \approx 6$, while up to ~30 PPSNe like 250-S would be detectable out to $z \approx 10$. Bottom: The 28.8 AB magnitude 3.6 μm limit taken in these panels again corresponds to 10 σ detections for a 10^4 second NIRCam integration. Moving to a longer wavelength results in a significant boost to both the number of detectable PPSNe and the maximum redshift at which they can be observed.

to a peak black-body wavelength of ≈8000 Å. This means that for all but the lowest redshifts, the majority of the emitted light is shifted substantially redward of the I band, which is centered at 9000 Å. Given the sensitivities described in Gardner et al. (2006), a 10^6 second NIRCam at 2.0 μm integration would be able to see a PPSN out to $z \approx 15$.

In fact, NIRCam will be so sensitive that using it to perform extremely long integrations will *not* be the best way to search for PPSNe. Rather, the key issue will be covering enough area to find them at the rates expected from theoretical models. As a single NIRCam pointing only covers 9.8 square arcminutes, any individual such field is not likely to host a PPSN, even at such very faint magnitudes. Rather a much more efficient method is to carry out a survey composed of roughly 100 pointings, each with a 10^4 second integration time. In this case, $\sim$10 PPSNe like 150-W or 200-I would be detectable out to $z \approx 6$, while up to $\sim$30 PPSNe like 250-S would be detectable out to $z \approx 10$. Furthermore, moving to slightly longer wavelengths increases the high-redshift sensitivity much in the same way as carrying out longer integrations at a fixed band. Thus the expected number of PPSNe/deg^2/dz above an AB magnitude limit of 28.8 at 3.6 μm is comparable to the number of PPSNe/deg^2/dz above an AB magnitude limit of 31.3 at 2.0 μm.

Taken together, these results imply that the optimal strategy for searching for PPSNe with *JWST* will be to carry out a two- or three-band NIRCam survey (to obtain color information on these objects), with an emphasis on longer wavelengths (to boost sensitivity), made up of $\approx$100 pointings with moderate integration times (to maximize sky coverage). Finally, as most of the features in PPSNe light curves are on the 30–100-day scale, this field should be revisited roughly once per 30 days $\times(1+z) \approx 1$ year, on three occasions. Although this would require about 1–2 weeks of dedicated time each year, clearly this program could be carried out in the context of a more general deep-field study, such as the present Supernova Cosmology Project in the context of the Great Observatories Origins Deep Survey with *HST*.

7. Modeling the Galactic descendants of metal-free stars

A secondary method of searching for PPSN is the detection of their nucleosynthetic products in present-day stars. However, interpretation of these abundances is complicated by the fact that searches for metal-poor stars are limited to the Galactic halo, where dust extinction and crowding are minimal. In fact, these analyses have already provided a number of intriguing constraints on the enrichment history of the halo (Freeman & Bland-Hawthorn 2002; Beers & Christlieb 2005), including uncovering the presence of extremely heavy-element deficient stars (Christlieb et al. 2002; Frebel et al. 2005). Yet, it is still unclear how the observed population of halo stars is related to Pop III star formation (White & Springel 2000; Diemand, Madau, & Moore 2005).

To quantify this, in Scannapieco et al. (2006), we combined a high-resolution N-body simulation of the formation of the Milky Way with a semi-analytical model of metal enrichment similar to that in SSF03. Our N-body simulation was carried out with the `GCD+` code (Kawata & Gibson 2003a), and it used a multi-resolution technique (Kawata & Gibson 2003b) to achieve high resolution within a 1 Mpc radius, while the outer regions exerting tidal forces were handled with lower resolution. In the high-resolution region, the dark matter particle mass was $M_{\rm vir} = 7.8 \times 10^5\, M_\odot$, compared to the final virial mass of $M_{\rm vir} = 7.7 \times 10^{11}\, M_\odot$. The simulation data was output every 0.11 Gyr, and at each output, we use a friend-of-friends (FOF) group finder to identify the virialized DM haloes, with a linking parameter of $b = 0.2$. As in Section 5, we then assumed star formation to occur in all haloes with virial temperatures above the atomic cooling limit of 10^4 K, which corresponds to a minimum mass of $M_{\rm min} \equiv 3.0 \times 10^9 (1+z)^{-3/2}\, M_\odot$. This means that even at a very high redshift of 20, all the haloes relevant to our study contained at least 50 particles, and are well identified by a FOF group finder.

Our next step was to use this accretion history to identify two types of objects: i) haloes that collapsed out of primordial gas, which we identified as Pop III objects containing "the first stars," and ii) haloes that collapsed from gas that has been enriched purely by Pop III objects, which we identified as second-generation objects containing "the second stars." Following our approach in SSF03, we adopted a model of outflows as spherical shells expanding into the Hubble flow (Ostriker & McKee 1988) for both Pop III and Pop II/I objects. These shells were assumed to be driven only by the internal hot gas pressure and decelerated by accreting material and gravitational drag. The evolution of each such bubble was then completely determined by the mechanical energy imparted to the outflow. In particular, the only difference between Pop III and Pop II/I outflows arises from the energy input per unit gas mass, $\mathcal{E}_{\rm g}^{III,II}$.

In the Pop II/I case, we calculated this assuming that 10% of the gas was converted into stars, that 10^{51} ergs of kinetic energy input 300 $M_\odot$ of stars formed, and we assumed an overall wind efficiency fit to results of Mori, Ferrara, & Madau (2002) and Ferrara, Pettini, & Shchekinov (2000). In the Pop III case, on the other hand, as there are no direct constraints, we varied $\mathcal{E}_{\rm g}^{III}$ over a large range as in Section 5. Finally, when outflows slowed down to the IGM sound speed, we assumed they fragmented and let them expand with the Hubble flow. For further details, the reader is referred to Scannapieco et al. (2006).

8. The spatial distribution of the Galactic descendants of metal-free stars

Figure 7 demonstrates the positions of the first and second stars at different redshifts in the Lagrangian model with $\mathcal{E}_{\rm g}^{III} = 10^{-2}$. At early times ($z = 9.84$), the first stars form close to the central density peak of the progenitor galaxy, due to the higher density peaks in this region. Second stars form in the halo in the neighborhood of first stars, because they condense from gas that is enriched by the material from the explosions of the first stars. At a later time ($z = 6.02$), new first stars are still forming, but now on the outer regions of the progenitor galaxy, because they are not yet affected by winds from the central region. The formation of the first and second stars is complete around $z = 3$, at which time the full region is enriched with metals. Also at this redshift, first and second stars start to be accreted into the assembling Galactic halo. This assembly is almost complete by $z = 1$, and thus the distribution at this redshift is similar to that at $z = 0$.

In Figure 8 we plot the radial mass density of first and second stars, as compared to the dark matter distribution in our simulation. In computing these masses, we assume that 10% of the gas mass in each halo is converted into stars and we intentionally make no attempt to account for mass loss due to stellar evolution. Furthermore, we plot only out to 250 kpc, which is 1/4 the size of the high-resolution region. Here we see that the density profiles of first and second stars are similar to the total dark matter density profile, although the second stars have a slightly shallower slope. As a result, the density of first stars at the center is 100 times higher than at the 8 kpc orbital radius of the Sun. However, the important number for developing observational strategies for finding the first stars is their relative density with respect to the field stars. While this is not directly computed in our simulation, the lower panels of Figure 8 show the local mass density of stars normalized by the local density of dark matter.

Amazingly, the mass fraction contained in the first stars varies only very weakly with radius. Moving from 1 to 100 kpc in the $\mathcal{E}_{\rm g}^{III} = 10^{-4}$ model, for example, the fraction of the mass in first stars decreases only by a factor ≈ 4. Furthermore, increasing the efficiency of Pop III winds to $\mathcal{E}_{\rm g}^{III} = 10^{-2}$ has the effect of decreasing the fraction of first stars without strongly affecting their radial distribution.

This is both good and bad news for PPSNe models. First of all, it means that observations of metal-poor stars in the Galactic halo should be taken as directly constraining the properties of primordial objects. Thus, the lack of metal-free stars observed in the halo is likely to imply a real lower mass limit in the metal-free initial mass function (IMF) of at least 0.8 $M_\odot$, the mass of a low-metallicity star with a lifetime comparable to the Hubble time (Fagotto et al. 1994). In fact the $\sim 10^{-6}\,M_\odot$ pc^{-3} value for Pop III stars in

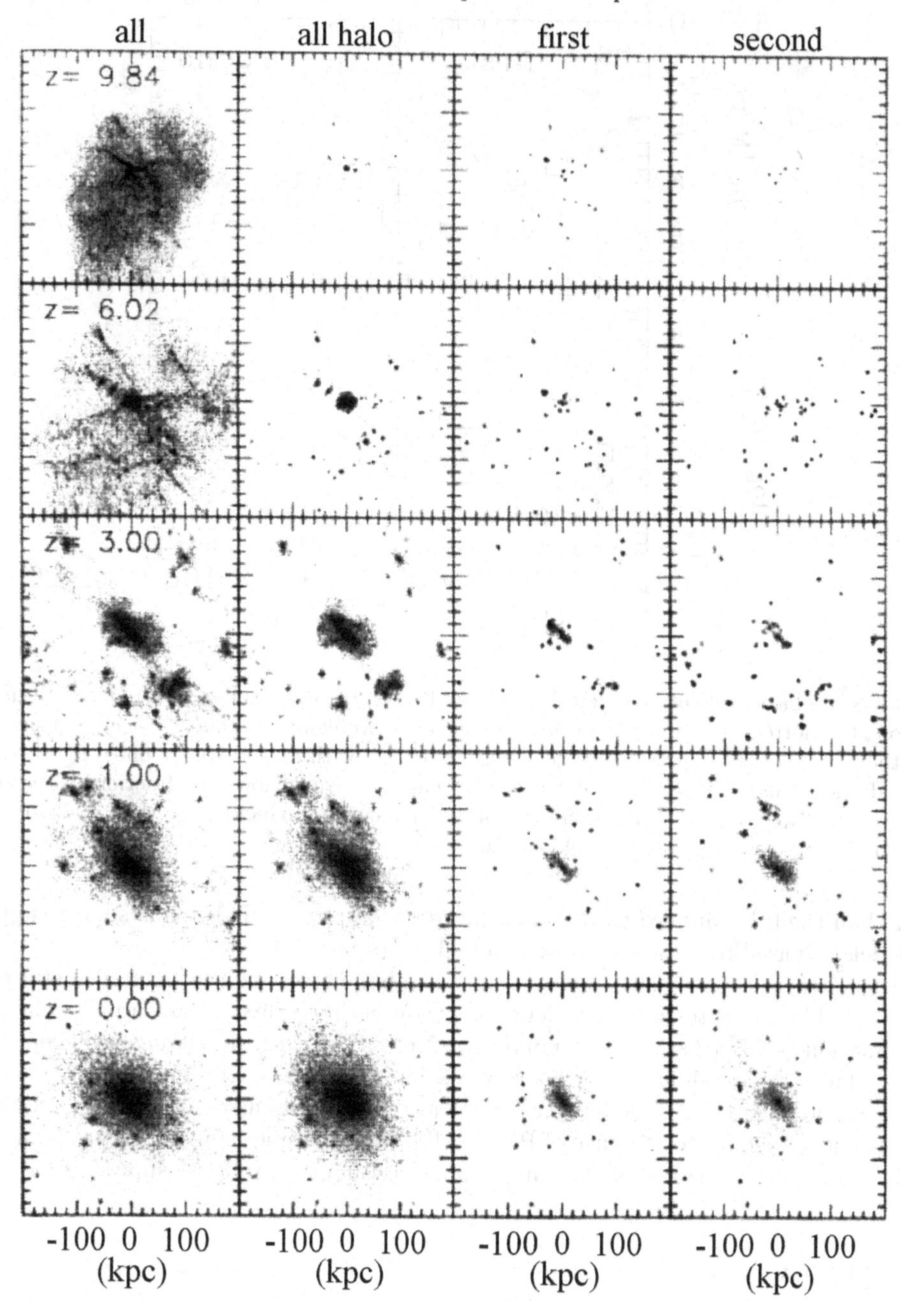

FIGURE 7. Distributions of first and second stars (3rd and 4th columns) in a 200 proper kpc^3 region, at various redshifts, for a model with $\mathcal{E}_{\rm g}^{III} = 10^{-2}$. For comparison, the 1st and 2nd columns show all the particles and all the particles that were ever within a halo with a virial temperature above the atomic cooling limit, respectively.

the solar neighborhood we compute is so high as compared to the observed stellar mass density of $5 \times 10^{-5}\, M_\odot$ pc^{-3} (Preston, Shectman, & Beers 1991), that several primordial $0.8\, M_\odot$ stars would have been observed even if f_*^{III} were over an order of magnitude

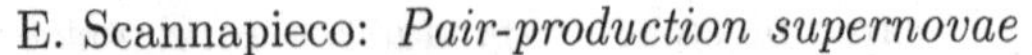

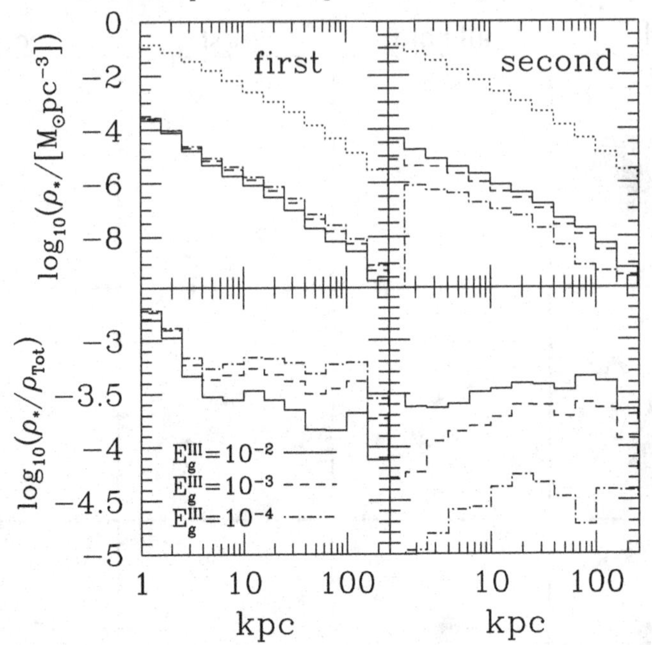

FIGURE 8. Radial profile of first (left column) and second stars (right column) at $z = 0$. In the top row, the dotted lines give the overall dark matter profile of the galaxy, which is compared with the radial density of first and second stars in models with $\mathcal{E}_g^{III} = 10^{-2}$ (solid), $\mathcal{E}_g^{III} = 10^{-3}$ (short-dashed), and $\mathcal{E}_g^{III} = 10^{-4}$ (dot-dashed). The lower row shows the fraction of the total density in first and second stars, with symbols as in the upper panels. Note that this plot does not include any mass loss due to stellar evolution.

lower than the 0.1 value we used to normalize our approach. This lends strong support to models of metal-free stars as biased to high masses.

On the other hand, it also means that the abundances we see in very metal-poor stars should be taken to constrain elements produced in the first stars. In particular, if a large fraction of Pop III star formation resulted in PPSNe, then the strong odd-even effect discussed in Section 2 should be measurable in a subset of stars with $[Fe/H] \leqslant -2$. To date these measurements have failed to uncover this signal. As discussed in Tumlinson (2006), this does not rule out the presence of PPSNe, but it does argue strongly that a significant number of Pop III stars ended their lives as more usual core-collapse supernovae.

9. Summary

Theoretically, PPSNe are the simplest of all supernovae. Driven by a well-understood dynamical instability, and leading to complete stellar disruption, they are the uniquely calculable result of nonrotating stars that end their lives in the 140–260 $M_\odot$ mass range (Heger & Woosley 2002). The issues are only when and where such stars existed. In the present enriched universe, the observed upper-mass limit of forming stars and the rate of mass loss in O stars argue strongly against these objects. However, in the primordial high-redshift universe, things are likely to have been very different. The typical fragmentation mass under these conditions is $\approx$1000 $M_\odot$ and stellar winds, at least of the line-driven type observed today, are expected to be negligible. This raises the real possibility that in the metal-free universe, a large fraction of stars generated PPSNe.

As metal enrichment is an intrinsically local process that proceeds over an extended redshift range, at each redshift, pockets of metal-free star formation are naturally confined to the lowest-mass galaxies, which are small enough not to be clustered near areas of previous star formation. As such faint galaxies are difficult to detect and even more difficult to confirm as metal free, the hosts of PPSNe could easily be lurking at the limits of present-day galaxy surveys.

In S05 we showed that the most important factors for modeling PPSNe light curves are the mass of the progenitor star and the efficiency of dredged-up carbon moving from the core into the envelope. In general, increasing the mass leads to greater ^{56}Ni production, which boosts the late-time SN luminosity. Mixing, on the other hand, has two major effects: it leads to a red-giant phase that increases the early-time SN luminosity, and it decreases the mass of the He core, consequently leading to a somewhat smaller mass of ^{56}Ni. Despite these uncertainties, PPSNe in general can be characterized by: (1) peak magnitudes that are brighter than Type II SNe and comparable or slightly brighter than typical SNe Type Ia; (2) very long decay times $\sim$1 year, which result from the large initial radii and large masses of material involved in the explosion; and (3) the presence of hydrogen lines, which are caused by the outer envelope.

The S05 light curves also allowed us to calculate the number of PPSNe detectable in current and planned supernova searches. Here the long lifetimes help to keep a substantial number of PPSNe visible at any given time, meaning that ongoing SN searches should be able to limit the contribution of VMS to $\lesssim$1% of the total star-formation–rate density out to a redshift of two, unless both mixing and ^{56}Ni production are absent for all PPSNe. Such constraints already place meaningful limits on the cosmological models.

The impact of future NIR searches is even more promising, as the majority of the PPSN light is emitted at rest-frame wavelengths longward of $\approx$8000 Å. In particular, *JWST* surveys with NIRCam have the potential to place fantastic constraints on PPSNe out to $z \approx 10$. In this case, the best approach will be a $\approx$0.3 deg^2 survey made up of $\approx$100 NIRCam pointings with $\approx 10^4$ sec integrations in two or three bands, with emphasis on the redder colors. Furthermore, this field should be revisited with a cadence of roughly once per year on three occasions. Although this would require about one to two weeks of dedicated time each year, clearly this program could be carried out in the context of a more general deep-field study with a much broader set of science goals.

Closer to home, we have also studied the final distribution of the elements synthesized in primordial stars. Despite the large uncertainties involved, all models generically predict significant Pop III star formation in what is now the Galactic halo. Thus, if they have sufficiently long lifetimes, a significant number of stars initially formed in primordial star clusters should be found in ongoing surveys for metal-poor halo stars. This is both good and bad news for PPSNe models. While it implies a real lower mass limit for the Pop III IMF of at least $0.8\,M_{\odot}$, it also suggests that the lack of an odd-even effect in the observed abundance ratios of metal-poor stars should be taken as evidence that a significant number of Pop III stars ended their lives as more common core-collapse supernovae.

However, definitive limits on PPSNe will only come from space-based NIR surveys. Should the surveys result in detections, they will open a new window on star formation and the history of cosmological chemical enrichment. If they result in upper limits, exquisite constraints will be placed on the presence of PPSN forming above the atomic cooling limit in primordial environments, but will leave open the question of their formation in the first "minihaloes" collapsing at extremely high redshift. While this would remain a possibility, the absence of detections from space-based searches would limit PPSN to the most remote and undetectable corners of the universe. Undaunted theorists might still wish to discuss them in workshops on the deepest depths of the cosmological dark ages. Observers may be reminded that "he who wishes to lie, should put the evidence far away" (Livio 2006).

I would like to thank my collaborators Chris Brook, Andrea Ferrara, Brad Gibson, Alexander Heger, Daisuke Kawata, Piero Madau, Raffaella Schneider, and Stan Woosley, for allowing me to present the results of our work together here. I am also thankful to Jonathan Gardner for providing detailed information on the planned capabilities of the *James Webb Space Telescope*. Finally, I would like to thank Mario Livio, Massimo Stiavelli, and the other members of the organizing committee, as well as the many excellent speakers, for a fun and informative symposium.

REFERENCES

ABEL, T., BRYAN, G. L., & NORMAN, M. L. 2002 *Science* **295**, 93.

BARBON, R., CIATTI, F., ROSINO, L., ORTOLANI, S., & RAFANELLI, P. 1982 *A&A* **116**, 43.

BARKAT, Z., RAKAVY, G., & SACK, N. 1967 *Phys. Rev. Lett.* **18**, 379.

BARRIS, B. J., ET AL. 2004 *ApJ* **602**, 571.

BEERS, T. C. & CHRISTLIEB, N. 2002 *ARA&A* **43**, 521.

BOND, J. R., ARNETT, W. D., & CARR, B. J. 1984 *ApJ* **280**, 825.

BOUWENS, R., ET AL. 2004 *ApJ* **616**, L79.

CAPPELLARO, E., MAZZALI, P. A., BENETTI, S., DANZIGER, I. J., TURATTO, M., DELLA VALLE, M., & PATAT, F. 1997 *A&A* **328**, 203.

CATCHPOLE, R. M., ET AL. 1988 *MNRAS* **231**, 75.

CIARDI, B., BIANCHI, S., & FERRARA, A. 2002 *MNRAS* **331**, 463.

CIARDI, B., FERRARA, A., & ABEL, T. 2000 *ApJ* **533**, 594.

CHRISTLIEB, N., ET AL. 2002 *Nature* **419**, 904.

DE VAUCOULEURS, G., DE VAUCOULEURS, A., BUTA, R., ABELS, H. D., & HEWITT, A. V. 1981 *PASP* **93**, 36.

DEKEL, A. & REES, M. J. 1987 *Nature* **326**, 455.

DIEMAND, J., MADAU, P., & MOORE, B. 2005 *MNRAS* **364**, 367.

EASTMAN, R. G., WOOSLEY, S. E., WEAVER, T. A., & PINTO, P. A. 1993 *BAAS* **25**, 836.

EASTMAN, R. G., WOOSLEY, S. E., WEAVER, T. A., & PINTO, P. A. 1994 *ApJ* **430**, 300.

ELMHAMDI, A., ET AL. 2003 *MNRAS* **338**, 939.

FAGOTTO, F. BRESSAN, A., BERTELLI, G., & CHIOSI, C. 1994, *A&AS* **104**, 365.

FERRARA, A., PETTINI, M., & SHCHEKINOV, Y. 2000 *MNRAS* **319**, 539.

FIGER, D. 2005 *Nature* **434**, 192.

FRALEY, G. S. 1968 *AP&SS* **2**, 96.

FREBEL, A., ET AL. 2005 *Nature* **434**, 871.

FREEMAN, K. & BLAND-HAWTHORN, J. 2002 *ARA&A* **40**, 487.

FRYER, C. L., WOOSLEY, S. E., & HEGER, A. 2001 *ApJ* **550**, 372.

GARDNER, J. P., ET AL. 2006 *Space Sci. Rev.* **123**, 485.

GIAVALISCO, M., ET AL. 2004 *ApJ* **600**, L103.

HAIMAN, Z., REES, M. J., & LOEB, A. 1997 *ApJ* **476**, 458.

HAMUY, M., SUNTZEFF, N. B., GONZALEZ, R., & MARTIN, G. 1988 *AJ* **95**, 63.

HEGER, A. & WOOSLEY, S. E. 2002 *ApJ* **567**, 532.

KAWATA, D. & GIBSON, B. K. 2003a *MNRAS* **340**, 908.

KAWATA, D. & GIBSON, B. K. 2003b *MNRAS* **346**, 135.

KUDRITZKI, R. 2000. In *The First Stars* (eds. A. Weiss, T. Abel, & V. Hill). p. 127. Springer.

KUDRITZKI, R. P. 2002 *ApJ* **577**, 389.

LIVIO, M. 2006 Free translation from Hebrew; *not so private communication.*

MORI, M., FERRARA, A., & MADAU, P. 2002 *ApJ* **571**, 40.

NAKAMURA, F. & UMEMURA, M. 1999 *ApJ* **515**, 239.

OEY, M. S. & CLARKE, C. J. 2005 *ApJ* **620**, L43.
OH, S. P. & HAIMAN, Z. 2002 *ApJ* **569**, 558.
O'SHEA, ABEL, T., WHALEN, D., & NORMAN, M. L. 2005 *ApJ* **628**, L5.
OSTRIKER, J. P. & MCKEE, C. F. 1988 *Rev. Mod. Phys.* **60**, 1.
PATAT, F., BENETTI, S., CAPPELLARO, E., DANZIGER, I. J., DELLA VALLE, M., MAZZALI, P. A., & TURATTO, M. 1996 *MNRAS* **278**, 111.
PRESTON, G. W., SHECTMAN, S. A., & BEERS, T. 1991 *ApJ* **375**, 121.
SCANNAPIECO, E., FERRARA, A., & MADAU, P. 2002 *ApJ* **574**, 590.
SCANNAPIECO, E., KAWATA, D., BROOK, C. B., SCHNEIDER, R. FERRARA, A., & GIBSON, B. K. 2006 *ApJ* **653**, 285.
SCANNAPIECO, E., MADAU, P., WOOSLEY, S., HEGER, A., & FERRARA, A. 2005 *ApJ* **663**, 1031.
SCANNAPIECO, E., SCHNEIDER, R., & FERRARA, A. 2003 *ApJ* **589**, 35.
SCHAERER, D. 2002 *A&A* **382**, 28.
SCHNEIDER, R., FERRARA, A., NATARAJAN, P., & OMUKAI, K. 2002 *ApJ* **571**, 30.
SMITH, N. & OWOCKI, S. P. 2006 *ApJ* **645**, L45.
SPERGEL, D. N., ET AL. 2003 *ApJS* **14**, 175.
TAN, J. C. & MCKEE, C. F. 2004 *ApJ* **603**, 383.
TUMLINSON, J. 2006 *ApJ* **641**, 1.
VINK, J. S., DE KOTER, A., & LAMERS, H. J. G. L. M. 2001 *A&A* **369**, 574.
WEAVER, T. A. & WOOSLEY, S. E. 1980. In *Ninth Texas Symposium on Relativistic Astrophysics*. p. 335. New York Academy of Sciences.
WEAVER, T. A., ZIMMERMAN, G. B., & WOOSLEY, S. E. 1978 *ApJ* **225**, 1021.
WEIDNER, C. & KROUPA, P. 2004 *MNRAS* **348**, 187.
WEINMANN, S. M. & LILLY, S. J. 2005 *ApJ* **624**, 526.
WHITE, S. D. M. & SPRINGEL, V. 2000. In *The First Stars* (eds. A. Weiss, T. Abel, & V. Hill). p. 327. Springer.

Cosmic infrared background and Population III: An overview

By A. KASHLINSKY

Code 665, Observational Cosmology Lab, Goddard Space Flight Center, Greenbelt, MD 20771, USA; kashlinsky@milkyway.gsfc.nasa.gov

We review the recent measurements on the cosmic infrared background (CIB) and their implications for the physics of the first-stars era, including Population III stars. The recently obtained CIB results range from the direct measurements of CIB fluctuations from distant sources using deep *Spitzer* data to strong upper limits on the near-IR CIB from blazar spectra. This allows us to compare the Population III models with the CIB data to gain direct insight into the era of the first stars, the formation and evolution of Population III stars, and the microphysics of the feedback processes in the first halos of collapsing material. We also discuss the cosmological confusion resulting from these CIB sources and the prospects for resolving them individually with NASA's upcoming space instruments, such as the *James Webb Space Telescope* (*JWST*).

1. Introduction

The very first stars to form in the universe, commonly called Population III stars, are now thought to have been very massive stars forming out of primordial metal-free gas at redshifts exceeding $z \sim 10$ (see review by Bromm & Larson 2004). Assuming that the density field responsible for structure formation is given by the Λ Cold Dark Matter (CDM) model, the first collapsing haloes hosting such stars may be too faint to be observed with present telescopes. Their studies may, however, be possible via the cumulative radiation emitted by the first luminous objects, most of which has by now been shifted into the near-IR wavelengths of ~1–10 μm. We review here the connection between the massive Population III stars, the cosmic infrared background measurements, and the implications of the latter for the duration, nature, and abundance of luminous objects that comprised the era of the first stars.

The CIB is a repository of emissions throughout the entire history of the universe. Cosmic expansion shifts photons emitted in the visible/UV bands at high z into the near-IR (NIR), and the high-z NIR photons appear today in mid- to far-IR. Consequently, the NIR part of the CIB spectrum ($1\ \mu m < \lambda < 10\ \mu m$) probes the history of direct stellar emissions from the early universe, and the longer wavelengths contain information about the early dust production and evolution. Recent years have seen significant progress in CIB studies, both in identifying and/or constraining its mean level (isotropic component) and fluctuations (see Kashlinsky 2005a for a review).

The CIB also contains emissions from objects inaccessible to current (or even future) telescopic studies and can, therefore, provide unique information on the history of the universe at very early times. One particularly important example of such objects are Population III (hereafter Pop III) stars—the still elusive zero-metallicity stars expected to have preceded the normal stellar populations seen in the farthest galaxies to date. Throughout this review we will use the term "era of the first stars," or "Pop III era," with the understanding that the actual era may be composed of objects of various natures from purely zero-metallicity stars, to low-metallicity stars, and even possibly to mini-quasars, whose contribution to the CIB is driven by energy released by gravitational accretion, as opposed to stellar nucleosynthesis.

Pop III stars are thought to have preceded the normal metal-enriched stellar populations, but even if massive and luminous, they are inaccessible to direct observations by current telescopes because of the high redshifts at which they are located. Extensive numerical investigations, as discussed by Bromm and Norman in this meeting, suggest that they had to be very massive, forming out of density fluctuations specified by the standard ΛCDM model. If massive, they are expected to have left a significant level of diffuse radiation which has today shifted into the IR, and it was suggested that the CIB contains a significant contribution from Pop III in the near-IR, manifest in both its mean level and its anisotropies (e.g., Santos et al. 2002; Salvaterra & Ferrara 2003; Cooray et al. 2004; Kashlinsky et al. 2004; see also review by Kashlinsky 2005a). This notion has recently received strong support from measurements of CIB anisotropies in deep *Spitzer*/Infrared Array Camera (IRAC) images (Kashlinsky et al. 2005), as will be discussed below.

The structure of this review is as follows: Section 2 summarizes the current measurements of the mean levels of the CIB and the contributions to them from the observed galaxy populations. Section 3 discusses the theoretical connection between the CIB and the first (massive) stars, and Section 4 gives the limits on the near-IR CIB from measurements of absorption in the high-energy spectra of cosmological sources. Section 5 reviews the recent *Spitzer*-based measurements of CIB fluctuations from early epochs and their interpretation in terms of the nature and the epochs of the contributing sources. Finally, the prospects for resolving these sources in future measurements are summarized in Section 6 followed with conclusions in Section 7.

The AB magnitude system is adopted throughout; the conversion to fluxes is simple, as zero AB magnitude corresponds to the flux of 3631 Jy.

2. Direct CIB measurements

Galactic and Solar System foregrounds are the major obstacles to space-based CIB measurements. Galactic stars are the main contributors at near-IR (less than a few μm), zodiacal light from the dust in the Solar System dominates between ~10 and ~50 μm, and Galactic cirrus emission produces most of the foreground at IR wavelengths of >50 μm. Accurately removing the foregrounds presents a challenge, and many techniques have been developed to do this as well as possible. Stars can be removed in surveys with fine angular resolution, or by statistically extrapolating the various stellar contributions to zero. Zodiacal light contributions usually are removed using the Diffuse Infrared Background Experiment (DIRBE) zodi model or its derivatives. Galactic cirrus and zodiacal light are both intrinsically diffuse, but are fairly homogeneous, adding to the effectiveness of CIB fluctuations studies at mid- to far-IR (Kashlinsky et al. 1996a,b).

In the near-IR CIB detections are difficult because of the substantial foreground by Galactic stars. Claims of detections of the mean isotropic part of the CIB are based on various analyses of DIRBE and IRTS data (Dwek & Arendt 1998; Matsumoto et al. 2005; Wright & Reese 2000; Gorjian et al. 2000; Cambresy et al. 2001). The measurements agree with each other, although the methods of analysis and foreground removal differ substantially. They also agree with the measured amplitude of CIB fluctuations using DIRBE data (Kashlinsky & Odenwald 2000). The results seem to indicate fluxes significantly exceeding those from observed galaxy populations. Figure 1 summarizes the current CIB measurements. Only the near-IR CIB at $\lambda < 10$ μm is relevant for the discussion that follows; we will not discuss here the possible dust contribution from the high z contributing to (re)emissions and to CIB at longer wavelengths.

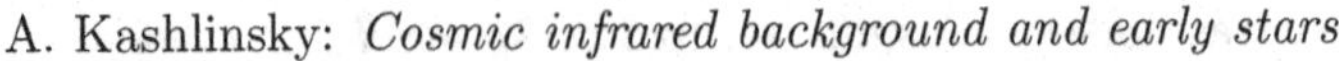

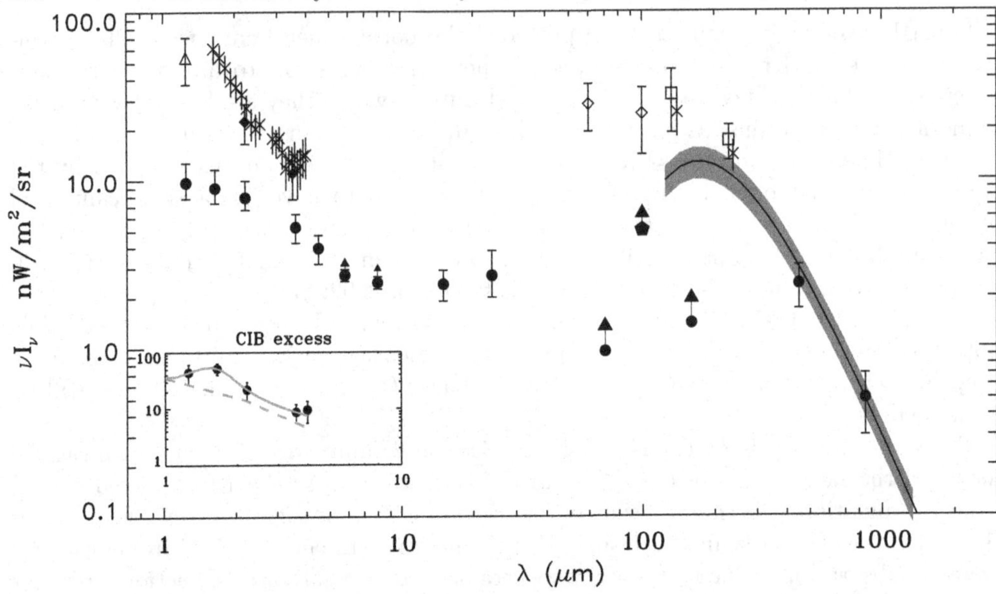

FIGURE 1. CIB and "ordinary" galaxy (OG) contributions (filled circles) vs. wavelength. The numbers are adopted from Figure 9 of Kashlinsky (2005a and references therein), and are discussed at length there. Briefly, the CIB fluxes from IRTS measurements are shown with crosses from Matsumoto et al. (2005), diamonds are from Gorjian et al. (2000) at 2.2 μm, and from Dwek & Arendt (1998) and Wright & Reese (2000) at 3.5 μm. The flux from ordinary galaxies is shown with filled circles and is taken from *HST* counts out to 2.2 μm (Madau & Pozzetti 2000) and from *Spitzer*/IRAC counts at 3.6 and 4.5 μm (Fazio et al. 2004). At $\lambda > 10$ μm no CIB excess was observed, and the levels of CIB are consistent with the net contribution from OG. The inset shows the NIRBE spectrum, νI_ν in $\mathrm{nW/m^2/sr}$ vs. λ in μm, from Kashlinsky (2005a). The two thick light-shaded lines show the ranges of the near-infrared background excess (NIRBE) spectrum suggested by the data: the solid line goes through the central points, while the dashed line grazes the lower edges of the data.

One can compare the claimed CIB levels to those obtained by integrating the contributions from galaxies seen in faint galaxy counts surveys. The net fluxes per magnitude interval dm from such counts are given by:

$$\frac{dF}{dm} = f(m)\frac{dN}{dm} \quad , \tag{2.1}$$

where dN/dm is the differential counts of galaxies per unit solid angle in the magnitude bin dm and $f(m)$ is the flux corresponding to m.

Figure 2 shows the near-IR CIB flux contributions per dm from the galaxy populations observed in various deep surveys. The total NIR fluxes by these OGs from the *HST* and *Spitzer* IRAC measurements are shown as filled circles in Figure 1; these total fluxes saturate at AB magnitude of about 20, and at IR wavelengths shorter than ~5–8 μm they appear to be lower by a significant factor than the detected CIB.

The difference between the claimed CIB levels and the total fluxes from OGs (i.e., not Pop III) are known as the near-IR background excess (hereafter NIRBE), whose integrated amount between 1 and 4 μm is (Kashlinsky 2005a):

$$F_{\mathrm{NIRBE}}(1\!-\!4\ \mu\mathrm{m}) = 29 \pm 13\ \ \mathrm{nW/m^2/sr} \quad . \tag{2.2}$$

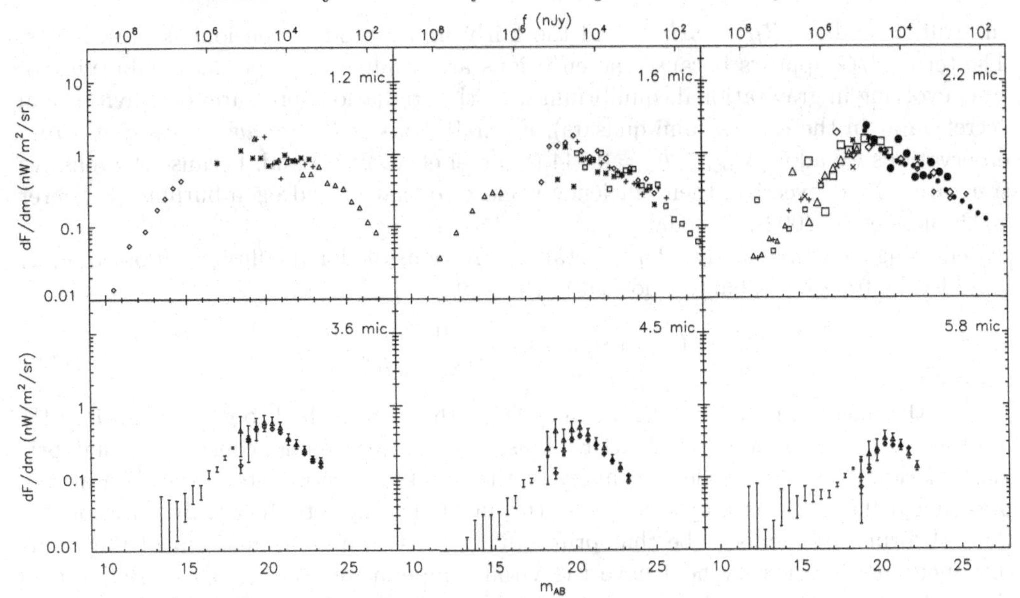

FIGURE 2. Cumulative flux in nW/m^2/sr contributed by galaxies from a narrow dm magnitude bin for *HST* (upper panels; Madau & Pozzetti 2000) and *Spitzer*-based (Fazio et al. 2004) counts data. The points are adopted from Kashlinsky (2005a; see Figures 15, 16).

3. Cosmic infrared background excess and Pop III

It was suggested that the NIR CIB excess is produced by massive Pop III stars at high z (>8–10; Santos et al. 2002; Salvaterra & Ferrara 2003; Magliochetti et al. 2003; Cooray et al. 2004; Kashlinsky et al. 2004). Because Pop III stars, if massive, would radiate at the Eddington limit, where $L \propto M$, the levels of the total flux produced by them are largely model independent (Rees 1978; Kashlinsky et al. 2004), leading to a robust prediction for the total bolometric flux from them. For completeness, we reproduce briefly the argument from Kashlinsky et al. (2004): Each star would produce flux

$$\frac{L}{4\pi d_L^2} ,$$

where d_L is the luminosity distance. Because for massive stars $L \propto M$, the total comoving luminosity density from Pop III is

$$\int n(L) L dL \propto \Omega_{\rm baryon} f_* \frac{3H_0^2}{8\pi G} ,$$

where $n(L)$ is their luminosity function and f_* is the mass fraction of baryons locked in Pop III stars at any given time. In the flat universe, the volume per unit solid angle subtended by cosmic time dt is $dV = c(1+z)d_L^2 dt$. Finally, these stars would radiate at efficiency ϵ ($\simeq$0.007 for hydrogen burning). This then leads to the closed expression for the total bolometric flux from these objects:

$$F_{\rm bol} = \frac{3}{8\pi} \frac{c^5/G}{4\pi R_H^2} \left\langle (1+z)^{-1} \right\rangle \epsilon f_3 \Omega_{\rm baryon} \simeq 4 \times 10^7 z_3 \epsilon f_3 \Omega_{\rm baryon} h^2 \frac{\rm nW}{\rm m^2 sr} . \tag{3.1}$$

Here f_3 is the mean mass fraction of baryons locked in Pop III stars and $z_3 \equiv \frac{1}{\langle (1+z)^{-1} \rangle}$ is a suitably averaged redshift over their era. The total flux is a product of the maximal luminosity produced by any gravitational process, c^5/G, distributed over the surface of the Hubble radius, $R_H = cH_0^{-1}$, and the fairly understood dimensionless parameters. The term c^5/G appears because the emissions are produced by the nuclear burning of stars evolving in gravitational equilibrium with the (radiation) pressure (or gravitational accretion as in the case of mini-quasars). From *Wilkinson Microwave Anisotropy Probe* observations we adopt $\Omega_{\rm baryon}h^2 = 0.044$ (Spergel et al. 2007) and, because the massive stars are fully convective, their efficiency is close to that of hydrogen burning (Schaerer 2002; Siess et al. 2002), $\epsilon = 0.007$.

Requiring that the massive Pop III stars are responsible for the flux given by eq. (2.2) leads to the fraction of baryons locked in them of:

$$f_3 = (4.2 \pm 1.9) \times 10^{-3} z_3 \frac{0.044}{\Omega_{\rm baryon}h^2} \frac{0.007}{\epsilon} \quad . \tag{3.2}$$

Within the uncertainty of eq. (2.2), only ~2% of the baryons had to go through Pop III in order to produce the entire NIRBE. This is not unreasonable, considering that primordial clouds are not subject to many of the effects inhibiting star formation at the present epochs, such as magnetic fields, turbulent heating, etc. The only criterion for Pop III formation seems to be that primordial clouds turning around out of the "concordance" ΛCDM density field have the virial temperature, $T_{\rm vir}$, that can enable efficient formation of and cooling by molecular hydrogen (Abel et al. 2002; Bromm et al. 1999). Assuming spherical collapse of Gaussian fluctuations and the ΛCDM model from WMAP observations (Spergel et al. 2007) the fraction of collapsed haloes at $z = 10$ with $T_{\rm vir} \geqslant (400, 2000)$ K is $(2.6, 5) \times 10^{-2}$, which is in good agreement with eq. (3.2). These stars would have to be dominated by masses >240 $M_\odot$ to be consistent with low metallicities observed in Population II (Heger et al. 2003).

The fraction given by eq. (3.2) was evaluated by using the bolometric CIB excess integrating from 1 to 4 μm. If there were significant amounts of CIB excess flux missed outside that range, the fraction f_3 would increase. However, at wavelengths $<0.1z_3$ μm the emission from the early times would be below the Lyman break and would be reprocessed to $\lambda > 1$ μm (Santos et al. 2002). If significant Pop III activity continued at $z_3 < 10$, which is unlikely, the rest-frame Lyman break may be redshifted to <1 μm, but the possible extra CIB excess from <1 μm will be compensated for by f_3 in eq. (3.2) decreasing with z_3. At longer wavelengths, the CIB excess given by eq. (2.2) can at most increase f_3 by ~30%. Thus the entire NIRBE can be explained if 2–4% of the baryons have been converted into massive Pop III stars at $z > 10$ (Madau & Silk 2005; Kashlinsky 2005b). This fraction of converted baryons scales linearly with the amplitude of $F_{\rm NIRBE}$.

4. CIB excess and high-energy gamma-ray absorption

An independent limit on the net amount of the isotropic component of the CIB comes from the measurements of absorption of high-energy cosmological sources. The latter results from the two-photon absorption, $\gamma\gamma_{\rm CIB} \rightarrow e^+e^-$, effective at high gamma-ray energies, $E_\gamma E_{\gamma_{\rm CIB}} > (m_e c^2)^2$. Because the reaction is due to electromagnetic interaction, its cross section is of order $\sim(e^2/m_e c^2)^2$ (the electron radius peaks at $\sim\frac{1}{4}\sigma_T$) which leads to significant absorption over cosmological distances.

The recent results from two $z \sim 0.18$ blazar spectra measurements at TeV energies by the HESS collaboration (Aharonian et al. 2006) and similar constraints using slightly lower z blazars (Dwek et al. 2005) indicate that the claimed NIRBE would lead to more attenuation at ~1–2 TeV than the known blazar physics would allow. The thick light-shaded lines in the left panel of Figure 3 show the attenuation due to NIRBE and galaxy counts fluxes for the CIB going through the central points of the near-IR CIB

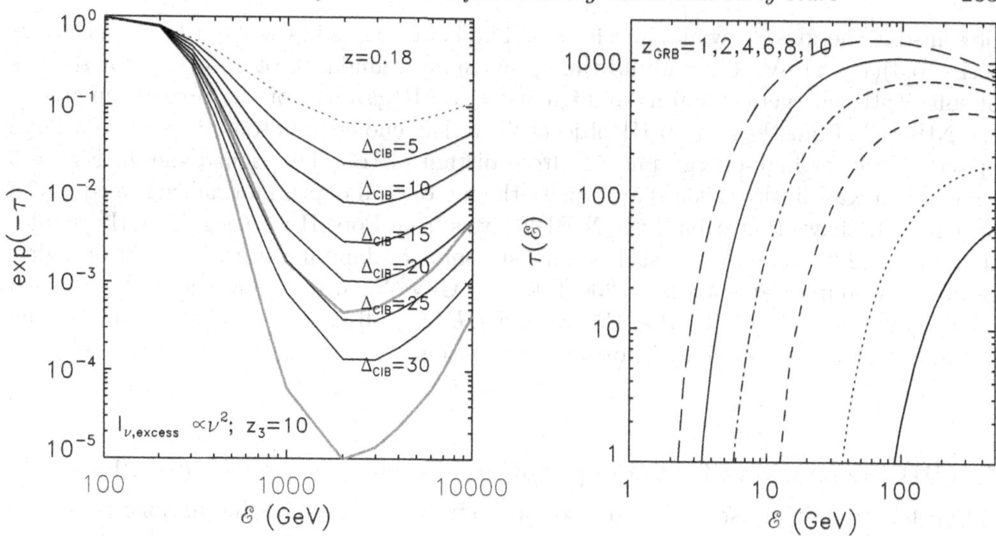

FIGURE 3. Left: Optical depth to photon-photon absorption for a source at $z = 0.18$. The convention for the thick light-shaded lines corresponds to the inset in Figure 1. Dotted line assumes only ordinary galaxies measured in deep counts and that their photons originated at $z \geqslant 0.18$. Solid lines correspond to the near-IR CIB excess from the Pop III era, assuming it ended at $z_3 = 10$. The values of the excess in nW/m^2/sr are marked near the lines. Right: The net τ vs. the GRB photon energy for the GRB redshifts shown in the panel. Solid, dotted, short-dashed, dash-dotted, dash-triple-dotted and long-dashed lines correspond to increasing order in z.

measurements (solid thick light-shaded line) and through the lower edges of the data (dashed thick light-shaded line). Such high attenuation factors (in excess of $\sim 10^3$) at $\sim$1.5 TeV have been argued against by the HESS results. This would indicate that—barring any changes in blazar physics—the NIRBE levels may be smaller than given by eq. (2.2).

However, it is important to emphasize that the HESS data still require significant CIB fluxes from the Pop III era. Indeed, this is also shown in Figure 3, where we computed the attenuation factors assuming that the NIRBE contribution from Pop III scales as $I_\nu \propto \nu^2$ with a Lyman-limit cutoff corresponding to the Pop III era ending at $z_3 = 10$ and normalized to the shown levels of the NIRBE flux, $\Delta_{\rm CIB}$ in nW/m^2/sr. The attenuation due to CIB from ordinary galaxies alone (filled circles in Figure 1) is shown as a dotted line and is *not enough* to account for the attenuation in the HESS blazars. The attenuation due to CIB levels claimed by the IRTS and DIRBE measurements is probably too strong. However, the figure shows that *significant amounts of NIRBE are still allowed and required by the data.* In particular, the HESS data requires the levels of NIRBE due to Pop III (i.e., with Lyman cutoff in the CIB at 1 μm) to be below ~15–20 nW/m^2/sr, still leaving up to a few percent of the baryons to have gone through Pop III.

This situation is expected to be resolved with the data from the upcoming *Gamma-ray Large Area Space Telescope* (*GLAST*) mission, which should measure spectra of high-z gamma-ray bursts (GRBs) and blazars out to 300 GeV (Kashlinsky 2005b). If Pop III stars at early epochs produced even a fraction of the claimed NIR CIB excess, they would provide a source of abundant photons at high z. The present-day value of $I_\nu = 1$ MJy/sr corresponds to the comoving number density of photons per logarithmic energy interval, $d \ln E$, of $\frac{4\pi}{c}\frac{I_\nu}{h_{\rm Planck}} = 0.6$ cm^{-3} and, if these photons come from high z, their number density would increase $\propto (1+z)^3$ at early times. These photons with the present-day energies, E, would also have had higher energies in the past: $E' = E(1+z) \simeq (0.1-0.3)(1+z)$ eV. They would thus provide an abundance of absorbers for sources of sufficiently energetic photons at high redshifts. Regardless of the precise amount of the NIR CIB from them, Pop III objects likely left enough photons to provide a large optical depth for high-energy photons from distant GRBs. The right panel in Figure 3 comes from Kashlinsky (2005b) and shows the net optical depth [normalized to eq. (2.2)] at high z. It shows that even if the NIRBE levels from Pop III were significantly smaller than in eq. (2.2) there should still be almost complete damping in the spectra of high-z gamma-ray sources at energies $<260(1+z)^{-2}$ GeV. Such damping should provide an unambiguous feature of the Pop III era and *GLAST* observations expected during the coming years would provide important information on the emissions from the Pop III era.

5. CIB fluctuations from deep *Spitzer* images and early populations

Pop III stars, if massive, should also have left significant CIB fluctuations providing a unique signature of their existence (Cooray et al. 2004; Kashlinsky et al. 2004). The reasons for strong CIB fluctuations are as follows:

- If massive, each unit of mass in Pop III stars would emit several orders of magnitude more luminosity than the present-day stellar populations.
- The Pop III era is expected to have spanned a relatively short cosmic time (<1 Gyr), leading to larger relative CIB fluctuations.
- Pop III stars are expected to have formed out of rare high-σ peaks of the density field which would amplify their clustering.

Kashlinsky et al. (2005) have attempted to measure these CIB fluctuations in deep *Spitzer*/IRAC data at 3.6 to 8 μm. The results of these measurements and their interpretation are discussed below. New and deeper measurements (Kashlinsky et al. 2007) support our earlier results.

5.1. *Observational results*

The main data used in Kashlinsky et al. (2005; hereafter KAMM) came from the IRAC guaranteed-time observations with ~10-hour integration of a field at high Galactic latitude. Additionally, the available data with shallower observations for two auxiliary fields were analyzed to test for isotropy of any cosmological signal. The datasets were assembled out of the individual frames, using the least-calibration method, which has advantages over the standard pipeline calibration of the data—the derived detector gains and offsets match the detector at the time of the observation, rather than at the time of the calibration observations.

For the final analysis, we selected a subfield of $\simeq 5' \times 10'$, with a fairly homogeneous coverage. Individual sources have been clipped iteratively. The images were left with >75% pixels for robust computation of the diffuse-flux Fourier transforms. CIB fluctuations from Pop III stars at high z should be independent of the clipping threshold, so the maps were also clipped progressively deeper to verify that our results are threshold independent. As more pixels are removed, it becomes impossible to evaluate robust Fourier transforms; then the diffuse-flux correlation function was calculated. In that analysis KAMM detected fluctuations significantly exceeding the instrument noise.

After the power of the instrument noise has been subtracted, Figure 4 shows the CIB fluctuation at 3.6 μm. The excess fluctuation on arcminute scales in the 3.6 μm channel is ~0.1 nW/m^2/sr; KAMM measure a similar amplitude in the longer IRAC bands, indicating that the energy spectrum of the arcminute scale fluctuations is flat to

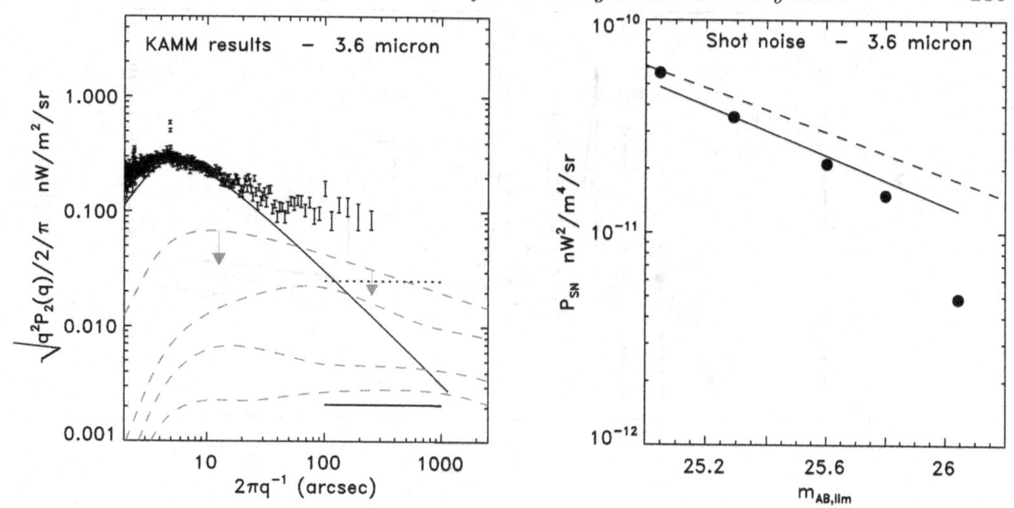

FIGURE 4. Left: the CIB fluctuations spectrum at 3.6 μm as measured by KAMM are shown with 1-σ error bars. Solid line shows the contribution from the shot noise of remaining OGs. The upper limits on the zodiacal light fluctuations are shown with the dashed-triple-dotted line, and on Galactic cirrus contribution with the dotted line. The upper limit of the contribution from OGs is shown with the upper dashed line, marked with down-pointing arrows. The lower dashed lines show more realistic contributions from OG populations as described in KAMM. Right: The amplitude of the shot-noise power spectrum from remaining OGs shown vs. the limiting magnitude of galaxy removal by KAMM. Solid line is from the KAMM fit to the *Spitzer* counts (Fazio et al. 2004); dashed line corresponds to counts analysis of Savage & Oliver (2005).

slowly rising with increasing wavelength—at least over the IRAC range of wavelengths. The detected signal is significantly higher than the instrument noise and the various systematics effects cannot account for it. There was a statistically significant correlation between the channels for the region of overlap, suggesting the presence of the same cosmic signal in all channels. When too few pixels were left for Fourier analysis, the correlation function at deeper clipping cuts remains roughly the same and is consistent with the power spectrum numbers.

Based on numerous tests, KAMM show that the signal comes from the sky. Below are the possible sources of the fluctuations; they include origin in the spacecraft, the Solar System, the Galaxy and the (cosmological) sources outside the Galaxy:

1. Instrument noise was estimated in KAMM, who show that it is too low and has a different pattern than the detected signal. Also, the cross correlation between the channels for the data's overlap region rules out significant instrumental contribution.

2. Residual wings of removed sources also appear unlikely to have contributed significantly to the detected signal. KAMM have done extensive analysis, and many tests to rule this out (e.g., the results remain the same for various clipping parameters, and for various masks of the clipped-out sources, as shown in their Supplementary Information).

3. Zodiacal light fluctuations were estimated by subtracting other data observed at two epochs separated by ~6 months. These are shown in Figure 4 and are very small; the amplitude at 8 μm is <0.1 nW/m^2/sr, and assuming normal zodi spectrum, the contribution to fluctuations would be negligible at shorter wavelengths.

4. Galactic cirrus is significant at channel 4 (8 μm), which may, in fact, be dominated by the cirrus component (diffuse-flux fluctuation ~0.2 nW/m^2/sr), but given the energy spectrum of cirrus emission the other channels, should have negligible cirrus. However,

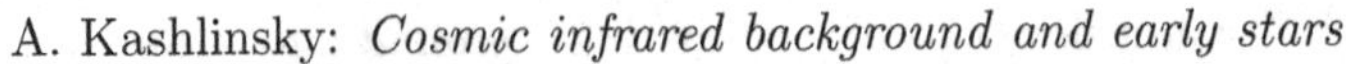

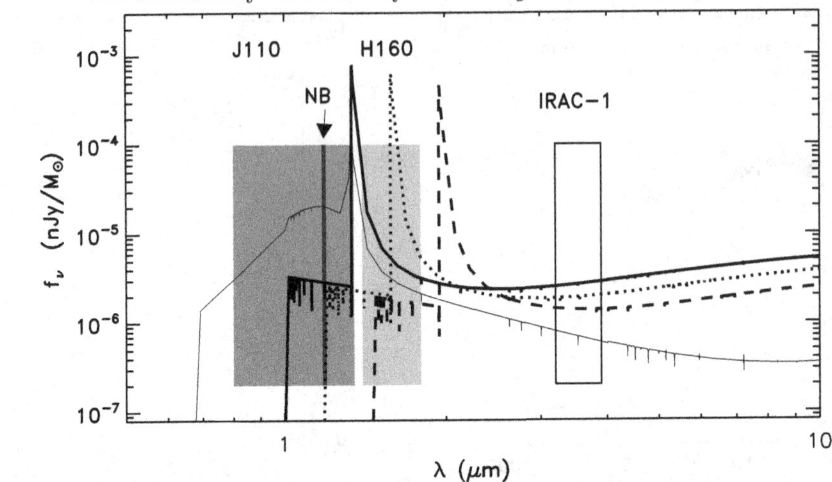

FIGURE 5. SED from Pop III system is shown for $z = 10$ (thick solid line), 12 (dotted) and 15 (dashed). The lines are drawn from Santos et al. (2002) for the case when the processing of the radiation takes place in the gas inside the nebula. The thin solid line shows the SED from Santos et al. for $z = 10$ when the nebula is assumed transparent and the processing takes place inside the intergalactic medium (IGM). The J, NB, H and IRAC-3.6 μm filters are shown. When the emission is reprocessed entirely by the IGM, the emission could extend below the Lyman limit at the rest frame of the emitters, in which case many sources could have existed even at $z = 10$, but escaped detection via the $J - NB > 0.3$ criterion of ZEN (Willis & Courbin 2005).

the map at Channel 4 has significant statistical correlation with the shorter wavelengths, suggesting the presence of the same cosmic signal at 8 μm.

5. Extragalactic sources thus seem to be the remaining logical explanation of the results. They include:

- Ordinary galaxies with "normal" stellar populations.
- Pop III with $M/L \ll (M/L)_\odot$ and located at high z.

We discuss the constraints the data impose on the two cosmological components. More discussion is given with new measurements in Kashlinsky et al. (2007).

5.2. *Interpretation*

Extragalactic contributions to the CIB fluctuations come in two flavors: 1) the shot noise, and 2) due to clustering out of the primordial density fields from which these sources trace. The shot-noise contribution, when evaluated directly from galaxy counts, gives a good fit to the fluctuations on smaller scales as shown in Figure 4. Ordinary galaxies have been eliminated from the maps down to very faint flux levels (~0.3 μJy in Channel 1). The remaining OGs' contribution to the CIB mean levels is small (<0.1–0.2 nW/m^2/sr), and its fluctuation levels are shown in Figure 4. On the other hand, the amplitude, power spectrum and the spectral energy dependence of the more than arcminute-scale fluctuation can be explained by emissions from Pop III.

Figure 5 shows the predicted flux due to spectral energy distribution (SED) of a Pop III stellar system at z = 12–15 and the region probed by the IRAC data. Although the *Spitzer* instruments cannot probe the peak of the emissions (due to Lyman emission), at the IRAC wavelengths we are still probing a region with substantial emissions by the Pop III sources.

Any interpretation of the KAMM results must reproduce three major aspects:

- The sources in the KAMM data were removed to a certain (faint) flux limit, so the CIB fluctuations arise in populations with magnitudes fainter than the corresponding magnitude limit, $m_{\rm lim}$.
- These sources must reproduce the excess CIB fluctuations by KAMM on scales $>0.5'$.
- Lastly, the populations fainter than the above magnitude limit must account not only for the correlated part of the CIB, but—equally importantly—they must reproduce the (low) shot-noise component of the KAMM signal, which dominates the power at $<0.5'$.

Below we briefly discuss the constraints on the populations contributing to the KAMM signal in the above order:

Magnitude limits. The nominal limit above which sources have been removed in the KAMM analysis is $m_{\rm AB} = 25.3$ at 3.6 μm, and this by itself implies that the detected CIB fluctuations arise from fainter systems. At this magnitude, one is already at the confusion limit for the IRAC beam at 3.6 μm, so fainter galaxies can be excised only by removing significantly more area in the map. The new mask then prevents reliably computing the power spectrum of the diffuse flux, which is why for deeper source removal KAMM presented its Fourier transform, the correlation function $C(\theta)$, in the Supplementary Information (SI). What is important in this context is that the $C(\theta)$, shown in Figure SI-4 of the KAMM SI, does not change in shape and amplitude with additional clipping. At 3.6 μm, the extra clipping goes to the limit of $\simeq$140 nJy for the faintest removed sources when KAMM stopped, as only 6% of the map pixels remained. Thus the KAMM signal comes from sources fainter than $m_{\rm AB} > 26.1$. For reference, this magnitude limit corresponds to $6 \times 10^8\ h^{-2}\ L_\odot$ emitted at 6000 Å at $z = 5$. A significant fraction of galaxies were thus removed from the data by KAMM even at $z \geqslant 5$, and the detected CIB fluctuations must be explained by still fainter and more distant systems.

Clustering component. From the remaining sources, the clustering component is given by the Limber equation, which we write in the form (Kashlinsky 2005a):

$$\left[\frac{q^2 P_2(q)}{2\pi}\right]^{1/2} = F_{\rm CIB}\Delta\left(q\left\langle d_A^{-1}\right\rangle\right) \quad ; \quad \Delta^2(k) \equiv \frac{1}{2\pi}\frac{k^2 P_3(k)}{c\Delta t} \ , \tag{5.1}$$

where $P_3(k)$ is the 3-D power spectrum of the sources' clustering, d_A is the comoving angular diameter distance, and Δt is the cosmic time interval spanned by the sources. Figure 4 shows that the clustering strength at $\geqslant 1'$ requires $\delta F_{\rm CIB} \sim 0.1$ nW/m^2/sr. The angle of $1'$ in the concordance cosmology subtends comoving scales of 2.2–3 Mpc at $5 \leqslant z \leqslant 20$. For ΛCDM density fields with reasonable biasing one can reach relative arcminute-scale fluctuations of $\leqslant$5–10%, meaning that the net CIB from sources contributing to the KAMM signal at 3.6 μm is $>$1–2 nW/m^2/sr.

Shot-noise constraints. The amplitude of the shot-noise power gives a particularly strong indication of the epochs of the sources contributing to the KAMM signal. This can be seen from the expressions for the shot noise (Kashlinsky 2005a):

$$P_{\rm SN} = \int_{m>m_{\rm lim}} f(m)dF(m) \equiv f(\bar{m})F_{\rm tot}(m>m_{\rm lim}) \ . \tag{5.2}$$

Figure 4 shows the shot-noise amplitude evaluated at 3.6 μm for the KAMM data as a function of the source removal threshold. Because the shot-noise sets important limits on the contributions to the CIB fluctuations from remaining ordinary galaxies, we have evaluated $P_{\rm SN}$ for the data at deeper clipping limits as follows: the maps analyzed and shown in KAMM's SI were clipped to progressively lower limits until $\simeq$6% of the map remained. At deeper clipping thresholds, where one cannot compute Fourier transforms reliably, we evaluated $P_{\rm SN}$ assuming it is proportional to the variance of the map, σ_F^2, minus that of the noise, σ_n^2, estimated by KAMM from the difference maps. (The noise, as expected, does not vary with the regions removed by deeper clipping and at 3.6 μm remains $\sigma_n \simeq 0.5$ nW/m^2/sr.) The last point may be an underestimation, as the precise value of the shot noise there may be influenced by not sufficiently well-determined (for that purpose) instrument-noise levels.

From Figure 4 we conservatively adopt the fiducial value of $P_{\rm SN} = 10^{-11}$ nW^2/m^4/sr as the *upper* limit to the shot-noise levels of the sources contributing to the measured CIB fluctuations. Above that level, it was shown that the sources contributing to the fluctuations must have CIB flux greater than a few nW/m^2/sr. Combining this with the values for $P_{\rm SN}$ shown leads via eq. (5.2) to these sources having typical magnitudes $m_{\rm AB} > 29$–30 or individual fluxes below a few nJy. *Such faint sources are expected to lie at $z > 10$, the epoch of the first stars.*

6. Resolving the sources of the CIB

That the CIB arises from emissions by discrete cosmological sources is therefore not really diffuse. It can and, of course, should eventually be resolved with faint/deep surveys which simultaneously have sufficiently fine angular resolution.

Salvaterra & Ferrara (2006) have recently raised an important question of whether the existence of the Pop III era is in conflict with the recent data on J-band dropouts (Bouwens et al. 2005, Willis & Corbin 2005). While the surveys are strongly incomplete at $H > 26$, at face value, the data indicate a paucity of $J - H > 1.8$ dropouts around $H_{\rm AB} \simeq 28$ of only at most a few per arcmin^2 (Bouwens et al. 2005) and a similarly low upper limit on the dropouts at $J - NB \geqslant 0.3$ in the ZEN survey of Willis & Corbin (2005). Is there necessarily a conflict, or do these data tell us something significant about the Pop III-era physics? Such theoretical computations of the counts of the Pop III sources are—for now—necessarily based on the assumptions listed here in decreasing order of their subjectively perceived likelihood:

1. the distribution of small-scale fluctuations of the primordial density field is Gaussian;
2. the small scale power is that of the concordance ΛCDM model;
3. the mass function of the first objects is described well by a Press-Schechter–type prescription, although the effective index of the ΛCDM power spectrum at these scales $n \to -3$ and pressure effects may be important at such low masses (cf. Springel et al. 2005);
4. that the redshift, z_3, out to which Pop III era extended is $z_3 < 9$;
5. the SED of Pop III systems is cut off at the Lyman-α frequency of the source epoch and the cutoff is not produced by the IGM at lower z (cf. Santos et al. 2002); and most crucially,
6. that the efficiency of Pop III formation inside the parental halos is the same for all masses and epochs. Dropping any of these assumptions would lead to significant changes in the predicted number counts.

The last assumption is particularly critical, because of the ease of both the destruction of H_2 molecules by Lyman-Werner–band photons (e.g., Haiman et al. 1997) and their creation following the ionization of the surrounding nebula (Ferrara 1998; see reviews by Barkana & Loeb 2001; Bromm & Larson 2004).

Figure 5 marks the filters around 1.1 and 1.6 μm used by Bouwens et al. (2005) and Willis & Corbin (2005) vs. an example of the emission template from Pop III stars at

$z = 10, 12, 15$. The figure shows that both surveys probe a very narrow range of redshifts: $z = 12$ by Bouwens et al. and $z = 10$ by ZEN. Indeed, only for $z = 12$ are sufficiently massive Pop III systems (the mass in Pop III stars $M_* > 10^6 M_\odot$) likely to produce $H \sim 27$–28 and satisfy $J - H > 1.8$. (The constraints from the ZEN survey are weaker). Thus the counts problem—even with the above assumptions—provides information only for these narrow ($\Delta z \lesssim 1$) epochs. The discussion in Section 5.2 shows that Pop III systems have the mean flux of less than ~5–10 nJy at 3.6 μm; such populations cannot be detected in the above surveys. Why Pop III stars form in such small systems is an interesting question, which is likely related to the variation of the efficiency of Pop III formation with epoch and mass.

In order to detect the faint sources responsible for the CIB fluctuations with fluxes below a few nJy, as discussed in Section 5.2, embedded in the underlying sea of emissions, their individual flux must exceed the confusion limit usually taken to be $\alpha \geqslant 5$ times the flux dispersion produced by these emissions (Condon 1974). If these sources do not have the necessary flux levels, they will be drowned in the confusion noise and will not be individually identifiable in galaxy surveys. Of course, this is precisely where CIB studies would take off. From observations one knows that confusion levels are not reached in J band until $m_{\rm AB} \sim 28$ (e.g., Gardner & Satyapal 2000). If such sources were to contribute to the CIB required by KAMM data, at 3.6 μm they had to have the average surface density of

$$\bar{n} \sim F_{\rm CIB}^2/P_{\rm SN} \sim 5\ \mathrm{arcsec}^{-2} \left(\frac{F_{\rm CIB}}{2\ \mathrm{nW/m^{-2}/sr^{-1}}}\right)^2 \left(\frac{P_{\rm SN}}{10^{-11}\ \mathrm{nW^2/m^{-4}/sr^{-1}}}\right)^{-1} . \quad (6.1)$$

In order to avoid the confusion limit and resolve these sources individually at, say, the 5-σ level ($\alpha = 5$), one would need a beam of the area

$$\omega_{\rm beam} \leqslant \alpha^{-2}/\bar{n} \sim 5 \times 10^{-3} \left(\frac{F_{\rm CIB}}{2\ \mathrm{nW/m^{-2}/sr^{-1}}}\right)^{-2} \mathrm{arcsec}^2 , \quad (6.2)$$

or of circular radius below ~0.04 $(F_{\rm CIB}/2\mathrm{nW/m^{-2}/sr^{-1}})^{-1}$ arcsec. This is clearly not in the realm of the currently operated instruments, but the *JWST* could be able to resolve these objects given its sensitivity and resolution.

7. Conclusions

Both the CIB and Pop III studies are rapidly evolving into observation-rooted, if not yet high-precision, disciplines. The current situation is therefore very dynamic, but one can summarize the up-to-date constraints on the Pop III era that emerge from the recent CIB measurements as follows:

• The measurements of the mean levels of the CIB indicate that about up to a few percent of baryons may have gone through Pop III. The new measurements of the $z \sim 0.1$–0.2 blazar spectra at a few TeV energies indicate that the net levels of the CIB may be not as high as the DIRBE- and IRTS-based analyses indicate. Nevertheless, these data show an evidence of absorption in excess of what can be provided by the observed galaxy populations, indicating additional fluxes from fainter (and likely more distant) sources. If real, these near-IR CIB fluxes should be detectable from the future *GLAST*-based observations of the high-z gamma-ray bursts.

• The recent measurements of the CIB fluctuations allow us to probe the emissions from high-z sources more directly. They indicate that the sources have fluxes below ~10 nJy at 3.6 μm, or $m_{\rm AB} > 28$–29. Such sources are likely to be located at $z > 10$, the era associated with the first stars. These sources must have provided CIB levels of at least a few nW/m^2/sr at 3.6 to 5 μm, and their clustering properties have to fit the measured CIB fluctuations of $\sim$0.1 nW/m^2/sr at arcmin scales, as well as have the SED that gives flat to slowly rising CIB fluctuations with increasing wavelength.

• The above constraint on the typical flux (a few nJy) of these sources suggests that they have less than $\sim 10^5\ M_\odot$ stars inside each of the haloes. This likely requires a halo-dependent efficiency if the parental haloes form out of the ΛCDM density field prior to $z \sim 10$.

• The sources contributing to the CIB fluctuations at the *Spitzer*-IRAC bands should be resolvable in surveys with sufficiently low noise (to measure the low fluxes) and high angular resolution (to overcome cosmological confusion). At 3.6 μm, this can probably be done with the future *JWST*.

I warmly thank my collaborators on the CIB fluctuations project, Rick Arendt, John Mather and Harvey Moseley, for many discussions and contributions. This work was supported by NSF and NASA/*Spitzer* grants.

REFERENCES

Abel, T., et al. 2002 *Science* **295**, 93.
Aharonian, F., et al. 2006 *Nature* **440**,1018.
Barkana, R. & Loeb, A. 2001 *Phys. Rep.* **349**, 125.
Bouwens, R J., Illingworth, G., Thompson, R., & Franx, M. 2005 *ApJ* **624**, L5.
Bromm, V., et al. 1999 *ApJ* **527**, L5.
Bromm, V. & Larson, R. 2004 *ARA&A* **42**, 79.
Cambresy, L., et al. 2001 *ApJ* **555**, 563.
Condon, J. 1974 *ApJ* **188**, 279.
Cooray, A., et al. 2004 *ApJ* **606**, 611.
Dwek, E. & Arendt, R. 1998 *ApJ* **508**, L9.
Dwek, E., Krennrich, F., & Arendt, R. 2005 *ApJ* **634**,155.
Fazio, G., et al. 2004 *ApJS* **154**, 39.
Ferrara, A. 1998 *ApJ* **499**, L17.
Gardner, J. P. & Satyapal, S. 2000 *AJ* **119**, 2589.
Gorjian, V., et al. 2000 *ApJ* **536**, 550.
Haiman, Z., Rees, M. J., & Loeb, A. 1997 *ApJ* **476**, 458.
Heger, A., et al. 2003 *ApJ* **591**, 288.
Kashlinsky, A. 2005a *Phys. Rep.* **409**, 361.
Kashlinsky, A. 2005b *ApJ* **633**, L5.
Kashlinsky, A., Arendt, R., Gardner, J. P., Mather, J. C., & Moseley, S. H. 2004 *ApJ* **608**, 1.
Kashlinsky, A., Arendt, R., Mather, J. C., & Moseley, S. H. 2005 *Nature* **438**, 45 (KAMM).
Kashlinsky, A., Arendt, R., Mather, J. C., & Moseley, S. H. 2007 *ApJ* **654**, L5; Erratum: 2007 *ApJ* **657**, L131.
Kashlinsky, A., Mather, J. C., & Odenwald, S. 1996a *ApJ* **479**, L9.
Kashlinsky, A., Mather, J. C., Odenwald, S., & Hauser, M. 1996b *ApJ* **470**, 681.
Kashlinsky, A. & Odenwald, S. 2000 *ApJ* **528**, 74.
Madau, P. & Pozzetti, L. 2000 *MNRAS* **312**, L9.
Madau, P. & Silk, J. 2005 *MNRAS* **359**, L37.
Magliochetti, M., Salvaterra, R., & Ferrara, A. 2003 *MNRAS* **342**, L25.
Matsumoto, M., et al. 2005 *ApJ* **626**, 31.
Rees, M. J. 1978 *Nature* **275**, 35.
Salvaterra, R. & Ferrara, A. 2003 *MNRAS* **339**, 973.

SALVATERRA, R. & FERRARA, A. 2006 *MNRAS* **367**, L11.
SANTOS, M. R., BROMM, V., & KAMIONKOWSKI, M. 2002 *MNRAS* **336**, 1082.
SAVAGE, R. S. & OLIVER, S. 2005; astro-ph/0511359.
SCHAERER, D. 2002 *A&A* **382**, 28.
SIESS, L., LIVIO, M., & LATTANZIO, J. 2002 *ApJ* **570**, 329.
SPERGEL, D., ET AL. 2007 *ApJS* **170**, 377.
WILLIS, J. P. & COURBIN, F. 2005 *MNRAS* **357**, 1348.
WRIGHT, E. L. & REESE, E. D. 2000 *ApJ* **545**, 43.

Printed in the United States
by Baker & Taylor Publisher Services

Printed in the United States
by Baker & Taylor Publisher Services